GREEN ENTREPRENEUR HANDBOOK

THE GUIDE TO BUILDING AND GROWING A GREEN AND CLEAN BUSINESS

WHAT EVERY ENGINEER SHOULD KNOW
A Series

Series Editor*

Phillip A. Laplante

Pennsylvania State University

*Founding Series Editor: **William H. Middendorf**

GREEN ENTREPRENEUR HANDBOOK

THE GUIDE TO BUILDING AND GROWING A GREEN AND CLEAN BUSINESS

ERIC KOESTER

CRC Press
Taylor & Francis Group
Boca Raton   London   New York

CRC Press is an imprint of the
Taylor & Francis Group, an **informa** business

CRC Press
Taylor & Francis Group
6000 Broken Sound Parkway NW, Suite 300
Boca Raton, FL 33487-2742

© 2011 by Taylor and Francis Group, LLC
CRC Press is an imprint of Taylor & Francis Group, an Informa business

No claim to original U.S. Government works

Printed in the United States of America on acid-free paper
10 9 8 7 6 5 4 3 2 1

International Standard Book Number: 978-1-4398-1729-2 (Hardback)

Visit the Taylor & Francis Web site at
http://www.taylorandfrancis.com

and the CRC Press Web site at
http://www.crcpress.com

Contents

Part II   The Green Startup

Part IV Green Progress (So Far)

Series Preface

What every engineer should know amounts to a bewildering array of knowledge. Regardless of the areas of expertise, engineering intersects with all the fields that constitute modern enterprises. Soon after graduation, the engineer discovers that the range of subjects covered in the engineering curriculum omits many of the most important problems encountered in the line of daily practice—problems concerning new technology, business, law, and related technical fields.

With this series of concise, easy-to-understand volumes, every engineer now has within reach a compact set of primers on important subjects such as patents, contracts, software, business communication, management science, and risk analysis, as well as more specific topics such as embedded systems design. These are books that require only a lay knowledge to understand properly, and no engineer can afford to remain uninformed about the fields involved.

Acknowledgments

Over the last two decades, I have had the great fortune to work with and learn from a number of outstanding individuals involved in the green business world. Many of those lessons and interactions have become a part of this book in some form or fashion. It would be impossible to name every person I have met along the journey, but I would like to thank a few people specifically for their involvement in this book.

My first exposure to the green business industry came at the urging of my father who decided I would get into too much trouble unless he put me to work during my summers off from school. During those teenage years, my father Larry hired me for a series of projects that gave me exposure and opportunities few teenagers have—specifically key lessons about global and U.S. environmental laws, recycling technologies, automobile recovery, plastics, waste-to-energy technology, tire recovery, green businesses, and entrepreneurship. With his help and support, we developed on a series of books and publications on these very topics that served as the initial foundation for my understanding of and passion for green business. This process also introduced me to a number of outstanding individuals involved in green business who offered invaluable insights into the field. Special thanks go to both my father and my mother Kathy for their support in all my endeavors. Love you both.

My law firm Cooley LLP provided a great deal of support and resources throughout this process. Specifically, I would like to thank John Robertson and Sonya Erickson for their continued support, each of the attorneys in our Seattle office, as well as the numerous attorneys on our clean energy and technologies practice group under the guidance of Jim Fulton and Tom Amis. The collaborative team of attorneys at Cooley continues to help outstanding companies across the globe.

I was also fortunate to have help from several outstanding research assistants and recent law school graduates, Brian Edstrom, Brittany Stevens, and Bill Dufour. Brian's help and efforts are on display in the chapters on government involvement in clean technology, specifically on regulations, laws, and certifications, as well as in the chapter on greening an existing business. Brittany provided outstanding research and preparation assistance for the chapter on utilities and selling to utilities. And finally, thanks to Bill for his help on the section dedicated to government contracting. The help from each of you was appreciated greatly, and certainly made this book much better!

I would like to give a special thank you to the team from the Taylor & Francis Group. Thanks are due to my acquiring editor Allison Shatkin who worked with me on this idea and project for the past year or more and to Kari Budyk who served as the project coordinator on the *Green Entrepreneur Handbook*. This book is my second publication with the Taylor & Francis team and I greatly appreciate their continued support and flexibility. Also, the outstanding drawings found throughout this book were done by Travis Fox from the Kansas City area, who is now my go-to artist for anything I need to have illustrated. Thanks again, Travis.

Finally and most importantly, my wife Allison continues to be my best friend and partner. Not only was she patient and understanding of all the time and energy the preparation of this book required, but she also helped me think about the structure and content of the book during the preparation process. Plus, I turned to Allison for help in the preparation of the chapter on taxes and incentive programs (saving my hide as usual!). Everyone always tells me how lucky I am and my response without fail is "you have no idea."

A project like this takes support from lots of talented and patient people including friends, family, colleagues, partners, advisors, and the like. So to everyone who helped with a few words of encouragement or more detailed critiques, thank you again.

The entrepreneurial community is alive and well—from the corner retail shop to the fast-growing startup and from a first-time entrepreneur located in Vietnam or Ethiopia to a startup team located in Silicon Valley. I truly believe that entrepreneurs, small businesses, and startups will continue to be an integral part of the worldwide economic engine, producing new technologies, employing more people and "flattening" our world more and more. This is truly a community made up of multiple people and organizations helping make small businesses and startups a success. To show my appreciation and support, I will be donating 100% of the proceeds from the sale of this book to two nonprofit organizations that focus on the worldwide entrepreneurial community (and that I personally believe in and support): Startup Weekend and Kiva.

Startup Weekend is a leading nonprofit providing experiential education for entrepreneurs. Startup Weekend has hosted their signature 54-hour event, which brings together various entrepreneurs and want-repreneurs to build communities, companies, and projects. Over 100 cities in 25 countries have hosted a Startup Weekend event impacting thousands of entrepreneurs. Learn more or donate at http://www.StartupWeekend.org/. Kiva is a nonprofit organization that combines microfinance with the Internet to create a global community of people connected through lending. They empower individuals to lend to an entrepreneur across the globe engaged in projects in countries, including Kenya, Mongolia, Cambodia, Peru, El Salvador, and the United States. Kiva's mission is to connect people, through lending, for the sake of alleviating poverty. Find out more or start lending at http://www.Kiva.org/. Both organizations support entrepreneurs involved in green activities, and are doing their part to make a difference. I urge you to take a minute and learn more about both organizations as their efforts are truly helping expand the scope of entrepreneurship, reaching from the remote ends of the world to the technology hubs and everywhere in between. I urge all readers to commit your time, your money, or your support—and just by buying this book, you are making an impact in these organizations as all royalties goes directly to help these outstanding organizations.

Eric Koester

Author

Eric Koester is a business attorney with Cooley LLP and has a practice focused on emerging technology companies, venture capital firms, and investment banks with particular emphases on venture capital and bank financings, corporate partnerships, commercial agreements, intellectual property licensing, public offerings, and mergers and acquisitions. He has started three businesses focused on environmental and technology consulting, Web site design, and mobile software and has a passion for working with entrepreneurs and technology-focused thinkers. Eric also volunteers his time at a number of nonprofit organizations that focus on entrepreneurship, microfinance, and education. Specifically, he is a board member of Startup Weekend (http://www.StartupWeekend.org) and an active supporter of Kiva (http://www.Kiva.org). All of the author's proceeds from the sale of this book will be donated to these outstanding organizations.

Eric published *What Every Engineer Should Know About Starting a High-Tech Business Venture* in 2009. While some information from that book is included in this book, *What Every Engineer Should Know About Starting a High-Tech Business Venture* it contains nearly 600 pages of more information on topics including formation matters, venture capital term sheets, compensation trends, board of directors, and various key insights into entrepreneurship and starting a business. More information on the book and related topics can be found on the blog: http://www.myhightechstartup.com.

Eric is a graduate of the business school at Marquette University and The George Washington University School of Law. He has also studied the intersection between the environment and leadership. Eric is a certified public accountant. Eric's wife Allison is a PhD candidate in business at the University of Washington's Foster School of Business in Seattle, Washington where they both reside. They have two pugs and enjoy travel, sports, and outdoor activities in their free time. Visit Eric's LinkedIn page at http://www.linkedin.com/in/erickoester or the blog and Web site for this book at http://www.greentrepreneur.org for more information.

Introduction

Green is an opportunity unlike anything we have seen before. I would argue it is bigger than medicine, broader than information technology and more global than the Internet. It is an opportunity as large as this third planet from the sun we call home—a planet that rotates in 24-hour increments and circles the sun in 365-day periods. In short, the green opportunity is about developing innovative solutions to advance our society while enhancing our most valuable asset: the Earth.

The best and most successful entrepreneurs recognize an opportunity and create products, services, and businesses that fill that unmet need. And with this green opportunity that I like to say is approximately 8000 miles in diameter, the challenges do not come much bigger than this one. That is why this book has been written, as a tool to assist entrepreneurs tackling the wide variety of opportunities to "go green."

Hopefully that is why you have picked up this book: to find *your* green opportunity. And let me tell you what, it may not be easy to find that perfect green opportunity, but now there is more support for your journey than ever before. And hopefully with a broader community behind it, smart, dedicated, and creative people will band together to support these goals.

What is so fascinating about green business is also its greatest challenge—its breadth. From cars to computers to buildings to clothing to travel to food to electronics, everything has the ability to be a part of the green movement. An entrepreneur building advanced solar arrays shares a connection with an all-natural cleaning company; a hybrid vehicle manufacturer is linked to an organic farmer; and a carbon trader is tied with the green building contractor. What do these disparate businesses and industries share? They each see an opportunity to reform and make an impact in the business world in ways that attempt to protect the Earth and our broader society. Separately, these greentrepreneurs may be tackling their own individual business opportunities, but together, change is occurring. And so greentrepreneurs are alike in that single regard.

But like any of you reading this book, I too have my doubts, my concerns, my fears, and my issues with green business. Is it a product of marketing? Is it a fad? Will the hype die out? Is it sustainable? Do consumers really care? Does it make business sense? We all understand that green business is not the answer to all our problems and will have challenges just like traditional sectors and industries.

In the end, businesses that make business sense—be they green, blue, red or otherwise—will survive. We hope your green business is one of those that not only survives but thrives. Some green businesses may fade as "hype" wanes, while still others may finally be able to grow once reality replaces the noise. I am betting on these challenges continuing to exist, and smart, savvy individuals tackling those opportunities arising from an Earth sometimes without committed stewards.

Inside This Book

This book is designed to be a one-stop resource for a green entrepreneur. Of course, it is simply impossible to cover everything an entrepreneur starting a green business would need, but we have done our darnedest anyway. Hopefully you can pick up the book, read and study portions of it, and feel you have a better sense for the unique challenges of building a green business. In the end, the same fundamentals exist as with any business—and the green angle provides a unique scenario that may not have existed in the past.

The book starts off looking at the broader green business marketplace in Part I (The Great Green Opportunity). This section is designed to lay the groundwork for any new entrepreneur to better understand the history of the environmental and clean technology movements (Chapter 1) and the drivers of the revolution into the future (Chapter 3). At the same time, as this book is aimed at entrepreneurs, we examine what motivates, drives, and inspires all entrepreneurs, but specifically greentrepreneurs (Chapter 2). And finally, we start where every entrepreneur should start: finding a market in need of your solution (Chapter 4).

Part II (The Green Startup) dives into the green entrepreneur's playground to examine a new business from initial idea to sales of your product or service. We address questions such as where greentrepreneurs can find ideas to build a business around (Chapter 5) and how to form a company to execute on that business concept (Chapter 6). Building any business, be it a green business or otherwise, ultimately requires people from the founders to the first employees to the advisors and directors (Chapter 7), and understanding how to find and retain those key contributors is a vital part of the success of the business. The second part of this book also examines two of the key challenges of many new businesses, raising money (Chapter 8) and making sales (Chapter 10). And we tackle the importance of your intellectual capital and assets (Chapter 9) that will drive broader adoption of your technologies and initiatives. Additionally, a case study is included in Part II that ties all these concepts together into an actual startup-like situation for a set of greentrepreneurs—hopefully, these lessons will help grasp much of the theory in the book and apply it to a real-world setting.

Part III (The Green Playing Field) emphasizes some of the most unique aspects of the green business environment. While it is important to not lose sight of the fact that you are starting a business that has similarities to other startups, green businesses today operate in a world that does have its own identity. Many green businesses touch on energy, from generation through consumption, and so Part III provides a greater understanding of utilities and energy generation and distribution (Chapter 11). Likewise, the potential size and scope of certain green businesses require funding that can only come through project finance (Chapter 12), which further distinguishes green business. But perhaps the most identifiable trait of green business is the connection to our governments. This portion of the book spends a series of chapters looking at the players and the process to sell to the government (Chapter 13), the federal, state, and local regulatory impacts (Chapter 14) as well as government incentives and tax programs designed to spur clean technology development (Chapter 16). Even outside of these tax programs, billions of dollars are being poured into green businesses through grants, loans, and other funds (Chapter 15), and can be a vital source of capital for your business.

Part IV of the book, Green Progress (So Far), tries to uncover some lessons learned and the challenges on the horizon as the green business sector grows. Today, all businesses are feeling the pressure to become greener at the hands of suppliers, customers, and even the government—and we try and provide some practical suggestions businesses can implement themselves in their aims to go green (Chapter 17). The fragmentation of the green business space has led to frustration and opportunity given that clear certification, labeling, and disclosure do not yet exist (Chapter 18), but we all know they is on the way and should be on the forefront of the minds of greentrepreneurs. Green business has also seen a new entrant into the field with venture capital and institutional investors (Chapter 19) playing a huge role in the green innovation game. And finally, Part IV examines the international trends in green, business (Chapter 20) as well as the potential for exit events including public offerings and mergers and acquisitions (Chapter 21).

The final part of the book (Part V) is aimed at the entrepreneur himself or herself—outside of even the green business context—dealing with *Green Business Fundamentals*. These are lessons, tools, resources, and fundamentals for any entrepreneur. And whether you are starting a software company, a retail store or a high-tech research company, this part of the book has a series of concise sections that can help you with common challenges you will face. Specifically, you will find information from market research and business planning (Chapter 22) to details of forming the business itself (Chapter 23). There are sections to help you with your founder team (Chapter 24) and the issues of employing people (Chapter 25). Many new entrepreneurs must also understand how to efficiently manage their intellectual property—oftentimes on a shoestring budget—and so there is an entire section dedicated to smart IP management (Chapter 29). We also recognize the challenges today's businesses face in a difficult fundraising climate and have devoted a substantial portion of Part V to these matters (Chapters 26 through 28). And finally, we have provided information on your partners (Chapter 30) and some additional background on various exits (Chapter 31).

The Future

The green opportunities remain substantial. Some pundits point to a broader green market that has not yet reached the mainstream—including a "saturated" solar market or a "depressed" biofuels market or a building freeze that has affected green building or a flat organics market. The reality is that solar generation remains a tiny, but rapidly expanding portion of worldwide generation; biofuels are becoming commercially viable in places like Brazil and research is identifying new non-food crop-feedstocks; green building has grown in the face of contracting slowdowns; and while organic products continue to be more expensive than conventional products, organics continue to see high levels of annual growth with new consumers.

Despite these challenges or critiques, green business remains a viable and booming sector. It is my sincere hope that this book can help, inspire, and keep you on track for your business. I would love to hear all about your green business and your path, so feel free to email me at Eric@greentrepreneur.org.

To a future full of green opportunities ...

A Web site has been created for this book to provide updates, address new topics, and answer your questions. Visit http://www.greentrepreneur.org for more information.

Part I

The Great Green Opportunity

1

Why Are We Going Green?

Welcome to the Green Business Revolution. That's right: everyone really *is* going green these days. It's "Green Week" on your television station. Cities promote car-free commuting options. Universities are shutting off the lights to save energy. Politicians are wearing green pins. Earth day has turned into Earth Year. Entrepreneurs are solving environmental challenges.

Countless magazines, news stories, popular television, dinner conversations, and even religious services have focused on today's green movement. People are trading in their gas guzzlers for hybrids and Teslas, retailers are putting solar panels and wind turbines on their roofs, and groceries are selling more organics and natural products. Green is the new black, to take from the popular fashion phrase. In fact, according to the Global Language Monitor from 2000 to 2009, the top words of the decade of were "Global Warming," and the top phrase of the decade was "Climate Change." So unless you are living under a rock, you're probably aware that there is this thing called a "Green Revolution" going on.

But like any revolution (or whatever you'd like to call the green movement), its roots reach farther back than this decade, and the hope is that its impact lasts well into the future. This chapter looks at the green movement—where did it come from and where is it going? We are going green and are all along for the ride.

Green: The Big Opportunity for Green (as in Money)

Let's start with the basics. You want to start a green business. Or perhaps you've decided to transition the focus of your company to become a "green" business. Or maybe you are an entrepreneur, a business owner or a manager who has an existing green business and are trying to gain a better understanding of the next steps ahead. Or you are just curious about green or clean businesses and want to know more. Either way, you've probably thought about the opportunities of "going green."

Beyond the social impacts of the green movement, there has sprung up a growing space for green businesses. The global recession of 2008–2009 brought with it worldwide public sector spending to generate "green jobs" and the businesses that employ those green-collar workers. Venture capital has invested heavily into solar, fuels, building, transportation, and efficiency technologies. And, new crops of entrepreneurs have graduated from universities and left their jobs in traditional sectors to build businesses to meet the demand for green products, services, and technologies.

According to Ira Ehrenpreis, a General Partner at Cleantech VC Technology Partners, a venture capital firm focused on investments in clean technologies, "Green is the New Green (as in Money)." And that mentality is based on good reasons. An aging power system, transportation industry, and general infrastructure system coupled with emerging regulatory trends and public support lead to the perfect opportunity storm. To put things into perspective, in the United States, energy is an estimated $2.1 trillion sector and transportation a $1.5 trillion sector. Chemicals represent a $3 trillion sector globally. And these are just a few of the key industries that are currently the focus of the clean technology movement. Overall, the green industry is estimated to be a $200 billion sector today and expected to experience substantial growth.

Market for Green Business

Experts estimate that the marketplace for green businesses is more than $200 billion worldwide and growing.

Entrepreneurs, investors, and various other business people see those numbers and understand the sizable opportunities they present. Think about it this way—the Internet boom in the late 1990s and even in much of the 2000s was built around a market that didn't exist—there were no online book retailers before Amazon, no search engines before Yahoo! and Google, no online auction sites before eBay, no webmail platforms before Hotmail, and no online pet supply stores before pets.com (Not all the Internet darlings could be big successes, right?). In the sectors targeted by green businesses, there are already customers paying for electricity, filling up their cars with gasoline and buying consumer products. All green businesses have to do is come along and offer a cleaner, greener, or more environmentally friendly option. (Well, as we'll discuss later, there is more to it than just that.)

But there is more to it than just only "green" (a.k.a. the money). Many green entrepreneurs look at opportunities to build green businesses as a way to build a company that does more than just create a profit. Said one entrepreneur who had started a small biofuels business, "I'm willing to do this because my kids deserve a world at least as good as what I had … and sadly my generation hasn't done much to give them that." And that attitude is one shared by many in the green movement.

This book is not meant to wade into the debate as to global warming, climate change, resource loss, water contamination, and so on. Simply put, we assume that in a world of limited resources, resource conservation will be a business driver for the future. To the extent it is determined that global warming is overstated or perhaps even a complete myth (which sentiment seems like a stretch, but the reality is we just don't know) resource management remains an important part of the future. There is more to green than just helping the environment. Going green can save money, provide health benefits, and help sell more products. Even the United States government has been able to expand its argument for investments in the green economy by promoting the duel benefits of both limiting climate change and developing additional energy security.

THE MEANING OF CLEAN

What key challenges are clean technology companies attempting to address:

- *Dirty industry modification.* Technologies that clean up previously dirty industries where pollution is already released. For example, technologies that remediate contaminated lead.
- *End of pipe.* Technologies that reduce or control environmental harm or externalities associated with industrial manufacturing. Examples include filters or scrubbers on smoke stacks or catalytic converters on car exhausts.

- *Clean substitutes.* Provide cleaner substitutes to existing technologies or materials, often using the same infrastructure.
- *Efficiency.* Enhance efficiency of existing processes—so that fewer inputs used leads to reduced outputs. Examples include energy efficient lighting and building materials that enhance thermal efficiency.
- *Pollution prevention.* Eliminate pollution—for example using sensors and monitors to optimize process inputs in order to reduce NO_x emissions.
- *Industrial ecology.* Models of efficient use of resources, energy and waste in a system-setting using closed-loop design. An example of this would be taking waste, energy or other materials and turning them into feedstock.

Source: SVB Financial Group. April 2007. *Earth, Wind, and Fire: A Cleantech Perspective.*

How Did We Get Here: The Roots of the Environmental Movement

The environmental movement did not begin with Al Gore's 2006 breakout documentary "An Inconvenient Truth" (although some argue that the movie did "kick start" the clean technology movement from just those "dark green," highly passionate consumers to every day Americans). In fact, the environmental movement did not start even in the 1970s with the enactment of numerous environmental regulations by the U.S. government. The true pioneers in the environmental movement within the United States had roots in the late 1800s.

The Industrial Revolution of the nineteenth century brought with it unprecedented growth and change to the world. And with that growth came increasing detachment from nature among the world's industrializing population. Before the Industrial Revolution and the movement of the population to urban centers, most of the world's societies held a deep connection to nature. The transition of the industrialized world from the agrarian countryside into urban centers brought with it a sense of greater detachment. In 1790, only 5% of Americans lived in urban settings. By 1850, that number was up to just over 15% of the population and continued to increase during the beginnings of the industrialization of America—increasing to 35% by 1890 and 40% by 1900. Today, over 80% of Americans live in an urban center. With that transition in the late 1800s, we can find the seeds of today's environmental movement.

With populations moving from the countryside into urban centers and more of America's great national resources being commercialized or exploited, certain leaders began to take steps to slow urbanization. Many point to president Theodore Roosevelt's love of nature and the outdoors as a key point in the environmental movement. In 1902, Roosevelt established the first national park at Crater Lake, Oregon. During his presidency, he created four additional national parks and 51 wildlife refuges, passed the Antiquities Act (which led to the creation of 18 national monuments), and created the National Park Service. In addition, the pioneers of the environmental movement in the United States include John Muir (founder of the Sierra Club), Henry David Thoreau (author of *Main Woods*), and George Perkins Marsh.

In the middle of the twentieth century, the concept of environmental activism began to grow. Organizations such as the Sierra Club began as groups dedicated to the protection

The Environmental Defense Fund was founded by a group of activists in response to the work of Rachel Carson to seek a ban on the use of DDT in Suffolk County, Long Island, New York.

of wildlife and the protection of wilderness. By the 1950s, those same organizations started to expand their message to other perceived environmental issues such as water pollution, air quality, population growth, and the reduction of the exploitation of natural resources. Many historians of the environmental movement point to a book by biologist Rachel Carson as a seminal work in the legislative reforms of the 1970s and the roots of today's green movement. Carson (1962) wrote *Silent Spring* detailing the impact of the use of DDT as a pesticide. Her research examined the potential link to cancer and harms to wildlife caused by the use of a chemical that had already been shown to harm birds. Eight years later saw the creation of the U.S. Environmental Protection Agency and, just two years later, a ban of the use of DDT as a pesticide. And more than the impact on the use of DDT, Carson's book led to an increase in the research and study of the actions of humans on the environment. New groups such as Friends of the Earth and Greenpeace sprung up and examined emerging issues such as water contamination, air pollution, and oil spills.

The 1970s ushered in the rise of environmental legislation, most notably the creation of the first agency dedicated to oversee environmental protection: the EPA. Congress passed the Clean Air Act, the Water Pollution Control Act, the National Environmental Policy Act, and the Endangered Species Act. The 1970s saw the first Earth Day and the rise of protests aimed at environmental ills. A series of environmental disasters including Three Mile Island in 1979 and serious oil spills such as the Exxon Valdez in 1989 further cemented the risks of human damage to the environment. And, although numerous attempts were made to overturn many of the environmental regulations enacted in the 1970s, the courts have continued to rule that these regulations are permissible. Some began to dismiss the environmental movement in the 1980s and 1990s as a fringe, radical movement—with groups such as PETA, Earth First, and ELF at the center of media firestorms.

But then, much like the mythological Phoenix, the environmental, green, clean technology, or eco-movement resurfaced again to the forefront of consumer's minds. Some point to the impact of the "The Inconvenient Truth." Others point to gasoline and oil prices rising to levels not seen since the 1970s oil embargo, and still others point to natural disasters such as Hurricane Katrina. Whatever the exact reasoning, green has again come to the forefront of society and has led to a new series of opportunities for smart, savvy business people and entrepreneurs to build businesses to capitalize on the "new" environmental movement.

SO IS THIS *REALLY* A CLEAN TECHNOLOGY?

"Going green" is not without its fair share of critics. Those critics note that our choices to "go green" may only be smoke and mirrors. Are the changes we are making really having an impact on our environment? Even the most ardent supporters of the green movement will agree that in some cases those critics make a fair point.

Let's take a simple example: *Cars.* How can we make cars better for our environment?

How about we try for the electric car? Governments around the globe have pumped gobs of money into the development of an all electric-powered car—which will take gasoline-powered vehicles off the road. Great progress ... removing thousands of tons of carbon dioxide and other harmful compounds from the atmosphere produced by those gasoline-burning cars. The problem lies in the fact that the electricity is being generated by dirty utilities using coal or natural gas. So you've just removed gasoline and replaced it with more coal power.

What about the hybrid vehicle—gas-electric combination vehicles? Critics are quick to point out that these vehicles are simply increasing mileage somewhat and not actually making a material difference to emissions. And better yet, what good comes from making your SUV a hybrid—now you've just upped the efficiency from 9 miles per gallon up to 14 miles per gallon?

Okay, let's go for biofuels—keep the same car, but just use a fuel from plants? Ethanol produced from corn was all the rage in 2006 and 2007 ... yet some scientists were quick to note that the actual environmental costs (the outputs) from corn-based ethanol was actually worse than using oil.

You've gotta ask the question ... are we really going green or are we just trading in one problem for another?

Obviously, this is an extreme example here, but certainly there may be cases when our desire to be greener does not synch up with our ability to be greener. The hope is through new technologies and additional research that businesses can address these critics—but be aware that they do exist and oftentimes do have a legitimate concern. "Going green" isn't an absolute—electric cars, hybrid vehicles, and biofuels may just be more efficient choices, but not perfectly efficient choices.

In the end, "going green" is a process—it starts with products that are better than the current alternative and leads to products that will be *much* better than current alternatives.

What Is Clean Tech, Green Business, Eco-Tech, EnviroTech or Just Plain Green?

Terminology in the world of "green" is not something easy to wrap your head around. Joel Makower, noted author and founder of the Web site Greenbiz.com, says of the green movement, "if you've spent any time tracking the green marketplace, there's a reasonable chance that you've emerged with your head spinning." Words like sustainable, triple bottom line, green, clean, environmental, eco-friendly, compostable, recyclable, renewable, natural, organic (did I say "green"?), and dozens more can all be applied to the concept of green businesses. And right or wrong, all of these words can accurately (or worse, inaccurately) describe aspects of an environmentally conscious business.

In order to start or build your green business, the first step is to decide what the heck a green business really is. That question is, unfortunately, a difficult one to define. As we'll discuss a bit later, the lack of any uniform definition around "green," "green business," "sustainability," and those numerous other words and phrases referring to this green movement offers some challenges for green entrepreneurs.

For the purposes of this book, we will use the term "green business" to describe the broad category of businesses that highlight sustainable, green, clean, or other similar attributes.

We'll use the term "clean technology" to broadly describe the various environmental, green, and other similar technologies.

In general, green businesses are just like any other business in that they must create sufficient profits to continue to operate. The difference lies in what else green businesses concern themselves with—weighing the value of sustainability and human capital, for instance. For the purpose of this book, we offer our own definition of a green business:

> A green business requires a balanced commitment to profitability, sustainability and humanity.

And while I happen to like the book's definition (unbiased as that may be), it is by no means the only one (nor perhaps even the best definition). Here are a few of the other interesting definitions for green businesses and clean technology:

- *The Green Times* uses the following definition: "Green is being concerned with and supporting environmentalism and tending to preserve environmental quality."

- Croston (2008, 2009), the author of *75 Green Businesses* and *Starting Green* offers the following definition: "Green Businesses have more sustainable business practices than competitors, benefiting natural systems and helping people live well today and tomorrow while making money and contributing to the economy."

- In his book, *Build a Green Small Business*, Cooney (2009) defines a green business by four criteria: (1) it incorporates principles of sustainability into each of its business decisions; (2) it supplies environmentally friendly products or services that replaces demand for nongreen products and/or services; (3) it is greener than traditional competition; and/or (4) it has made an enduring commitment to environmental principles in its business operations.

- *Clean Edge* describes clean technology as "a diverse range of products, services, and processes that harness renewable materials and energy sources, dramatically reduce the use of natural resources, and cut or eliminate emissions and wastes."

- Pernick and Wilder (2007) in their book *The Clean Tech Revolution* describe clean technology as "any product, service, or process that delivers value using limited or zero nonrenewable resources and/or creates significantly less waste than conventional offerings." Pernick and Wilder highlight eight major clean technology sectors: solar power, wind power, biofuels, green buildings, personal transportation, the smart grid, mobile applications, and water filtration.

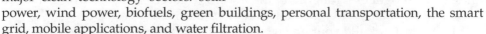

- According to the Cleantech Group, "Cleantech represents a diverse range of products, services, and processes, all intended to: (1) provide superior performance at lower costs, while (2) greatly reducing or eliminating negative ecological impact, at the same time as; and (3) improving the productive and responsible use of natural resources."

The Cleantech Group has gone one step further and classified green businesses into 11 categories, and a series of subcategories:

1. Energy generation	Wind, solar, hydro/marine, biofuels, geothermal, and others
2. Energy storage	Fuel cells, advanced batteries, and hybrid systems
3. Energy infrastructure	Management and transmission
4. Energy efficiency	Lighting, buildings, glass, and others
5. Transportation	Vehicles, logistics, structures, and fuels
6. Water and wastewater	Water treatment, water conservation, and wastewater treatment
7. Air and environment	Cleanup/safety, emissions control, monitoring/compliance and trading and offsets
8. Materials	Nano, bio, chemical, and others
9. Manufacturing/industrial	Advanced packaging, monitoring and control, and smart production
10. Agriculture	Natural pesticides, land management, and aquaculture
11. Recycling and waste	Recycling and waste treatment

DOES THE LACK OF "GREEN" STANDARDS HURT GREEN BUSINESS?

What is obvious from the variance and the number of different definitions of a green business is that there isn't a clear picture of what truly is green or clean. Makower (2009) suggests that the lack of "green" standards has hurt the broader movement. Said Makower in his book, *Strategies for the Green Economy*:

> The lack of a uniform standard, or set of standards, defining environmentally responsible companies means that anyone can make green claims, regardless of whether their actions are substantive, comprehensive, or even true. Want to put solar panels on the roof of your toxics-spewing chemical company? You can be a green business!

While some industries, products and sectors have instituted standards (LEED, organic-certified, ISO 14001, eco-label, and others) and the Federal Trade Commission (FTC) has begun to regulate eco-claims, green labeling remains a work in progress. Without a clear set of standards for green companies and products, consumers may not be able to differentiate between green and "greenwashing." One effort to develop a more comprehensive system is the Sustainable Business Achievement Rating system (SBAR). Others point to the Walmart Sustainability Index (discussed in Chapter 18) as a potential retail labeling schema. But without a recognized system in place, green businesses need to be aware of the risks and challenges in going green.

You can read more in Chapter 18 on emerging green certifications.

The "Green" Horizon

The "green" revolution represents one of the most exciting opportunities of the twenty-first century. Governments across the globe have developed incentive programs, directed research dollars, and emphasized the creation of "green collar" jobs. Investors have already poured billions of dollars into companies poised to capture the substantial opportunities of clean technologies. And people just like you are considering the possibility of building a green business.

Chapter 3 of this book will highlight 15 drivers of the green movement—drivers that have led us to our current focus and will lead the movement forward. Few experts predict that green will remain the word and phrase of the next decade, but who could have predicted the current state of play. Instead, much of the next wave of the green movement will likely be built on business fundamentals—growing businesses that are not only "green" but also represent a more efficient and better technology option.

The future of green business will move from early adopters of green technologies to mass adoption. Governments will continue to play a major role as the green movement impacts sectors from energy to transportation to water to waste recovery.

WHAT GREEN BUSINESS IS *NOT* …

In technology, it is easy to try and compare other sectors to clean technology. However, for all the similarities to other spaces and sectors, it also has substantial differences:

- *It is* not *a single sector.* It is easy to lump all things "green" into a category for reporting and study purposes. However, in the long run, expect this sector to be divided into categories such as energy generation and biofuels, or perhaps even subsectors such as solar and wind (in energy generation) and biodiesel and ethanol (in biofuels).
- *It is* not *biotechnology.* Biotechnology is built on the concept of development of a drug or other pharmaceutical that makes it through a long clinical study process and then can be sold at scale with large margins. While clean technology shares a longer time horizon, the comparisons tend to lose steam after that. However, some sectors (bioplastics or "green" chemistry) are more in line with the biotechnology sector.
- *It is* not *the Internet.* The Internet growth began with large investment dollars to capture new users (or eyeballs). With each new user, the return would increase. Clean technology will involve a different adopter (utilities, larger market-players, government) and will be judged on the ability to capture sales and dollars from those adopters. Again, some sectors (smart grid) would tend to align more with the Internet sector than the bulk of clean technology companies.

2

Becoming a Greentrepreneur (a.k.a. Green Entrepreneur)

Green entrepreneurs (greentrepreneurs) have become one of the fastest-growing groups of entrepreneurs. These entrepreneurs share a similar focus on products, services, technologies, and opportunities that have the duel benefit of providing economic value and sustainability. Is being a greentrepreneur really different from being other entrepreneurs? In some sense no, as any entrepreneur is recognizing an opportunity and building a business designed to fill some unmet need. But in another sense, greentrepreneurs represent a new type of entrepreneurs tackling challenges unlike others.

As a dedicated sector of entrepreneurship and startup businesses, green business is a relatively young field. Sure, there have always been green businesses out there, but most often they were more closely associated with another sector or industry. (Historically, for instance, green building practices were just part of the larger building sector, and it wasn't until recently that green building is its own sector and even industry.) Today, green business represents a $200 billion market with thousands of companies being built as dedicated green businesses. At the same time, many established businesses from Walmart and General Electric to Toyota and British Petroleum have undertaken substantial efforts to be seen as the ones with green practices.

Putting the *Green* in Greentrepreneur

As you make the decision to begin or expand a green business, it is important to note what motivates you and what has influenced your personal journey into the green business sector. Understanding the influences and motivation can help you focus on building a type of business that will fit your goals.

So just who are these greentrepreneurs? Research in the United Kingdom by David Taylor and Liz Walley examined the different types of greentrepreneurs by segmenting the entrepreneurs into categories based on the entrepreneur's personal motives and the influences for starting the business. By looking at why the entrepreneur was "inspired" or "motivated" to start a green business and overlaying it with the influences the entrepreneur

What type of greentrepreneur are you?

Innovative opportunist
Visionary champion
Ad hoc greentrepreneur
Ethical maverick

had in selecting the business type will give you a category of greentrepreneurs.

The entrepreneur's motivation would range from pure economic opportunity (whereby the green nature is merely the best opportunity rather than a motive itself) to a simple desire to help or change the environment (whereby the entrepreneur could well be starting an environmental nonprofit). The entrepreneur's influences will tend to come from a mixture of places, but may be based more on hard or soft sources. For example, hard or structural influences would be things like environmental regulations, a rise in funding for green businesses, increases in green consumers, growing market opportunities, and greater influence in the mass media of green themes. On the other hand, soft or socio-cultural influences are things like the prior personal experiences of the entrepreneur, his family and friends, his education, the organizations he belongs to, and the people in his personal network. These tend to create almost a passion for environmental or social causes, whereas the hard influences are much more data and factually driven.

Based on this research, Taylor and Walley identified four main types of greentrepreneurs: innovative opportunist, visionary champion, ethical maverick, and ad hoc greentrepreneurs.

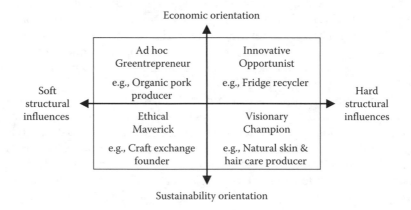

The first two types of greentrepreneurs are motivated by hard structural influences such as market regulation or the growth of certain green consumers. The *innovative opportunist* is an entrepreneur that spots a business opportunity in the marketplace that happens to be green. The researchers give the example of an entrepreneur that developed a recycling operation for refrigerators after the EU passed strict legislation on recovery of CFCs. Other examples include companies creating wind farms, new energy-efficient lighting, and hybrid and alternative fuel vehicles—all of which are likely motivated by new market opportunities from regulations, consumers, or similar market opportunities. The *visionary champion* shares the innovative opportunist's identification of a market opportunity, but likely has set out to change the world and founded a business on broader sustainable principles. The researchers point to Anita Roddick of the Body Shop and the founders of Ben & Jerry's Ice Cream.

The last two types of greentrepreneurs tend to build green businesses based on more on soft influences in sectors that may not have the benefit of government regulation, subsidies, or a mainstream market "pull." Instead, these businesses could well be in areas that could be described as "niches" and perhaps succeed because rather than in spite of that fact. The *ad hoc greentrepreneur* is referred to as an "accidental green entrepreneur," someone motivated by finance and not values. The researchers use the example of an organic pork producer that has a family motivated by sustainability, but this producer views this as an opportunity that just happens to be green rather than seek out a niche in the green space and is not a market that has been driven by government subsidies or environmental regulations. The *ethical maverick* is more likely to be influenced by friends, family, and his social circle to develop a sustainable business. The researchers highlight the "On the

8th Day" vegetarian café and health food shop setup in the height of the hippie era in the 1970s that set out to be ethical, environmental and support the friends and family who shared those values.

About Green Business Entrepreneurs

So what really makes a greentrepreneur different from other entrepreneurs? There truly is no single type of green business entrepreneur. In fact, greentrepreneurs may be defined by their diversity as much as anything else. Unlike software or biotechnology entrepreneurs who are likely to have "cut their teeth" on perhaps one of a couple dozen companies, greentrepreneurs hail from a broad range of backgrounds and businesses. Greentrepreneurs come from companies such as GE, 3M, Honeywell, Ford, General Motors, BASF, DuPont, Bank of America, and countless others. And greentrepreneurs hail from a diverse range of industries, including utilities, lighting companies, chemical companies, oil businesses, automobile suppliers, and many others.

This diversity has also led new crops of first-time entrepreneurs into green business, many starting their first company or joining new startups from large, multinational businesses. While this infusion of new talent is a welcome one, it had also led some experienced investors to suggest that greentrepreneurs are more "raw" and may lack some of the business and startup savvy of the other traditional entrepreneurial sectors. These facts may encourage these new greentrepreneurs to acquire more training or guidance from successful entrepreneurs in other sectors.

Likewise, greentrepreneurs hail from cities and states across the country. In sectors such as software, computer hardware, biotechnology, and medical devices, there are very concentrated populations of entrepreneurs in areas such as Silicon Valley, New England, Southern California, the Mid Atlantic, and the Pacific Northwest. Greentrepreneurs are springing up in locations such as the Corn Belt (for biofuels and bioplastics), the steel belt (for transportation-related technologies), the desert regions (for solar energy), and Texas and the Great Plains (for wind energy). You may be just as likely to find an entrepreneur with a green business located near a national laboratory or the headquarters of a major industrial company as you would in the typical hotspots for startups.

GREENTREPRENEURS: WHO ARE THEY?

Scott Cooney interviewed nearly 200 green entrepreneurs to see if he could identify any key traits or backgrounds that led them to undertake the creation of a green business. While he found all types of individuals tackled the challenges, perhaps the unifying feature he identified was a passion for environmentalism and building a sustainable future.

- *New to entrepreneurship.* Majority had been involved in a green business for less than 4 years; average was for 2.2 years.
- *Mostly men.* More men than women were green entrepreneurs, by a split of 60% men and 40% women.
- *At various points in their life/career.* The age range of green entrepreneurs was between 21 and 72—without a clear "normal" or median age.

- *Highly educated.* Most (over 90%) had a college degree, and many (over 40%) had a Masters or other advanced degrees.
- *Lean to the left (politically).* On a scale of Far Left, Left, Center, Right, and Far Right, most (>85%) ecopreneurs were either Left or Far Left. None reported being Far Right or Right.

From: Cooney, S. 2009. *Build a Green Small Business: Profitable Ways to Become an Ecopreneur.* New York, NY: McGraw-Hill.

About the Green Business Sector

What makes the green business sector different from some of the traditional sectors for entrepreneurs such as software, biotechnology, computer software, and medical devices? First things first; it is a bit of an oversimplification to label green business or clean technology as a single sector or industry. An entrepreneur in green chemistry or biofuels is very different from an entrepreneur creating wind farms or electric vehicles. Greentrepreneurs are more likely to fit with both a "traditional" business sector and the green business sector. So, while the green chemistry entrepreneur would likely associate with entrepreneurs and business owners developing chemicals and the wind farm developer with other entrepreneurs engaged in project finance, both would do so with other green business owners as well.

Still, the green business sector is characterized by certain general themes. Much of the green business sector is characterized by substantial regulation and government involvement. As you'll note in Chapter 14, there are hundreds of laws and regulations implicated in the various clean technology businesses, many more than in the traditional technology businesses. In addition, billions of dollars are being poured into green businesses by the government, from tax credits to grants to government contracts. Many believe that the ultimate success of the adoption of clean technologies will come from the government's willingness to regulate and fund many of the technologies that still are years away from successful commercialization.

The green business sector also has a heavy emphasis on energy. This includes businesses focusing on generating electricity from renewable resources and on energy efficiency, as well as those involved in the smart grid, alternative fuels and transportation, green plastics, and countless others. In fact, many of these must work with the utilities in order to succeed.

Green businesses are also characterized by large-scale projects. There certainly are smaller businesses and ideas in the green space, but more than in other sectors, the green business sector involves building large renewable energy generation plants, building alternative vehicle-manufacturing facilities, creating biofuels refineries and numerous other large-scale project businesses. For this reason, many businesses will ultimately require project finance to fund and develop those projects for mass adoption.

SOLAR PANELS: FROM THE GROUND TO YOUR ROOF

What does it take to create a solar panel and get it installed on your roof (or your business' roof)? Numerous companies are tackling one of the key sectors of the solar production value chain—and experts note that opportunities exist for a major player or players to tackle each aspect of the solar value chain.

- *Silicon production.* Raw silicon is processed into refined silicon of 99.9999% purity (estimated 11–15 participants according to a research report by Q-Cells).
- *Ingot/wafer production.* Purified silicon is melted and poured into molds to form ingots, which are then cut and sectioned into wafers (16–25 participants).
- *Solar cell production.* Silicon wafers are further processed to remove any damage from the cutting process in the wafer-creation and to refine the surface, followed by a series of treatments which allow for the generation of the electric field and increase in the absorption of the cells (30–40 participants).
- *Module production.* The solar cells are then combined into a module by attaching each of the cells to a frame and protecting the unit with glass or other covering. In addition, some modules have tracking systems or other reflective coatings applied to increase efficiencies (400+ participants).
- *Module installation.* Much of the local installation and mounting process has been outsourced to local companies and individuals working on local sites. However, there are now some companies that are exploring opportunities to provide local leasing or other pricing that allows for a sharing of the costs and benefits of the solar modules (numerous participants).

Putting the *Entrepreneur* in Greentrepreneur

While greentrepreneurs do have their share of differences from other business sector entrepreneurs, there may be more commonalities among all entrepreneurs in that they all share the same common goal: to build a profitable business. That means the entrepreneur must take on numerous roles outside their comfort zone. And nearly every entrepreneur will share a similar rollercoaster ride along the startup journey. Intuit asked more than a thousand small business owners what they felt was the most important characteristic to be a successful entrepreneur. The respondents identified the following traits as the top three most important:

- Hard work (37%)
- Visionary (19%)
- Good people skills (18%)

Traits of Successful Entrepreneurs

So what do most entrepreneurs share in common? Researchers such as George Solomon found that successful entrepreneurs tend to be focused on action over introspection, are inventive and innovative, and operate best when they are in charge. Norris Kreuger found that entrepreneurs were actually highly influenced by their environment, suggesting that entrepreneurial characteristics can be learned and heightened through education and experience. Alan Jacobowitz interviewed a large number of entrepreneurs in his research and concluded that nearly all entrepreneurs share certain personality traits including restlessness, sense of independence, a certain level of isolation, innovation and action orientation, and a higher level of self-confidence than their peers.

Below is a list of other skills that have been identified in successful founders and leaders of startup businesses:

- Instills a vision (and reminds others when they forget)
- Goal-oriented to keep the business on track and on target
- Practical and focused on solving problems efficiently rather than being "right"
- Leads by example and isn't afraid to answer the phones, make the coffee, or call customers
- Effective brainstormer who gathers information to develop the best solutions
- A good listener who really *hears* what others say
- Strategic thinker focused on making decisions for tomorrow
- Decisive when the time for discussion has ended
- Acts with integrity to maintain the reputation of the organization
- Good communicator who knows when a mass message won't do or when information has to be provided
- Genuine curiosity about all aspects of the business and all people involved
- Risk-taker who knows that a startup won't succeed by choosing to always play it safe
- Stubborn when the business needs stability and support
- Resilient when negative news hits or criticism arises
- Responsible for the actions of the company and isn't afraid to admit fault when needed
- Has an objective view of the company and isn't afraid to actively examine the business from an outsider's perspective

A very common question by first-time entrepreneurs is whether an entrepreneurial personality can be taught or an entrepreneur is just born with the right combination of inherent traits. The experts have mixed opinion and say that certain personality traits tend to be found in entrepreneurs, but that many other traits are learned or taught. What this information ultimately means is up to you to figure out—but it certainly seems to suggest that there are certain skills found in most entrepreneurs and, whether one is born with them or learn them, a prospective entrepreneur is bound to be more successful with these skills. So if you are looking to leave school directly into a green startup or following a successful career working for a large corporation, the government, or in academia, focus on building the set of skills likely to be required for a new business founder.

ARE YOU CUT OUT FOR ENTREPRENEURSHIP?

StrengthsFinder 2.0: A New and Upgraded Edition of the Online Test from Gallup's Now, Discover Your Strengths by Tom Rath.

This book is based on the popular personality assessment test from Gallup used by individuals to identify their strengths and weaknesses, and find careers that most closely match these talents. Check out the book or take the assessment online at www.strengthsfinder.com.

What Color Is Your Parachute? A Practical Manual for Job-Hunters and Career-Changers by Richard Nelson Bolles.

If you are still wondering whether starting your own business is for you, consider this book. It identifies two types of job hunts: a traditional job hunt and a life-changing job hunt. If you are considering starting your own business, check out the book or visit www.JobHuntersBible.com for more information.

Challenges of Building a Green Business

Starting a business is a risky endeavor, likely much riskier than a job at a large multinational corporation. And green businesses are no more of a "sure thing" than other businesses. In fact, some may argue that since the sector is built on political sentiment, changing regulations, and numerous first-time entrepreneurs the risks could be greater. That said, experts continue to point to the extensive green opportunities and predict growth well into the future.

But like any new business, green businesses are not without their share of risk and potential for business failure. According to a report by the U.S. Bureau of Labor Statistics, approximately one-third of new businesses fail within two years and more than half go out of business within four years. For a new entrepreneur, starting a new business brings with it unique challenges.

The U.S. Small Business Administration released a report studying the reasons small firms had been forced to declare bankruptcy. The reasons for filing were broken down into the following categories: (1) outside business conditions (mentioned by 39% of filers), (2) financing problems (28%), (3) inside business conditions (27%), (4) tax-related reasons (20%), (5) dispute with a particular creditor (19%), (6) personal problems (17%), and (7) calamities (10%). Outside business conditions included such factors as new competition, increases in rent, insurance costs, or declining real estate values. Inside business conditions included a bad location, inability to manage people, loss of major clients, or inability to collect accounts receivable. Personal problems often included divorce and health problems. One-third of the bankrupt businesses had less than $100,000 in debts, and 79% had less than $500,000 in debts. Mean assets were $841,000 and median assets $94,700.

And even businesses that don't go out of business have their share of challenges. In his research Roberts (1991) asked entrepreneurs to identify the primary problem for their business in its first six months of operations. Over 50% of respondents listed sales as the primary or secondary problem. The next two highest categories included personnel issues and personality conflicts. Entrepreneur Magazine and PricewaterhouseCoopers in their

"2006 Entrepreneurial Challenges Survey" asked CEOs of privately held, high-technology businesses about their most pressing challenges that they are focusing on in their businesses:

- Retaining key employees (74%)
- Developing new products/services (47%)
- Creating business alliances (27%)
- Expansion to markets outside the United States (21%)
- Finding new financing (15%)

And while these challenges may seem daunting and the number of businesses going out of business within four years (50%) may seem overwhelming, entrepreneurship still continues to grow. With the challenges comes the potential for rewards including independence, financial gain, personal and professional satisfaction, and others.

THE TOP REASONS WHY SMALL BUSINESSES FAIL

For every Google, Microsoft, Amgen and Home Depot, there are likely hundreds of failed businesses that were not able to make it. Understanding why businesses fail can help you anticipate where problems are likely to occur and address them before they become catastrophic to your business.

The following are reasons why business owners listed that their business had failed, which oftentimes included a number of factors for the failure, and the percentage of business owners that listed those factors as a reason for the failure.

General business factors
- Lack of a well-developed business plan, including insufficient research on the business before starting it (78%).
- Being overly optimistic about achievable sales, money required, and what needs to be done to be successful (73%).
- Not recognizing, or ignoring, what they don't do well and not seeking help from those who do (70%).
- Insufficient relevant and applicable business experience (63%).

Financial factors
- Poor cash flow management skills/poor understanding of cash flow (82%).
- Starting out with too little money (79%).
- Not pricing properly—failure to include all necessary items when setting prices (77%).

Marketing factors
- Minimizing the importance of promoting the business properly (64%).
- Not understanding who your competitor is or ignoring competition (55%).
- Too much focus and reliance on one customer/client (47%).

Human resource factors

- Inability to delegate properly—micromanaging work given to others or over-delegating and abdicating important management responsibilities (58%).
- Hiring the wrong people—clones of themselves and not people with complementary skills, or hiring friends and relatives (56%).

Source: Jessie Hagen of U.S. Bank cited on the SCORE—Counselors to America's Small Business Web site.

3

Drivers of the Green Revolution

Most successful entrepreneurs will tell you that success in their business was due to countless reasons—the people, the idea, the market, the partners, the competition, and just plain hard work. And so it goes in green businesses: there is no magic formula or approach that will make your business a success.

That said, one of the key things nearly everyone agrees on is that you need to identify the right opportunity for your business. This is described in numerous ways—the right "market" or the right "approach," or the right timing in the marketplace. We call it the opportunity—the place where you can fill a need that isn't being met elsewhere. But no matter how you describe it, the key is that you identify the opportunity out there for a need for your product or service by your end users.

For biotechnology, the opportunity has comes from the ability to heal people, extend life, and improve the quality of life. For personal computers, the opportunity has come from a tool that could lead to incredible gains in efficiency and for Internet companies, from a tool that could create a global marketplace where everyone from every corner of the globe could quickly participate in commerce. And companies that were able to identify their slice of that opportunity built successful businesses from Dell to Apple to Amgen.

And green is no different than these prior "boom" sectors. There are fundamental drivers that will push the green movement and accelerate clean technologies into businesses, homes, and countless new markets. For a green entrepreneur, the trick is to identify the opportunity and build a product or service that can capitalize on it.

This sounds a bit like gazing into a crystal ball, huh? Predict where the growth will occur, what the consumers will buy, and where the opportunities will be. And that's certainly true to an extent. However, this book doesn't intend to predict where the opportunity will lie for you. Instead, this chapter will look at what has been a driver of the green movement thus far and what is likely to drive the movement in the future. The green entrepreneur can use these opportunities to help as you gaze into your own crystal ball and decide where the opportunities lie.

This chapter lists 15 key drivers (or potential drivers) that have helped create opportunities in the clean technology sector and those that will continue to create opportunities going forward.

WHAT'S PUSHING CLEAN TECHNOLOGY?

Limited resources

- Growing population
- Decreasing profit margins
- International economic development
- Growing demand for commodities
- Increasing price and volatility of oil

Business trends

- Market liberalization
- Privatization
- Outsourcing
- Growing competition

Global drivers

- Energy security risks
- International political instability
- Pricing and markets for externalities (GHGs and Kyoto/Copenhagen)
- Climate change

Increasing demand

- Increased creation of corporate "green" initiatives
- Consumer preferences for faster, cheaper, lighter, cleaner products
- Demands from investors
- Need for infrastructure investment
- Demands for safe, reliable and clean energy, water, and air

Availability of capital

- Government funding programs
- Public markets supporting clean technology
- Growth in "green" entrepreneurs
- Venture capital funds focused on clean technology
- Active M&A market for clean technology

Stakeholder support

- Socially conscious investors
- Shareholder and shareholder advocacy group support
- Regulatory and environmental reforms
- Trade and industry organizations
- Environmental legislation

Driver #1: Green ($)

We started out the book with a quote by Ira Ehrenpreis that bears repeating. "Green is the New Green (as in Money)." Billions of dollars of private and public funds have flooded the clean technology marketplace, not only because green business makes sense for the long term sustainability of the planet, but because it represents a huge opportunity for wealth creation. Entrepreneurs have left sectors such as biotech, software, and hardware to look for new opportunities to build green businesses. Venture capitalists have repositioned their investments into the clean technology sector. Governments around the world hope to develop the next Silicon Valley of sun, wind, water, or fuel—and look to the growth of jobs, taxes, and opportunities those businesses can build.

Companies such as Walmart, BP, Google, and others aren't turning to clean technologies out of the goodness of their own hearts. Walmart sees the opportunity to reduce its costs by developing a more efficient and cost-effective supply chain. BP recognizes that while fossil fuels represent the bulk of its revenue stream, it knows that its knowledge of the energy sector positions it for the huge opportunities for new renewable energy. And Google understands that its cost structure is tied to energy, as some experts estimate that Google uses more electricity as a company than 140 countries of the world.

Look for dollars to continue to flow into the green business sector—driven by opportunities, consumer preferences and regulatory requirements. Yes, one of the key drivers of green business is and stands to be green—and the opportunity to create it.

Driver #2: The "Gore" Effect

It is without question that Al Gore has had substantial impact on the green movement. His film "An Inconvenient Truth" won numerous awards and was seen by millions around the globe. While some may question certain claims in the movie or may doubt the science Gore uses in his presentations, no one would question the importance of Vice President Gore on bringing awareness to the masses. In 2007, Al Gore was awarded the Nobel Peace Prize.

Gore has long been an advocate of enacting tougher environmental regulations and improving the U.S. position on clean technology. Today, former Vice President Gore has become the public face of the environmental movement. His efforts have been a key driver in regulatory reform, in consumers' exposure to the concept of climate change, and he continues to take an active role in the greening of media, investments, and politics. Awareness of the impact of climate change remains high. As the public persona of environmental activists around the globe, expect Gore to continue to find ways to drive adoption of clean technology.

Driver #3: Europe

Can you name the country with the greatest amount of solar energy production in the world? How about the country with the largest amount of wind production annually? (See Table 3.1.) Which country is the world's leading producer of tidal power? (The answers are Spain, Germany, and France, respectively.)

Where do greenhouse gases (GHGs) come from?

- Electricity (33%)
- Transportation (28%)
- Industry (20%)
- Agriculture (7%)
- Other (12%)

While the United States and China, the world's two largest greenhouse gas (GHG) emitters have yet to agree to binding reduction levels in GHG emissions, Europe has agreed to reduce the emission of carbon by 20% of 1990 levels by 2020. Europe stands to play a major role in the development of green business from the buying patterns of its consumers, spending of European governments, development of new technologies, and promotion of regulatory schemes that favor green

TABLE 3.1

Largest Wind Developers Worldwide

Developer	Country	Installed Capacity (MW)[a]
Iberdrola	Spain	7362
Acciona Windpower	Spain	5300
Florida Power and Light	United States	5077
Electricidad de Portugal	Portugal	3696
Babcock and Brown Wind	Australia	1859
Long Yuan Electric Power	China	1620
Eurus Energy Holding	Japan	1385
NRG Energy Inc.	United States	1300
EDF Energies Nouvelles	France	1218
Cielo Wind Power	United States	1148
Total[b]		29,965

Source: Data from BTM Consult.
[a] Estimated as of 2008.
[b] As a percentage of total installed capacity: 31.8%.

business. For businesses in the United States, China, and the rest of the world to tap into the huge markets of the EU, they will need to play by the rules of the EU.

Driver #4: Energy Security

Four of the top ten producers of oil (as of 2008) are Russia (#2), Iran (#4), China (#5), and Venezuela (#10)—nations that the U.S. government has publicly critiqued for their political and human rights policies. In addition, Saudi Arabia (#1), United Arab Emirates (#8), and Kuwait (#9) hail from politically charged regions of the world (Table 3.2).

TABLE 3.2

Top World Oil Producers (2008)

Rank	Country	Production[a]
1	Saudi Arabia	10,782
2	Russia	9790
3	United States	8514
4	Iran	4174
5	China	3973
6	Canada	3350
7	Mexico	3186
8	United Arab Emirates	3046
9	Kuwait	2741
10	Venezuela	2643

Source: Data from U.S. Energy Information Administration. Available at http://www.eia.doe.gov/country/ index.cfm
[a] In thousands of barrels per day.

Energy security has become a major driver of the emphasis on developing sustainable electricity generation and transportation alternatives. This has helped move the development of renewable energy technologies out of the "liberal" political base and into the hands of a bipartisan political base.

In addition, most experts will tell you that the energy prices in 2009 are unlikely to remain depressed against the 2007 and 2008 levels. Instead, the world is likely to see $5.00 per gallon gasoline and the price of natural gas continue to rise. With increasing prices coupled with a supply filled with political risks, the competitive landscape for renewable energy is likely to come into balance.

Driver #5: The First "Green" Bubble

As discussed in Chapter 1, much of the foundation of today's green revolution was laid in the 1970s. Legislation, government spending, consumer activism, and technology development all came out of this period of time. But unfortunately, initiatives like the electric car were killed (sorry); solar panels couldn't match costs with efficiency and consumer sentiment shifted away from the environment. Some experts point to this period as a bubble, whereby results never kept up with hype and enthusiasm.

Rewind to the beginning of the twenty-first century and the dawn of the second great green period. Have we learned our lessons or will we again kill the electric car such as the Tesla Roadster; will solar panels be too expensive and will consumers forgo the compact fluorescent light bulb (CFL) and the energy-efficient washer/dryer

Compact fluorescent light bulbs (CFLs)

In 2007, about 20% of the bulbs purchased in the United States were CFLs, representing a growth of over 300% in just 10 years.

set? It seems to be that we may just have learned our lessons from that first bubble. Tesla launched its Roadster as a high performance vehicle that just happens to be electric-powered. Solar panels are being installed on business rooftops with purchase power agreements (PPAs) in place, and CFLs are marketed as cost-efficient lighting (and selling faster than ever before).

These lessons from the first green bubble may not prevent some overenthusiasm and hype, but it has led businesses to examine fundamentals rather than buy into the green sentiment alone. The first green bubble also led to a great deal of technological development that has been refined and enhanced in the last 30 years—leading to a more efficient solar panel, better wind turbines, enhancements in wave energy, and improved recycling technologies. These enhancements to basic technologies have allowed for mass adoption of solar, wind, biofuels, and more technologies that simply were too early and too unproven in the 1970s and 1980s.

Driver #6: Health and Wellness

What is oftentimes overlooked in the green revolution is the connection between health and the environment. Many of today's eco-friendly businesses have developed products that offer health and wellness benefits—from chemical-free cleaning supplies and naturally dyed fabrics to organic foods and interior air pollution reduction. For instance, sales of organic foods are expected to grow 11% annually between 2007 and 2011 according to the Organic Trade Association's 2006 Manufacturers' Survey. The link between healthy living and products that are natural, nontoxic, nonemitting, and otherwise safer than those utilizing common chemicals will continue to drive the adoption of green products.

Driver #7: The "Dark Greens" and the LOHAS

As discussed later in this book, a segment of the population now exists that may have never existed before in our country's history—the Greenest Americans, who believe that their daily actions have a tangible and specific impact on the environment. Research suggests that these individuals make up as much as 9% of the U.S. population and may well be one of the only segments of society that consistently match their talk with their action when it comes to the purchase of green products. Consumer categories such as LOHAS (or Lifestyles Of Health And Sustainability) and other classes of green consumers make it possible for niche and not-so-niche products to have opportunities to gain initial adoption and expand into mainstream consumers.

LOHAS buying power

Estimated show that LOHAS consumers will spend between $200 and $300 billon annually.

One such example of a product initially launching targeting green consumers before expanding and going "mainstream" is the hybrid vehicle. First introduced by Toyota in 2001 with the launch of the Prius, the initial models emphasized the environmental benefits of the cars. Since the initial launch aimed primarily at the Greenest Americans, more American consumers have begun to purchase hybrid vehicles offered by various automakers, which now emphasize potential savings at the pump. More than one million Priuses have been sold worldwide to date.

Driver #8: Worldwide Stimulus Funds

Experts suggest that more than $400 billion of worldwide government stimulus funding has been allocated to clean technology and related green programs. While some of these funds may never be allocated to green projects or may only be tangentially related to green programs, the visible worldwide commitment has spurred private investors and companies. In particular, for large-scale projects including energy generation, energy efficiency, energy transmission, fuels, transportation, and green chemistry, the availability of financial capital around the world is likely to drive adoption of these new technologies. In addition, given the size and scope of these investments, it seems likely that more funds will be made available to fund green businesses and technologies in the event more is needed to avoid any political fallout from the failure of these investment dollars.

Driver #9: Venture Capital

Venture capital is a class of investment that looks for "home run" businesses—that means they are looking for a business that can take a small technology or concept, and scale into

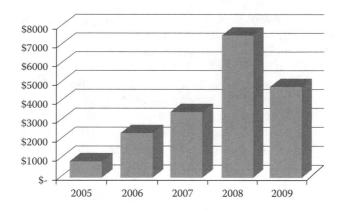

FIGURE 3.1
Clean technology venture capital investment (in billions). (Courtesy of Greentech Media.)

a market leader. Prime examples include Google, Yahoo!, Home Depot, Genetech, and countless others. And venture capitalists have set their sights on clean and green technology as an opportunity to identify those "home runs."

What is it about clean technology? Some venture capitalists point to the huge amount of dollars in the worldwide economy devoted to things like energy, transportation, buildings, and more (see Figure 3.1). Others point to the looming mandates and regulations coming from the government. Still others note that the old-world companies from utilities and oil companies to automakers and chemical companies are flatfooted when it comes to innovation. Whatever the reason, interest from venture capitalists grew clean technology to the largest investment class in 2009. This had spurred a number of entrepreneurs to move from the software, biotechnology, or other sectors and set their sights on clean technology. This flow of new money that is focused on high risk-high reward companies will continue to drive the clean technology sector forward.

Driver #10: The Internet and the PC

While some will look back at the technology growth of the late 1990s as nothing but a bubble, others will point to the slew of successful companies that grew to prominence in that period—Oracle, Intel, Sun Microsystems, Microsoft, Netscape, Dell, eBay, Amazon, Yahoo!, and countless others. These companies produced substantial wealth and returns for shareholders (some of which was lost in the popping of the bubble, but much remains). It also created larger-than-life personalities for the leaders of those companies, from Michael Dell to Bill Gates to Larry Ellison.

As recently as the 1980s, only 1% to 2% of graduating MBA students had the goal to become an enterpreneur at graduation, according to research by the University of California at Berkeley. Today, these numbers have jumped to between 10% and 20% of graduating MBAs, a tenfold increase over that period.

But why would the Internet and the PC be a driver of clean technology? The reason isn't the Internet or the PC themselves, but instead what they represent—the ability of an entrepreneur to build a business into a worldwide powerhouse in just a few short years. The 1990s and 2000s have incubated a rapid growth of entrepreneurship in the United States and around the world. Governments are now investing in fostering entrepreneurship (including tax credits, facilities, education, and more).

The growth of entrepreneurship (some of which can be traced back to the success stories of the Internet and PC entrepreneurs) has led more individuals to take a hard look at opportunities in green technology. As more MBAs and experienced executives look for the next Oracle or Microsoft, they are increasingly looking at clean technology as that opportunity. Look no further than the executives of many of the emerging clean technology businesses to see backgrounds from semiconductors, computer hardware, software, biotechnology, and medical devices that have made the transition into the world of green business.

Driver #11: Bush-nomics and Obama-nomics

The democrats and republicans may differ on many things, but both parties have recently been behind the green business movement. For example, former President George W. Bush famously spurred investment dollars into the biofuels sector during his 2007 State of the Union address. Likewise, President Obama campaigned on green jobs and a green economy, pushing for upwards of $80 billion of spending related to clean technology in the stimulus bill.

The political winds remain firmly behind the green movement. Despite differences on many fronts, politicians continue to see opportunities to promote green jobs, green innovation, green energy, and support of innovation. This bipartisan support will continue to drive innovation—and while it is quite uncertain as to the enactment of climate change legislation or a national energy policy, there continues to be support for many aspects of green.

Driver #12: Compact Fluorescent Light Bulbs

If you had to point to a single product that qualifies as the flagship of the green revolution from a consumer perspective, it would have to be the CFLs—with all due respect to the hybrid, the wind turbine, the solar panel, and the cloth grocery bag. In Chapter 4, we examine the story behind the growth of CFLs in the past 10 years, which some attribute to a refocused marketing effort (touting cost savings over environmental savings). Today a number of products can serve as the face of today's green revolution—perhaps none better positioned in homes and offices around the globe than the CFL.

Driver #13: Localization

Starbucks recognized a growing trend among consumers—an increasing preference for local brands and nonmass market products. Just what did the world's largest coffee retailer do about it? In 2009, the Seattle coffee company rolled out three pilot coffee shops that were meant to be the "corner coffee shops," without any Starbucks logos or products stocked on its shelves. And the trend isn't just in your local coffee shop. HSBC saw the growing backlash by some consumers against multinational banks and released a new marketing line—"the world's local bank." Frito-Lay is now attempting to localize its products to highlight the local markets where its products are manufactured and consumed.

An increased emphasis on localization comes as more consumers are beginning to recognize the environmental harms of shipping, transportation, and mass production. Localization will represent a substantial driver for smaller, customized, efficient, and targeted businesses—particularly those that can offer added environmental benefits by being local.

Driver #14: China

In a *New York Times* Op-Ed, Thomas Friedman summarized the role of China in the green business movement. Said Friedman, "C. H. Tung, the first Chinese-appointed chief executive of Hong Kong after the handover in 1997, offered me a three-sentence summary the other day of China's modern economic history: 'China was asleep during the Industrial Revolution. She was just waking during the Information Technology Revolution. She intends to participate fully in the Green Revolution.'" Not only will the influence of China be felt as a major player in the debate over global climate change, GHG emissions and the growing consumption of energy, China intends to play a major role as a developer of clean technologies, a consumer of green products and services, and a leader in the creation of green technologists and greentrepreneurs.

Driver #15: Carbon

While Kyoto, Copenhagen, and perhaps even Cancun did not create a global agreement on the reduction of carbon, it seems to be a matter of when the world will reach a consensus rather than if. In 2005, the Kyoto Protocol took full effect with 185 countries as signatories. In 2009, major principles were agreed to by key nations in connection with the Copenhagen meetings.

Regulation of carbon represents both a huge opportunity for green businesses, a major challenge of infrastructure for the management of carbon assets and credits, and a unique regulatory burden for the world to manage. Expect carbon to continue to be a major driver impacting green business now and into the future.

CARBON FOOTPRINTS, CONTENTS, AND EMBEDDED CARBON

As the world has become increasingly focused on CO_2 levels and emissions, new pushes have grown for product labeling with the amount of carbon produced when manufacturing or producing this product or service. A broad labeling scheme has not yet been accepted (the UK's Carbon Trust labeling scheme has developed a program, but has not yet gained mass adoption), but companies and regulators are trying to develop a system to define carbon content, carbon footprints, or the embedded carbon. Each of these concepts is basically a way to describe the aggregate amount of CO_2 emitted from every stage of its production and distribution, from the initial source to the final product purchased on a store shelf.

TAKING ADVANTAGE OF THE GREEN BUSINESS DRIVERS

EXAMPLES OF TECHNOLOGIES IN VARIOUS CLEAN TECHNOLOGY SEGMENTS

In each of the 11 segments of Clean Tech, the Cleantech Group has provided examples of technologies that are being developed or have been developed:

Air and environment

- Air purification and filtration products
- Multipollutant controls (e.g., sorbents)
- Catalytic converters
- Fuel additives to reduce toxic emissions
- Remediation
- Leak detection
- Pollution sensors and gas detectors

Agriculture

- Natural pesticides and herbicides (e.g., organic fungicides, beneficial insects, anti-microbial)
- Natural fertilizers (e.g., organic fertilizers)
- Farm efficiency technologies (e.g., sensors and monitoring of controlled insecticides and fertilizer use)
- Micro-irrigation systems (e.g., drip irrigation)
- Erosion control
- Crop yield improvements

Energy generation

- Renewable energy conversion (marine, tidal, solar, wind, biomass)
- Geothermal heat and electricity generation
- Waste to energy generation
- Cogeneration (combined heat and power units)
- Biofuel technologies (cellulosic fermentation, ethanol)
- Clean coal technologies
- Micro-power generators (e.g., vibrational energy)
- Electrotextiles

Energy infrastructure

- Power conservation
- Power quality monitoring and outage management
- Power monitoring and control

Energy efficiency

- Smart metering, sensors and control systems in applications
- Energy efficient applications (i.e., LED lighting)
- Chemical and electronic glass
- Energy efficient building materials (i.e., windows, insulation)
- Smart and efficient heating, ventilation and air conditioning systems (HVAC)
- Building automation and smart controls
- Automated energy conservation networks

Manufacturing and industrial

- Chemical management services
- Sensors for industrial controls and automation
- Advanced packaging (e.g., packing and containers)
- Precision manufacturing instruments and fault detectors
- Process intensification

Materials

- Green chemistry
- Advanced and composite materials (e.g., electrochromatic glass, thermoelectric materials)
- Bio-materials (e.g., biopolymers, catalysts)
- Nano-materials with clean technology applications (e.g., nanopowders, adhesives, gels, coatings, additives)
- Thermal regulating fibers and fabrics
- Environmentally friendly solvents

Transportation

- Different modes of transport (e.g., electric and battery vehicles, hybrid vehicles)
- Efficient engines

- Integrated electronic systems for the management of distributed power
- Demand response and energy management software
- Advanced metering and sensors for power (e.g., using active RFID networks, WiFi, mesh networks)

Energy storage

- Fuel cells for stationary and mobile storage
- Micro-fuel cells
- Advanced rechargeable batteries (NiMH, Li-ion, zinc air, thin-film, enzyme catalyzed)
- Heat storage
- Flywheels
- Super and ultra capacitors

Recycling and waste treatment

- Recycling technologies
- Waste exchanges and resource recovery
- Biomimetic technology for advanced metals separation and extraction
- Waste destruction

- Hybrid drive technologies
- Lightweight structures for vehicles
- Car-sharing tools
- Temperature pressure sensors to improve transportation efficiency
- Logistics management software for RFIDs
- Fleet tracking
- Traffic control and planning technology

Water and wastewater

- High purity water
- Desalination
- Filtration and purification
- Contaminate detection and monitoring
- Control systems and metering for water use
- Advanced sensors for water pollutants
- Separation of water into use-types
- Wastewater recycling and reuse
- Biological and mechanical wastewater treatment

4

Markets for Green Products

Type in the phrase "green marketing" into the Google search bar and it will give you over 126 million hits. Hundreds of companies now offer all varieties of green marketing services, information, support, and assistance. But green marketing is not a new phenomenon. In fact, some people have suggested that the overinflated green marketing claims of the 1970s and 1980s have today created skeptics out of many consumers for green claims, making the sales and marketing process for today's green companies all the more difficult.

Even so, today there are more consumers buying green products, more companies selling green products, and more marketing going on touting claims of health, safety, eco-friendliness, and happiness, all from going green. This chapter is designed to educate you on today's consumers and purchasers of green products.

Interpreting Green Sentiment

Eighty-two percent of Americans said they are buying green, according to a 2009 study by Green Seal and EnvioMedia Social Marketing. *82%, that's right!* While the number itself sounds amazing to anyone thinking about a green business, unfortunately the realities don't jive with those numbers. Green cleaning supplies make up a fraction of the overall cleaning supply market; organic food represents only 3–4% of the overall food market; and sales of hybrid vehicles represents less than 1% of global automobile sales in 2008. So, unless consumers are mistaken, green buying patterns are being drastically overestimated. We may think we are buying green or say we are buying green, but green products continue to be a niche buying decision.

Joel Makower (2009) dedicates an entire chapter in his book *Strategies for the Green Economy* to a central question of this chapter "Do Consumers Really Care?" Seems to be a fairly important question, right? For many individuals new to the green field, the response may be "Of course they care—aren't their dozens of studies out there where consumers have told us that they think about the environment when they purchase things and are even willing to pay a little more for things to help out Mother Earth?" The reality is that "caring" about the environment may not be enough to change buying habits and make consumers take action.

There have been hundreds of market research and consumer studies that have come out with fairly consistent messages: consumers say that they care about the environment. Makower (2009) lays out a number of studies that have highlighted findings showing that consumers are ready to "buy green." Here are just a few he points to:

- The Michael Peters Group put out a 1989 study that found that nearly nine in ten U.S. consumers were concerned about the environmental impacts of their purchases.

- In a study by GfK Roper Consulting, 79% of U.S. consumers say that a company's environmental practices influence what they recommend to friends and four in ten are willing to pay extra for a product that is perceived to be better for the environment.

- An Accenture study found that 64% of consumers are willing to pay a higher price—a premium of 11% on average—for products that produce lower greenhouse gas emissions.

Even Makower (2009) admits the research is pretty compelling, and may even sound compelling enough for a company to consider chasing these new, green markets. However, before you rush out and start pitching your green wares, there is more research data that seems to conflict with these findings:

- Seven in ten Americans and 64% of Canadians say that when companies call a product green or better for the environment, "it is usually just a marketing tactic," according to a study by Ipsos Reid.

- Seventy-five percent of consumers in the United States and the United Kingdom, although concerned about how their consumption affects climate change, feel paralyzed to act beyond small changes around the home, in a study by AccountAbility.

- Sixty-four percent of U.S. consumers can't name a single "green" brand, including 51% of those who consider themselves to be environmentally conscious, in a study by Landor Associates.

All these findings seem to conflict one another, with consumers saying they are concerned and willing to act to help the environment, but at the same time unaware of how to actually make changes to do so. Josh Dorfman, author of *The Lazy Environmentalist* suggests that the disparities can be attributed to convenience. Said Dorfman, "Green products are rarely on the shelves where consumers shop. People do not want to go out of their way or be hassled just to do the right thing for the planet."

Makower (2009) suggests two takeaways for these conflicting messages: "first, consumers are looking for ways to be more responsible in their lives, and they look to companies for solutions about what to do" and "second, being greener is not enough." The net result is that companies cannot be merely green to get a consumer's attention. Instead, a consumer is looking for a better product that has a byproduct of being greener and better for the environment.

Ultimately, the answer to Makower's initial question in this section is yes, consumers do seem to care about the environment. However, the challenge lies in marketing the green products as being better than the competing products, with the *added* benefit of helping the environment. Says Makower, "But there's a gulf between green *concern* and green *consumerism*. Bridging that gulf requires a deeper understanding of consumers' interests and motivations and the barriers that keep them from 'shopping their talk.'"

STRATEGIES FOR THE GREEN ECONOMY

For more information on green marketing, consumer preferences, marketing strategies, and strategies for green products, read *Strategies for the Green Economy* by Joel Makower and visit: http://www.greenbiz.com.

Understanding Green Consumers

While the prior section of this chapter detailed much of the differences between consumer's concern for the environment and making it part of their buying habits, it remains important to understand more about consumers and, in particular, individuals identified as green consumers.

While the general sentiment among consumers may overestimate the general ability to bank on most consumers to open their pocketbook to purchase green products, there is growing evidence that a committed group of consumers do exist who will support green businesses. Growing numbers of consumers shop at stores like Whole Foods, are avid recyclers, drive hybrid vehicles, will pay extra on their power bill to get "green" electricity and have replaced their light bulbs with energy-efficient bulbs. It is hard to estimate the number of these consumers, but it is safe to say they are still not the mainstream of America. One of the groups of consumers oftentimes associated with the proverbial "green consumer" is LOHAS, an acronym for consumers with *L*ifestyles *O*f *H*ealth *A*nd *S*ustainability. The Worldwatch Institute found that as many as 30% of Americans would fall into the LOHAS category and spent $200–$300 billion annually.

While it may seem the easiest to try to group green consumers by other typical marketing demographics such as age, geography, income, or ethnicity, other research has concluded that the green consumer may not easily fit within those groups. Seireeni and Fields (2008) in their book, *The Gort Cloud*, indicate that green is actually a psychographic defined by attitudes and values rather than age, gender, or ethnicity. This brings strange bedfellows together into the green consumer movement and may also make marketing to green consumers much more difficult.

For more information and details on market research tools, see

"Market Research" on page 271.

Another study by Cara Pike called the Ecological Roadmap found that there is a core group of Americans today that strongly hold the value of ecological concern—23% of Americans. Within that group, the Greenest Americans represent 9% of the population and closely tie their daily actions to the environment (Table 4.1). The groups that would together hold the value of ecological concern are the Greenest Americans, Postmodern Idealists, and Compassionate Caretakers—each have different primary reasons for supporting the environment and green living, but they all share a common value. The three groups in the middle as far as placing a value on ecological concern make up 44% of the U.S. population—the Proud Traditionalists, the Driven Independents, and the Murky Middles. And finally, 33% of the population believes day-to-day priorities and realities trump any environmental leanings for the Antiauthoritarian Materialists, the Cruel Worlders, the Borderline Fatalists, and the Ungreens.

TABLE 4.1

Ten Environmental Worldviews

Segment	U.S. Population (%)	Worldview in Brief
Greenest Americans	9	Everything is connected, and our daily actions have an impact on the environment.
Postmodern Idealists	3	Green lifestyles are part of a new way of being.
Compassionate Caretakers	24	Healthy families need a health environment.
Proud Traditionalists	20	Religion and morality dictate actions in a world where humans are superior to nature.
Driven Independents	7	Protecting the earth is fine as long as it doesn't get in the way of success.
Murky Middles	17	Indifferent to most everything, including the environment.
Antiauthoritarian Materialists	7	Little can be done to protect the environment, so why not get a piece of the pie.
Cruel Worlders	6	Resentment and isolation leave no room for environmental concerns.
Borderline Fatalists	5	Getting material and status needs met on a daily basis trumps worries about the planet.
Ungreens	3	Environmental degradation and pollution are inevitable in maintaining America's prosperity.

Source: Adapted from Makower, J., *Strategies for the Green Economy*. New York, NY, McGraw Hill, 2009, (Appendix: The Ecological Roadmap by Cara Pike).

Getting Green Consumers to Buy

While identifying these groups of green consumers is one thing, the next question is what determines whether or not individuals who value ecological concern actually spend their green on green. Why is one person in the LOHAS group or a particular Compassionate Caretaker more likely to buy green products than another individual? Aceti Associates identified the factors that were positively linked to buying green:

- *Perceived consumer effectiveness.* The more people believe that the efforts of individuals can make a difference in the solution to environmental problems, the greater their likelihood of buying green.

- *Perceived knowledge.* The more people believe they know about a "green" attribute (such as benefits of organics, the impact of buying goods from recycled content, or the impact on health of chemical-free cleaning supplies), the more likely are they to buy green.

- *Environmental concern.* Perhaps not surprisingly, the more consumers feel strongly about the environment, the more is the likelihood of their purchasing green products.

- *Feelings about environmental consumerism.* The more consumers tied the environment directly to purchasing decisions, the more likely were they to buy green.

At the same time, Aceti Associates identified the factors that were barriers to consumers buying green products:

- *Perceptions of inferior product quality.* Environmental benefits or advertising touting environmental claims have a diminished impact when the consumer perceives the replacement product is inferior in product performance.

- *Skepticism about "green" marketing claims.* Most consumers believe that environmental claims are overexaggerated and are unlikely to buy a product solely based on environmental claims without other corroborating evidence.

- *Difficulty in identifying green products.* If locating information on environmental benefits is difficult, it adversely affected consumer's desire to purchase a green product. In the absence of labeling of products, knowledge about environmental benefits of green products was relatively low even among environmentally concerned consumers (making the need for labeling more important).

- *Price sensitivity.* While survey data suggest that consumers are willing to purchase products that cost more and have environmental benefits, the data on actual willingness in the buying process show that consumers are not willing to pay more in most cases for environmental benefits.

The takeaway from this research suggests that yes, there are certainly viable markets of consumers who are concerned about the environment and willing to make buying decisions to reflect those concerns. However, it is important not to overestimate the numbers of those consumers or their innate willingness to buy green based solely or primarily on their concerns for the environment. In addition, green businesses should be aware that there are indeed barriers that can reduce a green consumer's willingness to buy a green product including a lack of knowledge, perceptions of inferiority, or skepticism of marketing claims.

WHAT YAHOO! KNOWS ABOUT GREEN CONSUMERS

Yahoo! has begun to look at the users of its site through a green lens to determine what green consumers search, click, and read. Erin Carlson, director of Yahoo's social responsibility department, discussed some of the key findings:

1. *Consumers don't want doom and gloom.* They want to hear about optimistic innovations—to hear about what's possible. For example, a story on an air-powered car proved a powerful draw.
2. *There's a lot of skepticism about celebrities' green endorsements.* "People want to know if there's been a backroom deal signed to promote that star's image," Carlson said. "Green and celebrities are not necessarily a good match." Imagery of real people making a difference is much more effective.
3. *Consumers love surprises.* Some of 2007s biggest clickthroughs? An article about a woman who lives in an 84-square-foot house, and a feature on the Pope adding environmental degradation to list of sins. "People want to be able to drop these tidbits at the next cocktail party," said Carlson.
4. *What's in it for me?* Consumers are interested in new gadgets that save money and products that offer health benefits.
5. *There's a shift from awareness to action.* Top-searched environmental term in 2006? "Climate change." Top-searched term in 2007? "Recycling."

Carlson's advice to online marketers? Piggyback green promos on traditionally high-interest categories (remember all those consumers that are curious about green products and services, but only if they're served up on a platter). For example, Yahoo's holiday gift guide last year featured a green product alongside the latest hot gadgets and toys—and traffic on those products went through the roof.

Source: http://ecopreneurist.com/2008/06/29/five-ways-to-attract-green-customers-from-yahoo-green/

Learning More about Green Markets and Consumer Attitudes

While there seems to be a number of mixed messages about green consumers and markets for green products, there has been a great deal of research in the past 40 years attempting to identify information about this growing market segment. As a result, there are numerous places where you can find out more about markets for green products, targeted customers, trade events for green products, and more.

A few of the resources available for researching green markets and green consumers include the following:

- Green business: http://www.greenbiz.org/
- Environmental e-market express: http://www.buyusa.gov/eme/enviro.html
- Talk the walk: http://www.talkthewalk.net/
- Eco labels: http://www.eco-labels.org/

CASE STUDY

INCANDESCENT LIGHT BULBS

The electricity used to provide lighting to homes and businesses consumes 22% of the electricity produced in the United States. Today, many consumers have made the switch from traditional incandescent light bulbs to the energy efficient compact florescent lights (CFLs). In fact, in 2007 about 20% of the bulbs purchased in the United States were CFLs, representing a growth of over 300% in just 10 years.

The story of CFLs is a unique case study in the green marketing phenomenon. CFLs were actually first developed in the mid-1970s in response to the energy crisis of 1973. The bulbs replaced the traditional bulbs and offered substantial savings in terms of energy consumption—so much so that the Earth Day Network stated on its Web site, "If every household replaced its most commonly used incandescent light bulbs with CFLs, electricity use for lighting could be cut in half ... and would lower our annual carbon dioxide emissions by almost 125 billion pounds."

The bulbs were launched in 1978, but received relatively low support among consumers. Philips branded its CFL as the "Earth Light" and touted the environmental benefits. Even as the bulbs came down in price and quality improved, consumer adoption was fairly slow.

What changed from the launch of the CFLs in the United States in the late 1970s until 2000s when sales exploded? Philips believes some of the change resulted in its approach to marketing the CFLs. In its research, Philips found that while consumers were generally sympathetic to environmental issues, it was actually the fourth or fifth most important reason for the purchase of the bulbs. The real benefit consumers would buy the bulbs for was the desire for the bulbs to last longer.

So by de-emphasizing the importance of the environmental benefits as more of a byproduct of the increased life of the bulbs, Philips and the other manufacturers of CFLs were able to encourage consumers to make the switch. And while the change in the marketing approach (coupled with a rising tide of environmentalism and the ease of the switch to CFLs) had led to dramatic growth in the purchase of CFLs, today some 95% of U.S. households lack a single CFL—meaning there is still a long way to go for mainstream consumer adoption.

Source: Adapted from Makower, J., *Strategies for the Green Economy*, New York, NY, McGraw-Hill, 2009.

Part II

The Green Startup

5

Green Ideas, Inventions, and Businesses

Think the twenty-first century was the start of "green ideas"? Think again. Today's increased emphasis on "going green" represents a recycling of many initiatives that in fact had their roots in the 1960s and 1970s—what many describe as the first real concerted social movement toward environmentalism. That first wave of environmentalism ushered in new regulations and also new spending on technologies for electricity generation from renewable sources, energy efficiency, alternative fuels for automobiles, and many ideas that have become en vogue again in the early twenty-first century. But, this century may represent for the first time that "green ideas" have been so successfully turned into businesses which can be self-sustaining around the globe.

So what does this mean for an aspiring green entrepreneur? For one thing, it says that you may not have been the first one to come up with your idea, concept, or initiative. For another, it says that a great deal of research and development dollars have already been invested by the government, universities, and other institutions that may serve as a source of these ideas, refinements, and technologies themselves. And, perhaps most importantly, it tells you that it takes more than just a "green" angle to make your business a success. There is no shortage of opportunities for today's greentrepreneur, but it still remains a "business-building" process before it can be a "world-saving" process.

For a new green entrepreneur, one of the first steps in the business-building process is identifying the opportunity for your business. That skill is known as opportunity recognition: the process whereby an entrepreneur identifies the need in the marketplace for a new technology, service, or product.

THE TOP 10 CLEAN TECHNOLOGY INVENTIONS

FROM *TIME MAGAZINE'S* 50 BEST INVENTIONS OF 2009

- *L Prize LED* by Philips Electronics (rank in the top 50, 3)—More energy-efficient bulb
- *Dashboard* by Energy Hub (4)—Dashboard for monitoring and adjusting home energy usage
- *Personal Carbon Footprinting* by Princeton University (12)—Research into personal carbon emission reporting
- *Powerhouse Solar Shingle* by Dow Chemical (13)—Roof shingle incorporating thin-film solar cells
- *Electric Bicycle* by YikeBike (15)—Foldable electric bike capable of reaching speeds up to 12 miles per hour
- *Hydroponic Vertical Farming* by Valcent (16)—High-density hydroponic farming system

- *Planetary Skin* by Cisco Systems and NASA (17)—Integrated sensor system aiding government and businesses in preventing and adapting to climate change
- *LEAF Electric Sedan* by Nissan, Tesla Motors, and Coda Automotive (25)—All-electric family vehicle
- *Packing Algorithm* by Johannes Schneider of the University of Mainz (37)—Algorithm to aid shipping companies in greater packing efficiency
- *Formula 3 Race Car* by University of Warwick (40)—Formula 3 race car that runs on a mix of chocolate and vegetable oil, converts ozone emissions into oxygen, and includes components made with carrot fibers, potato starch, and cashew shells

Recognizing Your "Green" Opportunity

You may already have that killer idea or you may be thinking about a series of ideas. Perhaps you are ready to start something, but aren't quite sure yet. So, what should that process of finding that green opportunity be like? Successful entrepreneurs usually describe the decision to found a business as a process, involving observations, numerous brainstorming sessions, detailed conversations, extensive research, and contemplation. There has been research into the process of opportunity recognition where two researchers from the University of Chicago and San Francisco State University (C. M. Gaglio and Richard Taub) explained the typical opportunity recognition process for entrepreneurs in four distinct steps:

- Prerecognition stew
- The Eureka! experience
- Further development of the idea
- The decision to proceed

For you, the actual business opportunity could have struck you quickly in a moment of clarity or could have been something that has developed over time. Some entrepreneurs are actively searching for a new business idea, while others aren't formally looking, but are open to possibilities. The economic uncertainties of 2009 and beyond have led a new crop of new entrepreneurs to consider building a green business—which has led some to actively examine opportunities in the green sector. No matter how you arrive at your idea or ideas, once the entrepreneur finds an interesting idea (or two ... or six) worth exploring further, one of the most important steps is the further development of the idea and the decision whether or not to proceed with this idea into a business concept.

What Comes First: The Business Idea or a Decision to Start a Business?

Entrepreneurs will often identify three different processes for coming up with their business ideas. In the first case, an entrepreneur will identify the opportunity and then will later decide to form their business. In the second case, an entrepreneur will decide they would like to start a business and begin to identify opportunities after that point. And finally, some entrepreneurs note that the idea or opportunity and the desire to start the business come simultaneously.

Researchers Gerald E. Hills and Robert P. Singh asked nearly 500 entrepreneurs to identify the process for recognizing their business opportunity. The research showed that entrepreneurs are quite mixed in how they identify their ideas:

- Business idea or opportunity came first—36.9%
- Desire to start business came first—42.1%
- Idea or opportunity and desire to have a business came at the same time—21.0%

Does it matter which path is chosen? Research by Teach, Schwartz, and Tarpley into idea generation by software firms suggests that the ideas "accidentally" discovered rather than those ideas identified through a formal screening process tend to achieve break-even sales faster. And, this result isn't entirely unexpected in the high-technology industry where rapid market entry is necessary. However, there are numerous examples of successful companies that had been formed following each of the paths discussed above.

For entrepreneurs who undertook a screening process for an appropriate idea (deciding to start the business first, then looking for the best idea), Hills and Singh asked the entrepreneurs the number of ideas they considered before selecting one. Over 65% of the respondents said they considered less than four ideas before settling on one.

- One—27.8%
- Two—18.4%
- Three—19.1%
- Four—10.0%
- Five—7.1%
- Six to nine—10.7%
- Ten to nineteen—4.4%
- Twenty to thirty-nine—0.7%
- More than forty—1.8%

Where Do Business Ideas Come from?

Business ideas tend to come from either a formal search or review process, or a "Eureka" moment. In either of those scenarios, where will the specific idea for a new business come from? Most surveys note that industry or market experience and discussions with individuals in your social or professional network are the primary places where ideas are generated.

Hills and Singh asked the entrepreneurs what had led to their business idea. They found that social and professional network contacts were one of the most important factors in idea identification, with 62% of the respondents noting that their idea had come from discussions

with business colleagues, friends, or family. Nearly 56% of the respondents pointed to their experience as an important factor in the development of their business ideas.

It developed from another idea I was considering	23.1%
My experience in a particular industry or market	55.9%
Thinking about solving a particular problem	22.9%
Discussions with my friends and family	42.4%
Discussions with potential or existing customers	30.9%
Discussions with existing suppliers or distributors	15.9%
Discussions with potential or existing investors/lenders	8.1%
Knowledge or expertise with technology	28.6%
Other	6.9%

Additional research by Chuck Eesley of the MIT Sloan School of Management looked at the sources of product or service ideas for alumni of MIT. According to Eesley, over 60% of the ideas for these products and services came from working in the industry.

While in school	14.4%
Discussions with social or professional acquaintances	13.0%
Research conference	0.7%
Working in the industry	61.9%
Working in the military	3.1%
Other	6.9%

The First Idea Might Not Be the Best Idea

Successful entrepreneurs note that the early stages of developing their business concepts are a struggle between obstinacy and flexibility. On one hand, the entrepreneur wants to stick to his or her vision *in spite of* the criticism and doubts. But, on the other hand, the entrepreneur recognizes the importance of adapting the initial idea as research, experience, and understanding increases.

Hills and Singh asked the entrepreneurs in their study about the change in their original business idea since its beginning. Over 50% of the entrepreneurs said that their idea is about the same, while the other half said the idea or opportunity has changed a "little" to a "great deal."

- Idea/opportunity has changed a great deal—13.2%
- Idea/opportunity has changed a little—36.4%
- Idea/opportunity is about the same—50.4%

According to the results of a survey by Launch Pad, a marketing consulting company, 44% of the technology companies they surveyed said the company had made a significant change in their business model in the prior year—changes ranging from a new product line or target market to pricing or sales model. The ultimate lesson for entrepreneurs is to recognize that there is a balance in play between obstinacy and flexibility. Recognizing when the idea, focus, model, or strategy needs to be changed, in any amount, is an important skill. However, the flipside of the discussion is that an entrepreneur also must recognize when to stick to their vision even in the face of outside pressures or scrutiny.

COMMON PROBLEMS WITH NEW BUSINESS IDEAS

Where do ideas go wrong? Oftentimes ideas go wrong when entrepreneurs start with their technology and try to sell it to consumers, rather than the other way around. The questions isn't: How can I get people to buy my invention? Instead, the question should be: What can I invent that people will buy?

Here are a few of the most common problems for entrepreneurs deciding what idea is best:

- *Marginal niche product.* You can make a successful small business by tackling a small niche problem. But, if you are looking to build a startup that will require outside investors (probably in the form of venture capital), you can't try and carve out a small niche just to avoid any competitors. In some sense, competition is good. It means that the end customer is actually willing to pay money for something. Outside investors rarely will fund a business that doesn't have a potential market of at least $500 million (this means if you had 100% of the sales in the market each year, your company would have sales of $500 million). It is okay to tackle a marginal niche market, but realize that it is more difficult to raise outside funds if that is the market you are set to pursue.
- *No specific user/customer.* This is the holy grail of problems with startup business ideas: not having a customer in mind before you build the product. Think like a sales person and consider who you would actually sell the product to. Before you go out and build any products, interview your likely customers and see if they would buy it and how much they would pay. In some cases, you've got a separate user and customer: Sure, it's the consumer who buys the product off the shelf, but you also need to sell the grocery store purchasing manager who would need to purchase the product and stock the shelves.
- *Derivative idea.* Many "new ideas" are actually just improvements or enhancements to existing products. This approach can work, but remember that being "just as good" or "just a bit better" is a difficult product to sell.
- *Willingness to pay for the product (or pay enough for your product).* Some ideas may seem to make sense and be an improvement over what is available, but if you can't make people pay money for the product or pay enough money for the product, then you may not have a viable business concept. It doesn't mean it is a bad idea or a bad product, but it may not qualify as a viable business concept to start a business around.

Finding "Green" Ideas

As previously discussed, many of today's successful clean technology companies have been enhancing, improving, or recycling some business and technology initiatives from the 1960s, 1970s, and 1980s. Solar power was the "future" in the 1970s, but could never quite compete effectively on price and was never able to gain significant traction (and even in the past several years with the "rebirth," solar has still had a small impact on overall energy generation). Recycling technology has seen a rebirth as more municipalities and cities enact recycling programs. And while GM may have "killed the electric car," we have seen a number of new electric and alternative fuel vehicles being developed.

GREEN BUSINESS IDEAS

There are literally thousands of businesses being formed with a green focus every year. Ideas range from the futuristic (high altitude wind, nuclear fission, and zero-energy urban transportation systems) to the realistic (eco-laundry services, green training centers, and solar installers) to the successful (utility-scale solar, wind farms, and hybrid vehicles) to hundreds more.

We've compiled below dozens of the new green business ideas the entrepreneurs are tackling—with many hundreds more out there:

- Drip irrigation systems
- Green on-line product marketplace
- Eco-landscaping
- Carbon dioxide capture
- Biodegradable plastic
- Waste/trash to energy
- Integrating sensors into agriculture
- Data center cooling
- Clean coal
- Advanced lighting systems
- Green cleaning supplies
- Green business grant writer
- Carbon offset exchange
- Green remodeler
- Green laundromat
- Electric car charging stations
- Sustainable fishery
- Eco-friendly pest control
- Home heating/cooling systems
- Water & electricity consumption tracking
- Green college and MBA programs
- Green chemistry
- Social financing

- Advanced thermostat systems
- Solar water heating
- Cool roofing installation
- Green building home supply store
- Green cement
- Modular home building
- Solar panel installer
- Green car dealership
- Biodiesel producer
- Ecotourism
- Organic farming
- Organic restaurateur
- Green builder
- Environmental accountant
- Green office advisor
- Recycling operator
- ISO 14001 consultant
- Wind turbine technician
- Heat pump installer
- Solar panel servicer
- Water purifying systems
- Alternative transportation and shipping services
- Home eco-gardening supplies

What a greentrepreneur should also realize is that there are various opportunities within each idea, field, industry, or sector. If you aren't ready to develop your own Advanced Lighting System from LEDs, you can create a reseller business and partner with several of the producers, be a low-cost supplier of LED parts, or become a repair technician for LED systems. If you don't have the resources to develop a carbon capture system, you can become a consultant and assist businesses that are looking for that support or develop tracking software for carbon saved.

For more ideas and examples, check out Chapter 3 on the Drivers of the Green Revolution.

So where do you find the next great green idea for your business? As discussed in the previous section, most entrepreneurs find their idea in the industry they are or have worked in. And, therefore, some solar companies have tapped utility executives, biofuels companies have been started by oil executives and some lighting companies have grown from semiconductor backgrounds or traditional lighting manufacturers.

RESOURCES TO FIND MORE GREEN BUSINESS IDEAS

75 Green Businesses You Can Start to Make Money and Make a Difference by Glenn Croston, Entrepreneur Press, 2008.

Build a Green Small Business by Scott Cooney, McGraw-Hill, 2009.

How Green Is Green Enough?

One of the questions greentrepreneurs struggle with is whether their green business is truly green enough. Perhaps your green business is an ISO 14001 Consulting Business which requires you to fly all over the country at the cost of releasing harmful greenhouse gases. Or what if your business is installing solar panels, but the construction of the panels utilizes harmful chemicals and metals? Building a "green" business may actually be shades of "gray."

Scott Cooney offers an example of a green cleaning business where three different cleaning businesses help understand the various levels of green in a business:

1. Business which offers home and office cleaning using only nontoxic products
2. Business which offers home and office cleaning using only homemade cleansers, which are packaged into renewable containers and spray bottles
3. Business offers home and office cleaning using only homemade cleansers, which are packaged into renewable containers and spray bottles, and limits its service within an area that can be reached on bicycle or by foot

While many individuals would consider the third example as the "most green" green business, the vast majority of average consumers would treat all three of these as viable green businesses. Determining if your business is truly green or truly green enough is a difficult determination. Most importantly, remember that being a green business may actually be a process where you continue to be greener and more efficient over time.

Development of Your Green Business Concept

Once a green entrepreneur identifies their business idea, the next stage of opportunity recognition involves further development and research into the business concept. Generally, there are several steps in moving an initial idea into a business concept. You may have heard of the "back of a napkin" as how the initial idea was first written down. The business concept usually involves taking that initial sketch of the idea and moving it into a business concept that can make money.

The development of your business concept typically involves deciding on two key things

Is this an idea that we can make money with?

AND

Am I interested in forming my own business to do it?

Once you've reached the idea stage, your work isn't done. Now you need to decide how to build a business based on the resources at hand. Do you have the right technical people, the necessary capital, and the right opportunity? This is the development of your business concept.

For example, think that the idea for your business is an online marketplace for refurbished solar panels. Developing that initial idea into a real business takes researching the market for competitors, suppliers, and pricing information. Using that information, you may quickly find that there isn't anyone providing that solution on the east coast in the United States, but there is a potential competitor offering this in Europe. Use that information to identify pricing, partners, and other key business concepts. Perhaps the CEO of this European company has made some statements about expansion plans into Asia, but not North America. Then, using some pricing information, you can develop a high-level model for expenses, sales, profits, and timing. With those tools in hand, you may have your business concept summary and be ready to start the details of business-building.

It may seem like an obvious step, but spending the time and energy researching the opportunity, the market, the capital requirements, and whether you are a good fit for this new business is crucial to make the decision whether to push the business forward.

To Business Plan or Not To Business Plan

Nearly everyone in the startup world (entrepreneurs, successful former entrepreneur, unsuccessful former entrepreneurs, investors, advisors, accountants, attorneys, writers, etc.) will have an opinion about preparing a business plan. Some people say business plans are a must while others think they stifle the flow of an early stage business. Some people say you should just focus on preparing an executive summary business plan or business concept spreadsheets, while others believe you should prepare full three to five year projections into a full business plan. Some will point to research that businesses which create business plans tend to be more successful, while others will hold out an example of a success story business that never had a business plan.

For more information on business planning, see

Basics of a Business Plan on page 274.
Template Executive Summary Business Plan on page 279.
Business Plan Software on page 280.

What is certain is that everyone has an opinion and they are just that—opinions. A greentrepreneur will need to determine what the business needs and develop that resource, be it a series of spreadsheets only or a 100 page business plan. To help you develop your business concept and potentially your business plan, we've included a number of resources in Part V of this book.

At a minimum, at the point when a greentrepreneur transitions from an idea into laying out the arguments for creating a business, you should begin to do some initial market research and set out your business concept summary. This is where you begin to analyze some of the projections, understand what competitors are out there, and determine what are the key milestones for the business to get a product developed and a business launched. Whether you decide to create some spreadsheets and models, conduct a series of interviews with potential customers, or begin getting quotes from manufacturing facilities on your product—all of these efforts should be aimed to determine how your business will take shape in the weeks and months ahead.

An Idea versus an Invention

As you've begun to flush out your idea and think about creating a business, you've likely also thought about the concept of intellectual property—those protections that can make sure your business has a limited monopoly on your technology or invention. You've got a great idea for a product that may be a major boon to the environment—everyone you've told about the idea really thinks you've found something with lots of potential. So, should you go out and get it patented?

For more information on intellectual property and startup businesses, see

- Green Intellectual Property on page 103.
- Strategies for Managing Startup Intellectual Property on page 365.

This is a fundamental question that many new entrepreneurs have when they have their business idea in hand and are ready to start their business. Unfortunately, some entrepreneurs fail to realize that an idea isn't an invention. And, ideas can't be patented; only inventions can be patented. So before you rush off to try and patent your new business idea, remember that you can't patent an idea—you'll have to "reduce it to practice," or build and design an invention based on that idea.

Deciding on a Startup or a Small Business

No two businesses are alike, but there are certain themes that apply to different categories of businesses. And when starting a green business, it is important to contemplate the future of your business as a small business or a startup business—which we'll discuss generally have different traits and objectives. (If you think these are the same things, don't worry, you aren't alone! However, there are important differences for the purpose of this book.)

What exactly is a small business or a startup you ask? Well, a few examples of **small green businesses** would be things like operating as a green consultant for managing office energy usage, developing an eco-friendly landscaping business, or operating as a solar or wind installer. These green small businesses tend to be more local, smaller in size (both revenues and employees), and less dependent on external financing or developing new technologies. On the other hand, a few examples of **green startups** would include developing green software to manage office energy usage sold to large companies, developing a line of products to be sold to eco-friendly landscaping businesses, or a company designing and selling a unique solar/wind device for use on rooftops. These green startups tend to require more capital, aim to employ a larger team, and usually have a more national reach and aggressive technology development plan. The differences largely come down to size, scope, reach, and focus. And in each case, the needs of the business will differ.

The terms "startup" and "small business" are sometimes used interchangeably, but in practice there is a difference between the two. According to the Career Action Center, "a start-up is a small company, most often with a high-tech focus, that is in the early stages of development, creating a product or service, or having a product or service needing manufacturing and/or marketing. They are looking to grow through possible venture capital funding, initial public offerings (IPOs) or acquisition by larger companies." On the other hand, a small business tends to have a more narrow focus and is not limited to high-tech applications. A small business oftentimes will not have growth goals similar to a startup and will not have plans to grow larger than its small business size.

Startup	Small Business
• Wealth generation	• Income substitution
• Importance of technology and proprietary I	• Broader range of businesses
• Broad markets	• Narrow markets (geographic or target audience)
• Goals of $10 million to $100 million in annual sales	• Goals of $500,000 to $10 million in annual sales
• Seeks angel and venture capital funding	• Relies on bootstrapping and bank loans
• Staff of 50 or more	• Staff of 20 or less

In the book *Engineering Your Startup*, James Swanson and Michael Baird separate the businesses into two categories: (i) income substitution businesses and (ii) wealth building businesses. Generally, a small business will fall in the category of income substitution, while wealth building entities are more likely to be startups. According to Swanson and Baird, "people who simply do not want to work for someone else can easily create a small income substitution business such as a one-man computer repair service shop." On the other hand, if you wanted to build a business that would be a national franchise or desired to create software monitoring energy usage, then this model would be a wealth building business.

These smaller, income substitution businesses are becoming more common and a viable option for individuals who are leaving corporate America for the independence of being self-employed. With the power of the Internet, the ability to quickly setup Web sites to promote and sell goods, and keep overhead costs low, many small businesses are being headquartered out of home offices—sometimes called "homepreneurs." In fact, today more than half of all U.S. businesses are based at home with an estimated 6.6 million home-based enterprises providing at least half of their owners' household income. These home-based businesses employ an average of two people, including the owners. According to a *Business Week* article on homepreneurs, home-based businesses tend to have fairly modest revenues with only about 35% of these businesses with revenues above $125,000. However, one of the most substantial financial benefits for home-based entrepreneurs is lower operating costs. According to the Small Business Administration, home businesses, on average, had lower sales and net profits than companies in commercial spaces, but profitable homepreneurs retained a greater share of their total receipts as net income: 36% vs. 21% for non-home-based businesses.

One of the first questions to ask yourself when determining which green business you are building is the ultimate size of the business. Do you envision your business as a small office or home-based business limited primarily to your own efforts and perhaps a couple of employees down the road? In some cases, the goal may be to build a small organization of 1–10 people that you hope would generate a hundred thousand dollars in profits and up to a million dollars in sales annually. On the other extreme, you may desire to create an organization with a national brand and operations across the country. In each case, the decision may be driven more by the type of business you want to build rather than your desire to operate a business of a particular size.

Does it matter what type of business you want to build? Yes, it does in fact. If you intend to build a small businesses with less than 20 people, most likely your concerns will be different than a green company intending to build its product or service into the market leader for a market or submarket. Small businesses tend to have different financing needs and usually will have a more narrow market, either geographically or in scope. Some portions of this book will focus on the subset of small businesses that intend to grow to a size where annual revenues exceed $20 million—which we will refer to as a startup. However, the most portions of this book are applicable across startups and small businesses.

The First Few Months Ahead …

Now that you've settled on your business idea, started thinking about the business concept and maybe even started mapping out your business plan, now you need to think about the first few months of actually building and running your new company. What can a greentrepreneur expect in those first few months and where should you focus your energies?

The first months of your new green business are oftentimes quite unique in the company's life. For many companies, the first six months to a year are run by a single entrepreneur or a small founder team. In his book *Entrepreneurs in High Technology*, Edward Roberts (1991) interviewed entrepreneurs to help identify issues that face a company in its first days. What he found is that entrepreneurs are forced to be "jack-of-all-trades" in the first six months of their venture, spending time on activities that may not be within the entrepreneur's comfort zone (see Figure 5.1).

Not surprisingly for the tech-focused startups in the Roberts' study, engineering-related activities required the most effort, with finance and administration requiring the smallest percentage of effort. Don't forget that sales and marketing efforts are still a key for businesses, even before their product is built and launched. The rest of Part II of this book will examine many of the key challenges faced by a new business in its initial forming and founding period.

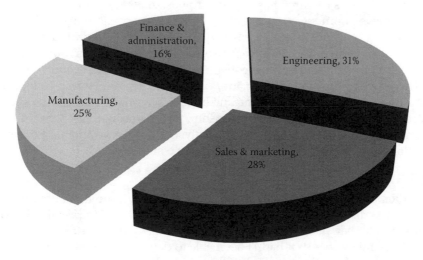

FIGURE 5.1
Founder time spent on various activities (initial six months).

ABOUT THIS CASE STUDY

In Part II of this book, we've included a case study that will take our fictional cofounders Teddy and Rebecca through many of the key decisions any new green business will face. Each chapter will examine another of their business challenges. Their kinetic energy capture technology may just be science fiction (although research is actually ongoing into kinetic energy capture), but the business challenges and decisions are real. While their specific business path may not match with every other green business, the decision-making process will be similar regardless of the end choices the business owner makes, including developing the idea and business concept, formation, stock issuance, hiring employees, selecting directors, raising money, protecting their intellectual property, marketing, and selling products.

GreenStartup, Inc.: A Case Study on Developing the Business Concept

The following is a case study detailing how to identify a business idea and turn it into a business concept for a green business. Here is a fictional case study of KinetiBaby.

Theodore Muir (who goes by Teddy) is an engineer's engineer, having focused his undergraduate, his masters and his PhD on electrical engineering—more specifically with his graduate research at University of North Dakota in the areas of kinetic energy capture. While Teddy's research has largely emphasized ways to increase efficiencies in existing engines, equipment, and machinery, he has also been looking at the developing field of capturing other forms of kinetic energy in the environment. Teddy is on track to finish his PhD and lead a life as a college professor in a prestigious university.

Rebecca Carson attended the same college as Teddy and they became friends during their freshman year studying introductory physics in the engineering college. Rebecca received a double-major in engineering and business, with her business studies largely in the areas of finance and entrepreneurship. After graduating, Rebecca stayed in California and has worked as a financial analyst, focusing on the marketing department of a large multinational consumer products company.

THE INSPIRATION

Teddy and Rebecca had been involved in their college environmental organization and shared countless discussions during and after college on ways to impact the world to help the environment. They had briefly worked on a business idea in college to try and sell solar panels to universities, but hadn't been able to develop a real market for their product. However, they knew they worked well together despite being a bit too early in their first business endeavor. Since graduation, they'd had a number of phone calls, discussion, and brain-storming talks about green business ideas. For the most part, the ideas either weren't unique enough, weren't large enough, or the business partners didn't feel they had the right background to pull it off. Yet, they both kept working and kept thinking.

One day, Teddy forwarded Rebecca an article from a conference he'd recently attended, detailing some efforts to create "smart clothing" that would contain sensor technology to help the wearer adjust to any environment. The sensors would be embedded in the clothing and help repel water, pull sweat from the body, generate heat, or countless other applications. In his email, Teddy had simply added the phrase, "Wouldn't this be interesting if clothing could capture our energy every day?"

It wasn't until a few weeks later that Rebecca finally connected the dots on how Teddy could be proven right. While visiting her sister who had two toddler-aged children, Rebecca realized that young children expended tons of energy—more than any adult she'd ever seen in their day-to-day life. After a half a day with her niece and nephew, Rebecca had been absolutely exhausted from chasing the kids around, pulling misbehavers off furniture and just generally playing "Auntie." And that's when it struck her—why not build smart clothing for kids that would harness all the kinetic energy of toddlers. If we could capture that kinetic energy, store it and use it to solve the energy crisis, then we may have something, she thought. Sure, kinetic energy was a bit difficult to capture and store, but Teddy was already thinking about that. Plus, parents were more than happy to dress their children up in silly looking outfits. By simply attaching a few electronic devices and batteries to a toddler's outfit for a full day of playing, Rebecca surmised you could power a household of electric devices.

THE BUSINESS CONCEPT

"Teddy. Listen, I think I've got something interesting," said Rebecca into her cell phone as she drove to the airport. Rebecca relayed her idea and waited for Teddy to start laughing at her.

"It's interesting," he replied. "I've heard of some researchers looking into this for some runners and in some other applications, but never for kids. Plus, there's this new area out there of making smart clothing too, so combining this could be really unique. It makes some sense actually. Plus, I've built some kinetic energy capture devices in my garage in the past that might be able to work as our starting point too."

"So you don't think this is crazy Ted?"

"Oh, it's totally crazy," he replied. "And it might just be crazy enough to work. But we won't know that until we dig into this a bit further."

Teddy told Rebecca he'd do a bit more research into these other potential kinetic energy harnessing competitors and see what they were working on. He also wanted to run some numbers to see if this was something that could theoretically work from an energy capture perspective. They agreed to meet for dinner next week when Teddy was back in California for a meeting with some of the former college instructors.

"I'll run some numbers on my end too," said Rebecca. "We sell a ton of baby products at my company, so I may be able to get a sense as far as pricing, sales channels, and distribution. If we crunch the numbers, we may know if this could even work."

GO OR NO GO

The following week, the two met for a working dinner to discuss what they'd found in the days after their initial phone call. Both were enthused by their findings and couldn't wait to tell the other the results.

Teddy began, "First things first, it doesn't look like anyone is publicly thinking about capturing kinetic energy through kids clothing. I saw about three or four companies doing work in the kinetic energy capture space, but two are primarily focused on military applications and the other two are looking solely at athletes. There is some intellectual property out there that we need to look more closely at, but nothing that said, 'don't do it.' We'll need to do some more digging there, but it doesn't look out of bounds."

"That's great news," replied Rebecca.

He went on, "And from a technical standpoint, I think it actually can work and I think I could build a working prototype with a bit of extra help. The numbers as far as energy capture and storage are way off the chart. Assuming we capture just 5% of the kinetic energy of a young, active child in a day, we can generate enough energy to power a refrigerator—and if we get 20%, the numbers start to get really interesting." Teddy talked about some of the challenges of embedding the electronics and sensors for kinetic energy capture, and the potential risks and liability issues of strapping your "pride and joy" with a battery pack. But he seemed to think he could build a working prototype in six months without too much trouble and without busting the bank. He'd even spoken to a few professors in the field who all seemed to indicate that they thought the technology was achievable.

"I've got some promising news on my end too," said Rebecca. "I ran the numbers and think we have a huge market to sell our products into. The baby clothing market is huge, and new parents, families, and relatives spend all kinds of money on new baby-related technology every day. I also spoke to a few recent parents who all seemed to think they'd be willing to buy it. The pricing may be the tricky part, but it seems like we've got a decent price range to work with." Rebecca pulled out a spreadsheet showing some potential costs to develop a product, as well as the sale price and cost of individual units. With a bit of help from investors, it looked like they could become huge a few years down the road if all went according to plans.

"The other option," Rebecca continued, "is to just see if we can partner with these other companies you found. Perhaps we just open a retail store and sell their products. Developing a line of clothing and this kinetic energy capture technology is going to be difficult. But we could also try something a bit smaller and more reasonable for the two of us."

LET'S START A BUSINESS

They sat in silence for a few minutes thinking about what they'd just discussed and seen. Finally Teddy broke the silence.

"I guess the next step here is deciding to do this. Both of us are in good paying jobs, have some college debt to pay off, and haven't built a successful business yet. That said, this sounds pretty interesting and I think we have the backgrounds and contacts to pull it off."

"I agree," answered Rebecca. "Personally, I'm leaning toward going full steam ahead with the idea. No one else seems to be developing kinetic energy capture technology embedded in baby clothing, so why not us? It's risky and all, but I think we've got a chance." They each discussed their current financial arrangements and willingness to leave their jobs. After they both flipped through the financial models Rebecca had prepared and relayed conversations with key contacts about the idea, they both knew what they had to do.

Teddy stuck his hand across the table. "Partners?" he asked.

"Partners," she replied. "Now the fun begins, huh? Let's have you start on the product development and I'll start on the administrative stuff and the marketing and sales."

The two agreed to spend the next month wrapping up at their respective jobs and working to lay out the next steps and stages.

And thus, KinetiBaby was born ...

6

Forming and Founding

Whether it is General Electric, Intel, Google, the small electronics shop at the mall, the restaurant across the street, or the consulting company working out of a basement, they all started at basically the same spot: that initial formation of their business.

And so as you begin to take the next steps in building your green business, you too will be faced with a number of questions: what entity to form, where to form the business, what to name your business, when to issue stock, and countless others. And while these questions are similar across businesses, the answers will vary depending on the business, your goals, and the opportunity ahead. This chapter is designed to help you think about these questions and give you some information to digest as you decide to move ahead.

This chapter is also designed to answer a few of the questions and give you an overview into the formation process. If you have already hired an attorney or an accountant, you should discuss the formation with him or her to see what advice they may have for you and to see what type of support they may be able to offer in the formation of your company.

The Basics

When should you form this new legal entity? How and under the laws of what state should it be organized? Once it is incorporated or formed, what are the next steps toward getting our business fully established to enter into contracts, hire employees, retain investors, issue stock, and do business?

Whether it is a corporation, or an LLC, whether you form a C-corp or S-corp for IRS purposes, whether you form the organization in Delaware, Nevada, California, New York, Massachusetts, Texas, Washington, or another state, and the numerous other choices you have to make during this period, it is important to note that one size does not fit all. Some green businesses may need to consider how to structure their business to take advantage of tax incentives or may consider adding subsidiaries for the protection of intellectual property. Structuring your business enterprise should be well researched and thought out. Remember that each decision should be made to fit your business and provide you the most optimal structure to succeed.

Thinking about the "formalities" of forming and founding a business can seem fairly overwhelming. You can spend days, weeks, even years reading books and the web, talking to people, and fretting over what is the "right" process for your business. For this reason, it is a useful process to talk to other entrepreneurs and advisors before taking steps to formally create the business. In most cities, there are organizations, meetings, and seminars designed to assist new entrepreneurs. Take advantage of those opportunities to get information from a variety of sources. They can help you to understand the process.

One entrepreneur referred to the process of incorporating, issuing founders stock, and the other formalities (founders agreements, employment agreements, intellectual property agreements, etc.) as "making our business 'real'." In some senses, that is correct. The entire formation process sets the framework for the business that you'll use going forward.

So what are the basic steps you'll need initially to form and found your business?

1. Getting a business entity
2. Issuing equity and dividing ownership
3. Agreements with founders, employees and consultants
4. All the other filings

If you think about the formation and founding process as those four major areas, it can help simplify things a great deal. You may have an attorney or accountant handle all of these items or some entrepreneurs may handle some or all of these steps on their own (if it is going to be more than just you in your business, it is worth at least having a conversation with an attorney and/or an accountant as you consider or take these steps).

For more information and details on founding and forming your business, see

- Selecting Your Company Name on page 283.
- Choosing an Entity on page 289.
- Initial Business Filings on page 291.
- Leaving Your Employer on page 296.

More information is available on details of forming and founding, including information on each of these four key steps in Chapters 24 through 26.

Why Do You Need to Think about These Choices?

The first thing to recognize about starting a business is that things like structure, issuing stock, and handling agreements among the founders are certainly *not* the most important things you need to think about. Sure, it can lead to headaches, hassles, and additional costs, but it is important not to lose focus on the real task at hand: building a business that can make money.

That said, it pays to spend some time up front researching and considering the right structure to your business. Some new entrepreneurs will rush to structure their business just like another entrepreneur or may decide to pick a structure that is used in their old company. Before you rush to follow the pack, take a moment to understand why it matters:

1. Your choice affects your *tax rates*.
2. There are varied *cost and fee structures* in each state.
3. It may be *costly to change* entities at a later point.
4. *Third parties* may prefer or require your company to be organized a certain way.
5. The *"Pain in the Butt"* factor—how easy and simple is it?

Whether you decide to consult with your attorney and accountant to determine the best form for your company or if you decide to forgo the advice and for go it alone, be sure you make a decision based on an understanding of how your choices affect the business and fit into your business strategy. Spending the time and money up front will help ensure that you make the right choices for your entity.

The Question of *When*: When to Legally Form Your Green Startup

You might think this is a simple question, right? However, there isn't a one-sized-fits-all approach to creating a green business. Some green businesses will choose to incorporate immediately after an idea is drawn on the back of a napkin. Some companies are established while their founders continue in full-time jobs. Other green businesses will operate fairly loosely as they lay out the business plan, develop the idea, technology, or product (almost like a hobby for the founder or founders), and won't get officially incorporated and "founded" for a number of months or years.

It is important to think about the concept of "*When*" for a number of reasons. Why not just incorporate your business right away once you have your idea? The incorporation process itself is fairly simple—in fact, in many states, you can apply online for just a few dollars. The main reasons why you might consider waiting to form the business are

- *It is just an idea.* If you haven't decided whether or not this is just a harebrained idea, or if you can recruit an adequate team to execute on the idea, you may not be ready to incorporate or form the formal entity.
- *You are employed elsewhere.* To the extent forming a separate business could breach some of the obligations to your current employer or raise suspicions of your bosses or coworkers, you may decide to postpone incorporation. More details on this are discussed below.
- *You aren't sure the type of business you are forming.* Switching from one entity to another or one state of incorporation to another is not incredibly complicated, but it does involve some amount of work, and possibly some costs to the extent you must involve your accountants and lawyers. If you are unsure about these choices, you may wish to wait until you've settled on certain of these choices.

WHEN WILL WE KNOW WE NEED TO LEGALLY FORM THE BUSINESS?

This is one of those times when it may make sense to call in some advisors, mentors, partners, or your lawyer and accountant. Oftentimes those advisors can help you understand when it is a good time to legally form the business. But what are key considerations to help you to decide when you should incorporate your new business?

- *Is your business currently operating?* If your business is already operating and doing activities such as entering into contracts, providing services (even via a Web site), or hiring employees, you may want to incorporate or form an LLC sooner. If you do not operate under the protection of a corporation, LLC, LP, or LLP, your personal assets may be exposed in the event of liability.

- *Is your business* not *currently operating?* If you are not operating the business (and are primarily in research, recruitment, and prestartup phase), there is much less business risk in waiting to incorporate.

- *Do you want to reserve a particular name for your entity?* If there is a particular name you want for the company, you may consider incorporating to reserve that name. Otherwise, you may wish to simply reserve the name without incorporating.

- *Are you employed at another job?* In particular, if your potential new business will have some overlap with your current employer, you may want to wait until after terminating your employment before incorporating or forming an LLC.

- *Are you expecting tax losses in near the future?* If you are planning to form a corporation rather than an LLC or another pass-through entity, you will treat taxable losses differently. Losses will not "pass-through" a corporation for tax purposes.

- *Do you have a financing that will likely close in the near future?* Problems can arise for certain new companies that want to issue stock to their founders at low prices if they wait until the company is nearing a financing. The problem arises because the company is selling its stock to investors at a much higher price than it hopes to issue stock to its founders.

- *Are you applying for any grants, loans, or similar programs?* Some grant and loan programs will require you to have established your entity in order to participate in the program application process.

- *Are there cost implications?* If you are not operating, when you form the company you'll likely have additional costs to pay your accountant for filing your annual taxes and to pay your attorney to assist with formation matters.

- *Are there tax or fee implications?* Some states have minimum tax filing amounts for a corporation even if you are not active.

- *Are you currently paying self-employment tax?* If you are currently operating as a sole proprietor or in a partnership, you will likely be subject to self-employment tax. This may mean you are currently paying over 15% of your earnings for self-employment obligations to Medicare and Social Security. If you instead incorporated, you may not pay the self-employment tax on any profit that remains in the corporation.

- *Do you have any timing requirements?* Depending on the time of year, you may find that you have to wait longer to receive confirmations of filings. This is particularly true at the beginning of a year and at the beginning or end of a month.

If you are certain you are going to start your business, you shouldn't wait too long before taking steps to incorporate your business as a corporation or to form an LLC. The costs to form the entity are fairly low, and it is generally a good idea to protect your personal assets that may be exposed if you operate a business without proper protections of a legal entity.

Can You Wait *Too Long* to Incorporate?

Let's say you and your business partners have an eco-business idea and a business plan that is going to change the world. It's so great, in fact, that you've been able to convince some people to invest in the business … once you form it.

This is one of those times when you need to be careful about waiting too long. For many companies that are just an idea, you will want to issue some stock to the founders of the business at a low price (let's say just at a penny or at tenth of a penny per share). If you wait to issue that stock until you receive funding, you may create a "Cheap Stock" tax problem. Issuing stock at $0.001 per share to founders on the same day (or perhaps just a few days before) you sell stock to these excited investors at $1.00 per share will raise some flags with Uncle Sam. The IRS may consider this difference in stock price (between the price paid by the founder and the price paid by the investor) as income for the founder.

If you are in the process of raising money and have yet to form the business, be sure to talk to an accountant or lawyer about how to handle this timing issue.

Running a Startup "On the Side"

One of the common questions many entrepreneurs have is regarding being a part-time entrepreneur and business founder—oftentimes called "Moonlighting." Particularly when the business is just getting started, perhaps you intend to solely focus evenings and weekends on the business, while retaining your regular job (and the salary) at the same time.

Many successful businesses began as a side project or a part-time project—so being a part-time founder is not unheard of. That said, running (or starting up) a competing business with your current employer can possibly raise numerous issues. That means you need to be cautious and carefully follow the rules when engaging in potentially competing ventures. In fact, numerous court cases have ruled in favor of the employer in the scenario when an employee starts a competing business while still employed—especially if the employee was managing those operations or was a key employee. You could even be at risk if you incorporate the business while still employed.

Make sure that before you decide to start a new business or join an existing business part-time that you review your current company's policies and your employment documentation before acting formally on behalf of your new business venture. In some circumstances, employees have found that disclosing your business plans prior to beginning this venture has been met with greater support than expected (but use your judgment on whether or not you should disclose!). Perhaps your employer may be able to reassign you so as not to raise competition issues. No matter how you plan to handle the scenario, you should not misstate or misrepresent the activities of the new business.

Don't risk litigation for your new venture—when in doubt, check with the appropriate parties. Your attorney may be able to provide some guidance for this transition period.

CHOOSING AN "ALTERNATIVE" CORPORATE STRUCTURE

Many entrepreneurs starting green businesses have begun to look to certain new types of corporate entities, rather than traditional corporations, LLCs or partnerships. Historically, the traditional entities have each primarily (or purely) emphasized the creation of shareholder value—through a focus on the maximization of profits. Today, more businesses are being formed with the express purposes of a "triple-bottom line," emphasizing people, planet and profit, creating a new choice for today's greentrepreneur who may too be looking at an emphasis on the planet when choosing an entity.

In 2010, a new class of corporate entities have begun to be approved by state legislatures. Maryland and Vermont approved "Benefit Corporations" or "B Corporations" as a new class of entities in 2010 and a number of states are currently considering the B Corporation. The nonprofit behind the legislation, B Corporation (http://www. bcorporation.net/), is working to enact legislation in all states to provide entrepreneurs the option of forming a B Corporation. In California, legislation has been introduced to create a "flexible-purpose corporation." Under this legislation, existing California corporations could adopt the new model to formalize their social mission and emphasize purpose over profits, giving companies broader discretion of acting in the "best interests" of the corporation. These aren't the first entity types aimed at social entrepreneurs that have been approved. Vermont, Michigan, Illinois, and New York (among others) each have the Low-Profit Limited Liability Corporations (or L3Cs). Given the current focus on green businesses and the triple-bottom line, expect more states to consider these new entities.

So what does this mean for you as a greentrepreneur? On one hand, this certainly signals that triple-bottom line entrepreneurship is growing in popularity (B Corporation promotes that over 300 corporations representing more than $1 billion in revenues are now B Corporations). For a business that aims to promote its "greenness" and leverage the green movement, creating a B Corporation or flexible-purposes corporation certainly signals the entity's intentions. Certain investors may be particularly inclined to invest in B Corporations or flexible-purposes corporations over C Corporations (in fact, B Corporation lists nearly a dozen investments into and several sales of B Corporations). On the other hand, entrepreneurs should be cautious because of the newness of these entities. There is not a well-established set of corporate laws or court decisions that can help guide behavior. And, it still remains the case that sophisticated investors, including venture capital, public capital and many sophisticated angels, may only invest in classic C Corporations. If your strategy requires these investors (or is likely to need large amounts of capital), a B Corporation may not fit.

The most important lesson for today's greentrepreneurs is to become more educated about B Corps or flexible purpose corporations if you are seriously considering forming a B Corporation or flexible purpose corporation —"The Change We Seek," a blog run by B Corporation (blog.bcorporation.net) contains a great deal of information about these new entities. In addition, there are a number of entities that have already chosen the B Corporation that may have valuable insights to share. And before you make any choice, be sure to speak to your lawyer, accountant, banker, partners, and board members as well as likely fundraising sources (if you entity needs to raise funds). An increasing number of investors in sustainable or green businesses are looking at these alternative entity structures as appropriate investment vehicles, which increases the opportunities for green businesses to choose these corporate structures.

GreenStartup, Inc.: A Founding and Forming Case Study

The following is a case study detailing some of the formation and founding options for a green business. Here is a fictional case study of KinetiBaby:

A quick recap ...

While Teddy Muir was a researcher at the University of North Dakota, his friend and college lab partner, Rebecca Carson came to him with an idea to solve the world's energy troubles. Rebecca was working as a financial analyst and had been interested in starting a business since college. Now, five years after Teddy and Rebecca had graduated, they finally had their idea. At first, Teddy was skeptical, but after doing some research and analysis he figured it actually had a chance. They sat down over dinner one night and hatched a plan to leave their current employers and start their new green business: KinetiBaby. Rebecca had received her inspiration while visiting her sister who had two toddler-aged children. Sure, kinetic energy was a bit difficult to capture and store, but parents were more than happy to dress their children up in silly looking outfits. By simply attaching a few electronic devices and batteries to a toddler's outfit for a full day of playing, Rebecca surmised you could power a household of electric devices. And, based on Teddy's research and modeling, they were both convinced this seemingly crazy idea had legs. And thus, KinetiBaby was born ...

LET'S FORM A BUSINESS

With a handshake and a plan on a napkin, KinetiBaby was born with Teddy and Rebecca as cofounders. Their first step was to formalize it and create the business. They decided to find an attorney to help them with the process. Based on a bit of advice from a mutual friend who had recently built a clean tech business, they were introduced to Derek, an attorney who works with other startup companies. A few days later, Derek, Teddy, and Rebecca sat down to map out some of the key formation and founding matters for KinetiBaby.

TO PASS THROUGH OR NOT TO PASS THROUGH …

"Before we get started," began Derek, "I wanted to make sure that you guys had checked all your employment documents and to see if you'd spoken to your current employers about the new business. We certainly don't want to create any unneeded stress between your old employers and the new business."

"We did," replied Teddy. "I wanted to be sure that there were no issues with any of the intellectual property I'd created at the University of North Dakota, so I sat down and talked with their Office of Technology Transfer and mapped everything out. They are fully supportive of this and my department chair even offered to serve on our advisory board. I'm going to keep working part-time at the University initially, so we've talked about how to document my efforts and to be cautious about keeping these two jobs very separate. Rebecca's boss wasn't quite as enthused with her leaving mostly because they like her a lot and she's leaving to do this full-time. She had some clauses in her employment agreements about noncompetition, confidentiality, and nonsolicitation, so we'll need to be a bit careful there, but we don't think it is an issue since these are pretty different business ideas. Here are copies of those documents." And Teddy handed over a folder with their employment paperwork.

"That's great," said Derek. "The best course is for me to review those documents just like you did, to be sure that nothing will force us to put the brakes on things. Sounds like you guys are on top of that." He took the employment documents from Teddy.

"Now let's start out with a basic question about what kind of business you are hoping to start," said Derek. "Tell me about what you envision for the future of your company, KinetiBaby. Where do you see things in two years if all goes well?"

"That's an interesting question. I guess we've talked about it a bit as we thought about our business plan. Ultimately, we just didn't know the answer though. There are so many variables that we can't control," said Rebecca. "Can you explain why the future even matters at this point?"

"One of the important things to think about early on," said Derek, "is selecting a business entity that best fits your goals for your business. While you may not have it all figured out yet, we try and do the best we can at the outset."

Derek went on to explain some of the key differences between a limited liability company and a corporation. He described that it is more typical to see a company choose the limited liability company if the business was likely to stay moderately small that didn't intend to try to raise money from venture capital. It would be more appropriate to choose a c-corporation if Teddy and Rebecca believed they would need to raise a substantial amount of capital and build a more sizeable business for KinetiBaby to be a success. Ultimately, while it involved a bit of "crystal ball gazing," Teddy and Rebecca would need to think about the kind of business they envision two to three years down the road when selecting the entity.

"But this isn't something you should lose sleep over," said Derek. "If things change, you work with your attorney, accountant and advisors to change with it. If your consulting business suddenly creates a product and needs to IPO, then you find smart people to make that work. So it's an important decision when you start, but isn't 'unfixable' if you get it wrong."

"That's good to know since this is our first time doing this," said Rebecca.

Then Teddy asked, "Can you explain why an LLC would be more typical in a smaller business that requires less capital?"

"Generally, a limited liability company is selected because it has certain tax advantages," said Derek. "An LLC is a pass-through entity for tax purposes, meaning that the entity itself isn't taxed, but the owners are taxed. So each year the owners would pay taxes on the business' profits based on their ownership – meaning the taxes 'pass-through' the entity to the owners. So for a smaller business that isn't likely to have substantial capital needs, an LLC is fairly simple to setup and customize. If you were going to start a two-person consulting company or a small retail store selling kinetic energy capture devices built by other companies, other then an LLC is probably a good choice to consider. If, on the other hand, you want to build KinetiBaby devices to outfit every toddler across the country, then you'll probably need to raise quite a bit of money, likely from venture capital or other similar sources. And those investors tend to favor investing into a c-corporation. The downside of a c-corporation is that corporate earnings are 'double-taxed' by the IRS."

Derek went on to explain how the earnings in a c-corporation are first taxed at the corporate level. He explained that if KinetiBaby were to have profits of $1 million, the company would pay approximately 35% or $350,000 to the federal government. Then, if the company wanted to distribute the remaining $650,000 of profit to the shareholders of the business, those dividends would be taxed at approximately 15%, so approximately $100,000 of additional taxes—so from a starting point of $1 million to a net to the shareholders of $550,000 after all taxes are taken out. Therefore, the total tax rate on the profits of the corporation would be approximately 45%.

Net Income	=	$1 million
Corp. Tax Rate	=	35%
Profit	=	$650,000
Personal Tax Rate on Dividends	=	15%
Net to Shareholders		$550,000 (55% of Net Income)

If KinetiBaby were an LLC, with two owners, splitting the profits equally, the tax rate could be as high as 35%, but more likely around 30%. This means the owners of the LLC would likely receive net profits of around $700,000.

Net Income	=	$1 million
Personal Tax Rate on Income	=	Approx. 30%
Net to Shareholders		$700,000 (70% of Net Income)

"So, at a million in profits, the difference between an LLC and a c-corporation could be as much as $150,000 more in taxes," said Derek. "So that's the main reason why you'd choose a pass-through entity, if it made sense for your business."

After Teddy and Rebecca processed what they'd heard and thought a bit about it, they realized that to really have the impact on the environment that they hoped, KinetiBaby was going to need to grow quickly and would require more capital than they could currently bring to the table. It sounded like KinetiBaby was probably going to need to be a corporation.

THE BEST OF BOTH WORLDS

"I guess we would have to be a c-corporation then," said Rebecca. "But, what about if we don't think we'll need to raise money for a couple years? Do we then give up the tax benefits from having an LLC for those years?"

Teddy and Rebecca both agreed that long term they would definitely need to raise money, probably from venture capital or other corporate investors. However, they thought there was a period of several years of development before they'd need to make that decision. Before any venture capitalist was going to be looking at KinetiBaby, Teddy and Rebecca believed they had a good two to three years of research, development, and commercialization ahead.

"That's a common problem that tech entrepreneurs like you both have," said Derek. "The business will likely require capital down the road, but who knows how long or if it will ever happen. Does it mean you have to give up all those potential tax benefits just because you *might* raise venture capital? Fortunately, no. Some entrepreneurs decide to initially form as an LLC and then convert or merge their LLC into a c-corporation later. However, there is a hybrid approach that may also work for KinetiBaby—an S-corporation."

Derek explained that a corporation could make an S-election with the IRS and gain the benefits of the pass-through entity. You would incorporate as a c-corporation and then make a filing with the IRS, requesting to be treated as a pass-through entity. Therefore, if the corporation had any profits, they would "pass-through" to the shareholders just like they do for the owners of the LLC. There are some limitations to an S-corporation, including having less than 100 shareholders and having only individuals as shareholders (not corporations, LLCs, or other entities).

"By making the S-election for your corporation," Derek continued, "you keep the potential tax benefits, initially. Then if you are ready to raise venture capital or take money from another investor, you simply terminate the S-election, and become just a c-corporation. You may have to pay back some prior tax benefits at that time, but this is oftentimes simpler and cheaper than initially forming an LLC and converting down the road."

I'VE HEARD IT IS NICE IN DELAWARE

"That sounds like us," said Teddy. "I think this is all starting to come together. My next question is which state do we pick for our corporation? I've heard people say places like Delaware, Nevada, and Wyoming are the places to choose from. But I've never been to any of those places and we don't want to have to move our business to Las Vegas or Cheyenne." Rebecca interrupted, "Why can't we just pick right here in California?"

"Don't worry," replied Derek. "You've got lots of options and none of them require you to relocate Dover or Cheyenne."

Derek explained that choosing a state of incorporation or a state of formation if you are forming an LLC doesn't require moving to that state. Instead, it primarily determines which corporate laws the business is subject to and certain state corporate taxes, often called franchise taxes. This means that you could be incorporated in Delaware and never set foot in Delaware, having your corporate offices located elsewhere.

"Okay, I was thinking how I was going to explain the move to my wife!" said Teddy. "So how do we choose where to incorporate then? I was told that most Fortune 500 companies are incorporated in Delaware. Shouldn't we incorporate there too?"

"Lots of companies do choose Delaware," replied Derek, "but it certainly isn't the only option these days." Delaware has a history of being a very business friendly state, explained Derek, which explained the large numbers of companies that are incorporated there. Formation in Delaware is fairly inexpensive and fast, and franchise taxes are relatively low. Most corporate lawyers know Delaware law well, making it the "law of corporate governance" in many instances. States such as Nevada and Wyoming are trying to model their corporate laws on Delaware and lower their state tax rates to attract corporations to incorporate there.

"For a new corporation like KinetiBaby," said Derek, "the choice will probably come down to Delaware or wherever you think the headquarters will be. It used to be that Delaware law was the best place if you were forming a business with substantial growth potential, but these days the differences aren't as large and there is less of a stigma of incorporating in your home state." In the case of KinetiBaby, California likely represents a viable option for consideration, explained Derek. There are differences between the two states for some matters like dividends, shareholder voting, and directors, but the similarities likely outweighed those differences. California isn't likely to be as quick with processing filings and may have some more hoops to jump through, but in the end, there isn't anything that would necessarily make your attorney recommend that you definitely not incorporate there. From a cost perspective, the initial filing fees in each state are both around $100.

"The difference for some companies is that incorporating in Delaware actually requires to still register in the state where you are operating, here in California," said Derek. "This means that if you become a Delaware corporation, KinetiBaby will still need to register as a foreign corporation (meaning a corporation incorporated in another state) and pay a filing fee and annual filing costs. So for some companies these costs might lead you to choose just your home state and save the additional filing fees."

"Will it matter to our investors?" asked Rebecca. "I mean if we are going through all this trouble, I don't want us to pick a state that scares away investors."

"Choosing California over Delaware probably wouldn't scare any investors away," answered Derek. "However, if you asked most venture capitalists, my guess is most will prefer that you be a Delaware corporation, since that is the form they are most familiar working with. If you picked a state like North Dakota where Teddy is from, an investor may request you convert to another state since they may not be sure how North Dakota corporate laws work. But states like California, New York, Massachusetts, Washington, Illinois, and a few others probably won't cause anyone to bat an eyelash. Now, if your business is going to require you to talk to investors all over the place, including outside California and even outside of the United States, the safest choice may be Delaware, not by a mile, but by an eyelash," Derek said with a smile.

HOW ABOUT A NAME?

"Got it," replied Rebecca. "I think I'm with you then. Since the costs don't seem to be drastically different and I've heard green investors are located around the world, let's go with Delaware. Now, we really like the name KinetiBaby for our company. We've looked and there is a domain name available and Teddy checked the USPTO Web site and didn't see any trademark on the name. Can we pick that name?"

"Sounds like that should work," said Derek. "As long as that name isn't used in Delaware, we should be able to get it. In Delaware, you are usually required to add Corporation, Corp., Incorporated, or Inc. at the end. That isn't required typically in California. We can reserve the name in Delaware if you are worried someone may swoop in and take KinetiBaby, Inc., otherwise we should be able to file your Certificate of Incorporation in a few days and get the name outright."

"KinetiBaby, Inc. it is then," said Teddy.

IT CAN'T BE THAT SIMPLE, RIGHT?

Derek looked down at his notes and back to Teddy and Rebecca. "Well," he said, "with all that information, I think we're ready to get you guys incorporated. We will need to file a Certificate of Incorporation with the Secretary of State of Delaware and file a Statement and Designation by Foreign Corporation with the Secretary of State of California. There are a couple other items we'll need to take care of in order to get all the necessary licenses and permits in place. We'll also want to talk to your accountant about the S-election, and coordinate that filing."

Derek also explained the need to file for a Federal Employer Identification Number, as well as certain state permits. Each state, including California, usually has a business portal or licensing agency that lists all the required forms and paperwork to fill out. He suggested to spend some time on the state Web site and to call the office line of the agency and ask questions about what is required. "And be sure to check on whether there are any specific licenses, permits, or other filings you need to make due to the type of business you have. For example, there could be filings with the state environmental office if you are using certain metals or chemicals. Those are easy to miss, so be sure to understand all that is required."

"We will also have some corporate items to handle," said Derek. "This includes preparing your bylaws, appointing directors, issuing stock, and a host of other items that the corporation needs to approve. We'll discuss most of these soon, but for now, we'll get KinetiBaby incorporated."

HOW ABOUT AN IP HOLDING COMPANY?

"I've got one more question," asked Rebecca. "Someone told me we may want to set up a subsidiary for our intellectual property."

"That's a good question," said Derek. "Some entrepreneurs feel the need to try to protect themselves from 'giving away the farm' in their new business. So the entrepreneur will consider setting up a separate holding company for the intellectual property and create a contract between the startup and the holding company for the startup to use the IP for a fee. The theory is that this structure may allow the entrepreneur to form multiple startups based on this intellectual property. In some cases, that approach works, but in others it could discourage an investor from putting money into the business. If you have a legitimate reason for separating the intellectual property from your business, we can discuss it further, but if you are going to be developing the IP as you build the business, this may not give you much protection anyways."

"I see. Well, what about if we want to use project finance to build a manufacturing facility down the road?" asked Teddy. "Should we set up those subsidiaries now?"

"Usually those subsidiaries don't need to be set up until the business is ready to do a project finance transaction," said Derek. "There isn't any rush and when that time comes you'll have plenty of smart attorneys helping you structure this correctly."

Teddy and Rebecca thanked Derek for the time and headed off to start reading up on baby clothing fashions and kinetic energy capture (quite the odd combination). They had a business to build and now were on their way to having a real business entity formed.

For more information on some of the issues discussed in this case study, check out Chapter 23 "Forming the Business." This includes topics such as Selecting Your Company Name, Choosing an Entity, Initial Business Filings, and Leaving Your Employer.

7

Assembling Talent

People are the core behind a business, and success or failure of a startup or small business oftentimes is determined on the backs of a few key founders, employees, consultants, directors, and advisors. Identifying green collar individuals for your green business is about finding someone with the right "green" credentials mixed with the right personality, technical skills, and experiences needed to design and build your green product or service.

Founders

Generally, a venture begins with a small, core group of individuals who will serve as the backbone of the company until it can afford to hire additional employees. These people are oftentimes called the founders. Additional talent will be added and will supplement in key areas. Many successful entrepreneurs will tell you that these future hires will not supplant the need to bring the most outstanding collection of talent together for your founding team.

Research by David Hsu at the Wharton School of Business found that the ability of the founder or founders to recruit talented executives (other founders and management-level employees) for their organization from *within* their own social circles (as opposed to recruiting talent from within their investors' social circles) was positively associated with the resulting valuation of the business at a future fundraising event.

Making these top notch hires early on in your founding will add immediate value to the organization.

For more information and details on founders and founders agreements, see

- "Founder Decisions and Agreements" on page 299.
- "Founding Team Questionnaire" on page 304.

Famous Founders

Many of the success stories of the past three decades in entrepreneurial ventures have come from a team of dedicated founders: Bill Gates and Paul Allen founded Microsoft together; Larry Page and Sergey Brin cofounded Google; YouTube has its famous cofounders Chad Hurley and Steve Chen (the third founder, Jawed Karim, left the company to attend graduate school); and David Filo and Jerry Yang founded Yahoo! together in 1994. Other examples include Steve Jobs and Steve Wozniak of Apple, Oracle Corporation was founded in 1977 by Larry Ellison, Bob Miner, and Ed Oats (but Larry remains the central executive), and Marc Andreessen and Jim Clark of Netscape.

FAMOUS GREENTREPRENEURS

Environmental blog Grist.org profiled a group of famous entrepreneurs who are active in developing green businesses. To view the full profiles of these greentrepreneurs, visit: http://www.grist.org/article/bizfounders/.

- Yvon Chouinard, *Patagonia*
- Nell Newman, *Newman's Own Organics*
- Ray Anderson, *Interface*
- Janice Masoud, *Under the Nile*
- Jeffrey Hollender, *Seventh Generation*
- Jeff Lebesch and Kim Jordan, *New Belgium Brewing*
- Gary Hirshberg, *Stonyfield Farm*
- Michael Gordon and Vaughan Lazar, *Pizza Fusion*
- Colin Crooks, *Green-Works*
- Gary Erickson, *Clif Bar*
- Sean O'Hea, *Vehizero*
- Joseph Whinney, *Theo Chocolate*
- Jigar Shah, *SunEdison*
- Martin Ernegg, *Zelfo Australia*
- John Mackey, *Whole Foods Market*
- H. Harish Hande, *SELCO-India*
- Josh Dorfman, *Vivavi*
- Virginia Young and Janie Lowe, *YOLO Colorhouse*
- Mark Bent, *SunNight Solar*

There are also examples when a single founder was the central public figure in the company's initial success, including Jeff Bezos from Amazon, Michael Dell of Dell Computers, Mark Zuckerberg from Facebook, and Pierre Omidyar of eBay. And while these individuals may have been the initial driver of a business or technology; each would likely volunteer that there were always other key early stage partners, founders, and employees crucial to the long-term success of these businesses.

Size of Founder Teams

Chuck Eesley from the MIT Sloan School of Management studied early-stage businesses to identify what factors about their founding teams were integral to their success. Eesley found that technology startups with larger founding teams were more likely to be successful than those with smaller teams. As the number of cofounders increased in Eesley's study, so did the likelihood that the company would ultimately do an initial public offering (IPO). Likewise, Eesley identified that a founding team was more likely to be successful if the team had any prior experience at businesses that had previously done an IPO. Plus, the more connections to venture capital firms that a founding team had, the more likely that business would ultimately file for an IPO.

Many startups will begin with a three-person founding team including a president/ chief executive officer, a vice president of marketing/sales, and a vice president of engineering/chief technical officer. You will continue to build out the organization as you continue to raise funds, develop your product, and begin sales efforts. But ultimately, the core team will likely consist of some combination of these key business and technical skill sets.

Identifying Cofounders

Additional research by Eesley at MIT looked at where founders had met one another prior to forming the business. Eesley's research looked into founding teams from recent college graduates (less than five years following their graduation) and more established an alumni of MIT (more than five years following graduation). Among founding teams of the recent graduates that had formed companies since 2000, approximately 30% of the founding teams grew out of their MIT research, 20% grew out of work relationships, 20% from extracurricular activities, and 30% from social activities. For founding teams of established alumni that had formed companies since 2000, a larger number of founding teams grew out of work relationships. For these teams, less than 15% grew out of MIT research, 40% grew out of work relationships, 40% from social activities, and less than 5% from extracurricular activities. As you might expect, as graduates advance in their careers, a greater number of founding teams will grow out of work and social relationships, while for recent graduates more startup teams will be formed based on prior research and extracurricular activity relationships.

In many cases, a founding team will grow out of personal relationships or a working relationship (e.g., Paul Allen and Bill Gates of Microsoft, who became friends in high school in Seattle, and Larry Page and Sergey Brin of Google, who met as Stanford University graduate students). In these situations, you may have had a self-contained team in place ready to begin efforts to develop the organization. In other situations where you do not have a readily identified cofounder (e.g., Steve Jobs convinced an initially skeptical Steve Wozniak to join him after Jobs had proposed selling a computer as a fully assembled PC board), you may need to begin searching your social network to find additional key members to join you.

There are a number of things that can be identified as key traits to look for in any member of your founding team. However, researchers have identified six key features that an ideal team would have (but remember you may not find anyone who has all of these characteristics, so look for someone who combines the best mixture for your team):

- Compatibility
- Mixture of new and old team members
- Entrepreneurial experience
- Technical experience
- High integrity and ethics
- Peer/equal

USING YOUR NETWORK TO IDENTIFY A COFOUNDER

It may seem like a daunting task—finding the right cofounder. What you may not realize is that the best people might be right under your nose and you just need to identify and recruit them into your venture.

The best place to find a cofounder, business partner, advisor, or even an initial investor is someone you already know in your personal network: people you have worked with at prior and current employers (but, you should be careful with coworkers in your current employment situation if you have a nonsolicitation clause); people you know through volunteer, civic or other organizations; and friends and connections from your high school, college, or graduate school.

GOING ON THE "CIRCUIT"

Once you've identified people you know and think may have some interesting insights, connections or input for your new business concept, consider scheduling meetings with these individuals. At this stage, these meetings are nothing more than chances to reconnect and gain some additional information about your business concept from trusted individuals.

Your "meeting circuits" should not focus on a cofounder—instead you should be looking for conversations, meetings, and lunches. As a new business founder or potential founder, you'll need to begin developing a personal network to help in a variety of areas (e.g., finding a good lawyer, recruiting various types of talent, finding customers, soliciting investors). Ultimately, you may be surprised at what you are able to find from these initial conversations, including the amount of resources you had at your fingertips!

Even founders who believe they have a founding team in place already will likely benefit from connecting or reconnecting with their extended network. You may not even realize the value of a connection until you begin these initial discussions. Founder teams have found important members of their advisory boards, future CFOs, investors, customers, landlords, lawyers, accountants, and countless key contacts they've utilized in their business from these initial meetings. Each founder should begin, even at this early stage, having these initial dialogues with people in their network.

Initially, many entrepreneurs will not make an express mention of the fact that they are looking for cofounders. Instead, these meetings can serve as opportunities for feedback from interesting people.

IDENTIFYING INTERESTED CONNECTIONS

As you continue to have meetings with your initial contacts as well as meet with second-degree connections, hopefully you'll begin to get a sense of people who may be interested in the opportunity in some fashion. In some cases, a perfect candidate may simply raise his or her hand to join you as a cofounder. In others, you'll scratch your head after a series of meetings wondering if you are any closer to identifying a cofounder. But remember, this is just the first phase—developing a list of resources and contacts who you'd be interested in having play a role in your business.

You should search broadly at this stage: don't be afraid to meet with a CEO of a company who you are certain won't leave his company—oftentimes this person will be a great source of referrals and may be a perfect person to serve on your advisory board. You may not need a biologist or a financial controller, but you may be surprised at who they know and can connect to you.

Identifying Green Collar Talent

One of the most popular phrases of the United States Presidential Election in 2008 was "green collar jobs." What exactly these jobs are is somewhat hard to define (like some of the other definitional issues related to green business). What is clear is that there are more programs being developed to train green business executives and employees. While your business may not require a freshly minted MBA of sustainability, you may be surprised to find a recent undergraduate has taken a course of green business practices.

Government "Green" Training

The federal government and many states are also trying to spur the growth of green business through funding for training of "green collar workers," the popular phrase for jobs in green and clean sectors and industries.

In particular, the federal government has funded a series of Energy Training Partnership Grants. The education funded by these grants will be based on the needs by the local communities and will train workers for careers as hybrid/electric auto technicians, weatherization specialists, wind and energy auditors, and solar panel installers. Examples of recipients of these grants include the Oregon Manufacturing Extension Partnership in Oregon and Washington, the Blue Green Alliance in Minnesota and the United Auto Workers-Labor Employment and Training Corporation in Missouri. The aim is to provide training and retraining for individuals transitioning from careers in traditional industries into careers in various green business sectors.

Sustainable MBA Programs

As emphasis on green business and clean technology has grown, so too have educational opportunities for individuals wanting to join those fields. In 2009, there were 149 schools in 24 countries participating in the Beyond Grey Pinstrips organization which rates and tracks sustainability programs in global MBA programs. Between 2001 and 2009, the number of MBA programs that required students to take a course dedicated to business and society issues grew from 34% to 69%, with schools today offering an average of 19 courses that feature some degree of social, environmental, or ethical content.

For a ranking of MBA programs leading the way in the integration of issues concerning social and environmental stewardship into the curriculum visit

http://beyondgreypinstripes.org/rankings/

Employees and Consultants

Building the startup team is a crucial piece of the success or failure of your venture. Hiring well is not a one-time event, but represents an ongoing demand on the company to recruit, attract, hire, retain, and develop key talent for the organization.

Depending on the position that needs to be filled and the stage in the cycle of your companies, you will use a variety of methods to recruit new employees. For many positions, you may find that your personal network will be able to provide you a wealth of referrals to consider for positions. The downside may be that this referral network will consist of a number of people already in positions and convincing them to leave the position might be a challenge. For other positions, you'll likely find that job postings made on Web sites are a good source of applicants for a position. And for certain key positions, you may find that retaining a search firm is the best approach for attracting the right talent.

Early on, one of the key considerations a new company has is the timetable for filling its management team. Typically, a company will be formed by a single founder or a small group of founders, each filling a number of key roles within the company. As the venture continues to grow, the business will need to add key members to the management team to oversee areas of finance, marketing, sales, business development, manufacturing, and, in some cases, an experienced chief executive officer or chief operating officer.

For more information and details on employment matters, see

- Recruiting Employees on page 311.
- Hiring an Employee on page 313.
- Employee Compensation on page 316.

While your business may require a unique timetable for executive hiring, Nesheim (2000) in his book *High Tech Start-Up* noted a general pattern in the process by which founders added to the executive team. According to Nesheim, a company would typically be founded by the inventor and a core group, but begin by adding an experienced engineer, preferably someone with prior experience at a vice president level of a startup venture. Next, founding teams would add an experienced chief executive officer, particularly near to the time of fundraising. Following fundraising, companies often first hired a vice president of marketing to help with creation of the first products of the company, followed by a vice president of business development who may oversee sales or marketing early on in the life of the company. Tech startups would usually add a vice president of sales approximately two months prior to product launch date and a vice president of manufacturing or operations near the time of product launch. Nesheim found that companies were less consistent in their hiring of a chief financial officer or vice president of finance and administration. In some cases, this position would be filled just after financing while other times investors would encourage the company to wait, according to Nesheim, to give the investors greater input and oversight of the company's financial policies.

GREEN JOB SITES

The following sites are green and environmental job boards where you can identify job seekers or post your resume if you are seeking a green job.

- Clean Edge (http://jobs.cleanedge.com/)
- EcoEmploy.com (http://www.ecoemploy.com/)

- Envirolink (http://www.ecoemploy.com/)
- Environmental Career Opportunities (http://www.ecojobs.com/)
- EnvironmentalCAREER.com (http://www.environmentalcareer.com/)
- Ethical Jobs (http://www.ethicaljobs.net/)
- Great Green Careers (http://www.greatgreencareers.com/)
- Greenjobs (http://www.ethicaljobs.net/)
- Green Biz Job Listing (http://www.ecoemploy.com/)
- Green Career Central (http://www.greencareercentral.com/)
- Green Dream Jobs (http://www.sustainablebusiness.com/index.cfm/go/greendreamjobs.main)
- Idealist.org (http://www.idealist.org/en/career/index.html)
- The UK Green Directory (http://www.greendirectory.net/about.htm)

Board of Directors

According to investment banker Bill Vogelgesang, failing to focus some energy on developing a strong board represents a significant missed opportunity for any new company. Said Vogelgesang, a managing director and principal in the Cleveland office of Brown, Gibbons, Lang & Co., an investment banking firm, in an interview with Inc. magazine, "What entrepreneurs often don't realize is that a strong board can help a young or growing company build credibility in the outside world. A good board—especially one with

For more information and details on boards of directors, see

- Board of Directors on page 306.

heavy involvement from other CEOs and decision makers—reflects well upon a chief executive because it shows that he or she can take criticism and doesn't just want to impose one business vision on the company."

Recognizing the importance of these outside advisors, many startup companies choose to have a board of directors, as well as advisory boards or a scientific advisory board. These boards and the board members can represent important sources of strategic thinking, networking, and experience for a young company. Identifying board members is a difficult process for many startup companies. So what are the important things to consider when you are looking at a potential director?

- Relevant industry experience
- Ability to provide relationships and "open doors"
- Background in entrepreneurial activities, marketing, sales, or strategy
- Interpersonal and communications skills
- Ethical reputation
- Ability to work in a team
- Good relationship with the CEO or founders
- No conflicts with your business

- Adequate time to devote to your business
- Reasons for joining the board

Recruiting directors can be a tedious process, but references, contacts, and even search firms in some cases can be useful to find the right people. Allow plenty of time to find board members and don't hold up executing on your business plan because you may not have the right team in place right away. Look for references from your current board members, attorney, accountant, or other professional services providers, from other entrepreneurs, from local professional associations or groups, from directors and executives of successful companies in your area and from former colleagues or executives from prior employment situations. Finding the right team is important and deserves a substantial commitment of time and energy from the new business.

Selecting directors that can open doors for the company is an important characteristic. Research has shown that the number of contacts your business has with venture capitalists is correlated to your ability to get funded (proving that one aspect of getting financing is simply a numbers game—finding as many potential investors that you can get access to). Look for potential board members that can offer you access to a personal network for business opportunities, sales leads, potential employees, interesting investors, and other board members. Particularly for early stage companies, you will have the luxury to wait longer to find high impact board candidates—so set the bar high for your ideal candidates and work to fill the seats with top tier candidates.

POTENTIAL BOARD MEMBERS

- *Academia.* Former professors that may have expertise in a key aspect of your business.
- *Retired executives.* Executives, former partners of accounting or law firms, or other successful individuals who have or will soon retire.
- *Former employers.* Executives from former employers who will know you personally and may offer an outsider's perspective.
- *Board search firms.* Utilize a search firm to fill important board seats.
- *Board members of other companies.* Individuals currently serving on a board may be more willing to serve on another board.
- *Industry and trade associations.* Leaders and members of key trade associations.

Advisory Board

Many entrepreneurs wish to recruit an advisory board (also known as a strategic or technical advisory board) for their new business. An advisory board can serve to assist the business, provide credibility when reaching out to investors, and be a source of board members as the company continues to grow. In certain sectors, particularly clean technology, medical, high-tech, and software, the use of an advisory board is important for companies as a networking tool. Even so, many startups have found maintaining an effective and efficient advisory board to be a difficult proposition.

Sometimes, an advisor may be an individual you have targeted as a potential customer, a potential investor, or a potential executive. In the case of potential customers or investors, this can give that individual a way to get to know the business team before making a more substantial investment or partnership. Bringing on a prospective company executive as an advisor for a small amount of equity can be a "try-before-you-buy" approach to executive recruitment.

Bernard Moon identifies three different types of advisory board members for technology startups. Your advisors may be:

- *"Prestige" advisors.* These advisors are likely to be well-known public figures that give your company some public relations "buzz" and added visibility.

- *"Credibility" advisors.* These are advisors that are not necessarily as well known as "prestige" advisors, but will be well-respected academics or professionals, usually within the personal networks of the management team.

- *"Practical" advisors.* These advisors can provide assistance and guidance in areas where your management team may be lacking (accounting, finance, technical) or may have startup or other valuable industry experience.

ENGAGING EMPLOYEES IN SUSTAINABILITY

GreenImpact has developed a report to help businesses better engage their employees in sustainability. According to the report, *"The business value of integrating sustainability includes: cost savings by integrating energy efficiency into the workplace and products and services; attracting and retaining the best and brightest talent who want to work for companies with an authentic green commitment; and increased market share and revenues resulting from a stronger brand and new, innovative green products and services."*

The report highlights 10 best practices for green teams:

- Start with the visible and tangible: focus on internal operations
- Get senior management involved, but don't lose the grassroots energy
- Engage employees to capture ideas
- Communicate and share best practices
- Engage employees with their bellies: The low carbon diet campaign
- Engage employees in their personal lives
- Engage customers to be part of the solution
- Use art to raise awareness
- Create a toolkit to support and guide green teams
- Align green teams with corporate sustainability goals

For more information on Green Teams, go to http://www.greenimpact.com or to download the report, visit: http://www.greenbiz.com/sites/default/files/GreenBizReports-GreenTeams-final.pdf.

GreenStartup, Inc.: A Case Study about Building the Team

The following is a case study detailing the process a green business goes through in making hires, assembling an advisory board and completing its board of directors. Here is a fictional case study of KinetiBaby:

A quick recap …

While Teddy Muir was a researcher at the University of North Dakota, his friend and college lab partner, Rebecca Carson came to him with an idea to solve the world's energy troubles. Rebecca was working as a financial analyst and had been interested in starting a business since college. Now, five years after Teddy and Rebecca had graduated, they finally had their idea. At first Teddy was skeptical, but after doing some research and analysis he figured it actually had a chance. They sat down over dinner one night and hatched a plan to leave their current employers and start their new green business: KinetiBaby. Teddy and Rebecca, with the help of their attorney Derek, chose to incorporate their business in Delaware as a corporation and then make an S-election with the IRS in order to initially be treated as a pass-through entity. They incorporated the business as KinetiBaby, Inc. and registered as a foreign corporation in their home state of California. Now they were ready to start tackling some of the other decisions of the business including determining titles, issuing stock, determining the board of directors, and hiring some employees.

COFOUNDER STOCK SPLIT

With their handshake and the beginnings of a business plan on a napkin, Teddy and Rebecca had agreed to be cofounders of the business. Now that the two cofounders had incorporated their business, they were ready to issue the stock in the business and determine some of the other issues cofounders face.

The first issue that Teddy and Rebecca hoped to address was stock ownership. While both of them thought equal partners was probably the way to go, their attorney Derek encouraged them to write out a list of what each person was bringing to the business and their key role. They sat down to write out the list and came up with the following:

Teddy	Rebecca
CTO	CEO
Initially Part-time	Full-time
Contributing Patent	Startup Cash
	Came up with Idea

"Frankly, I don't see any good reason for us not to be equal partners," suggested Rebecca. "I feel like both of us are bringing equal things to the business. You are assigning the patent you own, I'm putting in some cash to get us started and we are both putting in our time."

"What about a doing 50-50 split?" asked Teddy. "Could that cause us to be deadlocked down the road on something?"

Derek had discussed the issue with them at their last meeting. Some founders who were contemplating a nearly equal split elect to do a 51-49 stock ownership split initially to solve for potential deadlock on a key issue down the road. While that approach certainly does work suggested Derek, he offered another potential solution to solve those issues. "I actually think the board of directors can provide a nice solution to deadlock. Since both of you will be directors initially, some companies will appoint a third director from the outset that both founders approve of and grant that director a small amount of stock—thus allowing that director to help break any ties." Derek had further explained that by bringing in another savvy individual initially who could offer some business expertise and experience and could break ties in the event of a conflict you may be able to create a practical solution to the potential issue (see Table 7.1). "It won't solve all problems," said Derek, "but that way you can remain equal partners with a mechanism to resolve conflicts down the road."

Teddy and Rebecca already knew who would be a great third director and minority stockholder. Alan Smith had been their professor while in college, but also owned his own engineering design company. They both recognized that his ties to the university and his experience operating a successful business would make him an ideal fit. Rebecca had already had lunch with Alan during the initial planning and Alan had already agreed to be an advisor, so they knew Alan would probably be happy to step in and serve as their third board member.

TABLE 7.1

Stock Ownership

Shareholder	Ownership (%)
Theodore Muir	49
Rebecca Carson	49
Alan Smith	2
Total	100

VESTING

The next thing on their list to tackle was discussing vesting on their stock. Derek had suggested that most cofounders will want to include vesting on at least some portion of their founders stock. Vesting ensured that if a cofounder left the company, he or she would only keep the portion of their stock that had vested. Vesting gives the company a buy-back right if the founder departs before reaching certain service time or performance milestones. So if Teddy got a terrific job offer at another university and couldn't continue with the business, he'd only be able to keep the portion of his stock that had vested. For example, if his stock was going to vest over a three-year period and Teddy left after a year, he would own 33% of his stock. Most companies would tie vesting to the expiration of time, but some companies tie it to milestones such as fundraising, product development, or other key business milestones.

Derek had suggested there were really four key things to consider with respect to vesting: length of vesting, vesting increments, if there was "cliff vesting," and if there was acceleration. "First, determine the length of time for vesting. For many high-tech or clean-tech companies, the vesting period is between three and four years. For your company, think about the amount of time it would take to get your company to reach its key goals. If you both think the business needs three years to launch the product and get sold in big-box retailers, then a three-year vesting may be the right length. If that time period is more like four or five, then a four-year vesting is probably the right length."

Both Rebecca and Teddy thought they were going to take at least four years to get their product developed and fully commercialized, so they settled on four years. They also decided to make vesting monthly so that each month over the four-year vesting period 1/48th of their stock would vest.

Derek had also offered some information on cliff vesting and acceleration. "Sometimes cofounders will want a trial period to lapse before any vesting happens—say three, six, or twelve months. That way, if things don't work out after the first few months and a founder departs, then there is no stock that would vest. Then after that trial period elapses, called the cliff, all stock that should have vested in that three, six, or twelve month period would automatically vest if the founder stuck around. For initial cofounders, you may decide you don't need it, but as you bring on new executive team members down the road, the cliff vesting concept is probably more appropriate. In addition, you may also determine that you want acceleration of vesting for certain events—usually things such as an acquisition or merger or sometimes for a termination of a founder."

Derek explained that acceleration could be for a percentage of the unvested stock—say 50% would accelerate on the acquisition—or for all of the unvested stock. In other cases, the acceleration would only occur if the founder was terminated in connection with or shortly after the merger. This can sometimes protect the founders in an acquisition. Some companies will also provide for acceleration of vesting if a cofounder is fired from the company.

"My vote is to skip the cliff vesting here since we both know one another so well, and to add in the acceleration of our stock if we are acquired," suggested Teddy. "If we get bought, there is no reason why we shouldn't get our full payout."

"I agree," said Rebecca. "When my dad's company was acquired, they came up with a package to keep him around and I'm sure they'd do the same if we ever got that far."

They agreed to give Alan the same deal with a four-year vesting on his stock, but decided to add a 12-month cliff since they weren't sure if he'd truly have time for the business as a board member and key advisor.

Rather than pay cash for their stock, Derek suggested that they exchange their intellectual property, including the patent currently in Teddy's name, for their stock. The stock would be sold to them at a very low price initially—$0.001 per share—and so rather than pay for that stock in cash, the intellectual property such as the business plan, their know-how and any tangible IP like the patent would be assigned to KinetiBaby in exchange for that stock.

OPTION PLAN

"Okay, what about our option plan then," asked Rebecca. "We know we are going to need to do some hiring in the next year, so let's be sure to reserve enough to make that possible."

Derek had suggested that most early stage companies would reserve approximately 10–30% (sometimes more and sometimes less depending on the specific circumstances) of the authorized stock of the company to be issued to future hires. Rebecca and Teddy agreed that a 20% reserve seemed about right given what they thought they'd need to give to future team members. They also settled on a four-year vesting with a one-year cliff for all optionees. When Derek had helped them to incorporate KinetiBaby, Inc., he had set up the company to initially authorize the company to issue up to 10 million shares of common stock. But both of them understood that as they took more investment in the future, KinetiBaby would authorize more common stock, which was why they didn't need to reserve more shares for their potential investors. Table 7.2 shows how their initial capitalization looks like.

TABLE 7.2

Initial Capitalization

Shareholder	Number of Shares	Ownership (%)
Theodore Muir	3,920,000	39.2
Rebecca Carson	3,920,000	39.2
Alan Smith	160,000	1.6
Option Pool	2,000,000	20.0
Total	10,000,000	100.0

Teddy and Rebecca were each appointed as directors of KinetiBaby, and Alan was appointed as the third director and issued his stock. Rebecca took the role of chief executive officer, while Teddy agreed to serve as chief technical officer. Teddy and Rebecca each receive their nearly four million shares in exchange for the assignment of any intellectual property they'd created related to the business (which included the initial research and development they'd each done as well as the "napkin" business plan they'd put together).

EMPLOYEE NUMBER ONE

Teddy and Rebecca had also agreed that they needed to hire an engineer to help build their prototype, especially since Teddy would only be working part-time initially on KinetiBaby. They'd posted the job on CraigsList and received an outstanding candidate, Julie Bowen who would be terrific to lead the development effort.

Julie understood that this was a startup company and was willing to decrease her salary demands to a level that KinetiBaby could afford, but expected a moderate stock grant to make up for the forgone salary. Teddy and Rebecca agreed to an annual salary of $40,000 and an initial option grant of 3% of the total stock authorized. Julie's options would vest over four years with a 12-month cliff. Julie signed an offer letter detailing this information and also executed an agreement assigning any inventions to the company, agreeing not to compete with KinetiBaby or solicit employees or customers for one year after she left KinetiBaby and maintaining confidentiality. Teddy, Rebecca, and Alan approved the option grant at their next board meeting.

To fund Julie's salary and some of the other startup expenses, Rebecca had agreed to put $20,000 of cash into the business. Derek suggested they issue Rebecca a simple promissory note that would earn a fair rate of interest that could either be paid off down the road or Rebecca and KinetiBaby could agree to convert that into the purchase of more stock.

ADVISORY BOARD

Teddy and Rebecca also set out to build an advisory board in order to tap into some valuable resources without having to hire any more employees before they had any money.

Teddy had identified his department chair at the University of North Dakota who had already agreed to serve as an advisor. His department chair had checked with the university and was unable to take any equity compensation for the job, but agreed to help out just the same. Teddy had also spoken to his father who was currently an executive at a local electric utility company and knew a great deal about the utility business, something both Teddy and Rebecca knew was crucial for the success of their business. His father was an easy sell—he'd been one of the first people Teddy had spoken to, so he knew he'd be a great advisor.

Rebecca had also been successful in identifying an individual to bring in as an advisor—a former client from her prior job, Mary Orosco. Mary was the purchasing manager for a large retail store. Before considering an offer to her to serve as an advisor, Rebecca first checked to see if her old employment paperwork prohibited asking a client to be an advisor. She didn't think it did, but she decided it was probably best if she gave a courtesy call to her former boss to make sure everything was clear there.

Teddy and Rebecca also knew they'd need to raise money in the near future in order to do some further testing on their prototype. So they decided to see about adding an advisor who they'd already targeted as an angel investor down the road. Tim Gotten had built and sold his business several years back and now was an active angel investor. Teddy and Rebecca had met Tim through Alan and asked if he would be a business advisor. Tim had agreed, and Rebecca and Teddy planned to use his role as an advisor to give them some key insights into the startup world as well as to help set the stage for making an investment pitch to Tim six months from now.

They had agreed to give each of their advisors who could accept options a grant of 25,000 or 0.25% of KinetiBaby, but decided to give Tim a larger grant of 50,000 since

TABLE 7.3

Capitalization

Shareholder	Number of Shares	Ownership (%)
Theodore Muir	3,920,000	39.2
Rebecca Carson	3,920,000	39.2
Alan Smith	160,000	1.5
Reserved Option Pool	1,600,000	16.0
Julie Bowen	*300,000*	*3.0*
John Muir	*25,000*	*0.3*
Mary Orosco	*25,000*	*0.3*
Tim Gotten	*50,000*	*0.5*
Total	10,000,000	100.0

they believed his role would be more substantial initially. Derek had suggested that a grant anywhere between 0.10% and 2% would be appropriate for an advisor (the more prestigious the advisor or more involved the advisor would be the higher the ownership stake) and said most advisors would receive a two-year vesting period. So Teddy and Rebecca had Derek draft up advisory board agreements for each advisor. They also approved the small option grants with Alan and issued the options to their advisors. Above is the resulting capitalization table (see Table 7.3).

For more information on some of the issues discussed in this case study, check out Chapter 24 "Founders" and Chapter 25 "Employees." These include topics such as Founder Decisions and Agreements, Founding Team Questionnaire, Board of Directors, Recruiting, Hiring an Employee, and Employee Compensation.

8

Raising Green (Money)

Money is oftentimes one of the largest struggles for a new business—getting it, managing it, keeping it, and raising more of it. Whether that money comes from sales, cash advances on the entrepreneur's credit cards, a bank loan or venture capital, money always seems to be in short supply for a startup business. This chapter is designed to help a greentrepreneur understand how much is required to get started, where to look for additional funding and how to raise it.

Starting Out

Starting a business takes money. You'll need to buy equipment (servers, beakers, or something else); you'll probably need to hire a few helping hands; and you'll usually need to handle some general administrative costs. That doesn't include things like building your product or going out and selling it. Oh yeah, and you certainly aren't getting paid yet either …

So what will it take to get things started? Much of that depends on the plan you set up for the business. Today, you can build a business for a couple hundred dollars if you make the company entirely virtual—say building a mobile phone application to track your CO_2 consumption and working out of your home. On the other hand, if you plan to develop a new solar panel or biofuel, you are likely to have a few more upfront costs needed to get a laboratory or manufacturing line up and running.

For more information and details on raising funds from various sources, see

- Raising Money from Friends and Family on page 329.
- Raising Money from Angels and Angel Groups on page 332.
- Raising Money from Venture Capital on page 336.
- Obtaining a Business Loan on page 339.
- Grants, Loans and Other Green Government Funds on page 199.

How Much Does It Take to Get Started?

While it is impossible to tell you exactly what it will cost you to get your business going, we do have some estimates to go by. Research by the Global Entrepreneurship Monitor (GEM) in 2004 suggested that the average amount needed to start a business was $53,673. That amount was less in cases where the business that was "necessity-pushed," meaning the business was created out of an unmet need in the market or industry. For the necessity-pushed businesses, those startups only required $24,467—less than half of the average business.

In his book *Entrepreneurs in High Technology*, Edward Roberts (1991) interviewed entrepreneurs who were building high-tech businesses (which would include things like information technology, biotechnology, clean technology, and other similar "tech" entrepreneurs).

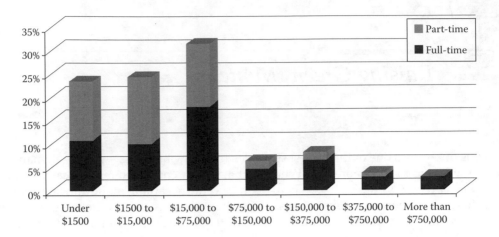

FIGURE 8.1
Part-time and full-time founder teams and the initial capitalization.

The goal of Roberts' research was to identify the key issues that face any startup in its first days. One of the areas of focus was on the initial capitalization of the business and where startups had obtained those funds. Nearly 80% of those companies surveyed said they had needed less than $75,000 in initial capitalization, with nearly 48% needing less than $15,000. And it isn't just part-time founder teams that can get started on less than $75,000. In fact, a full 71% of the full-time founders in Roberts' study were able to start their technology business on less than $75,000 of initial capital (Figure 8.1). Both the GEM and Roberts' research prove that you can in fact form and found a business without the need for substantial capital investment.

While these average figures will be helpful as you budget out initial costs and planning, you will still need to build a budget specific to your business (do you need to make paid hires, buy specific equipment, or outsource some development?) This process can be part of a business plan building exercise or can just be part of standard budgeting you go through. In any case, ensure that you are sufficiently capitalized initially to start the business—and build in a cushion as life in a startup business is rarely on schedule or budget given the challenges you'll face. This is where spending some time with Microsoft Excel (or another spreadsheet program) and a rough sketch of the business can really help you out and ensure you aren't left without money to get from formation to product launch.

Where Does the Initial Funding Come from?

If you expected to start a business and have money from outside investors pouring in over just your initial idea on a napkin, you may be quite disappointed. In fact, fewer than one in ten thousand startups have money from a venture capitalist in the bank when they open the doors for operations. Most startups start the old fashioned way—behind the passion of a founder or founders, and some money from the pockets of those same founders. Research finds that nearly 90% of company founders will invest some amount of personal money into the business at the early stages of the business, with an average investment by each founder of around $10,000. In Roberts' research, more than 74% of entrepreneurs said the primary source of their initial capitalization was personal savings of the founding team.

Personal savings	74%
Family and friends	5%
Private institutional investors	7%
Venture capital funds	5%
Nonfinancial corporations	6%
Commercial banks	0%
Public stock issuances	3%

Data from the GEM report confirms this information, finding that more than 65% of the total amount of the startup phase funding (which includes the entire startup phase, which would typically be more than just the initial three to six month period used in Roberts' research) comes from personal savings, credit, and through informal investors such as family and friends (not angels or venture capital).

And what are these initial funds used for? It somewhat depends on the industry, not surprisingly. According to Roberts, hardware companies stated that the primary purpose for their initial capital was for product development, followed by production facilities and working capital. On the other hand, software companies needed funds for working capital since software development requires programmers and technical personnel more than expensive equipment. Companies performing contract R&D listed the priority of their needs as lab equipment, product development, and working capital.

WHAT DO STAKEHOLDERS (INVESTORS/PARTNERS/ CUSTOMERS) WANT TO KNOW?

BIOFUELS COMPANIES

- Do you have long-term access to the feedstocks needed to produce your product?
- Do you have a distribution strategy and partners for your product?
- Can you succeed without incentives or government subsidies?
- How can you manage volatile commodities prices?

"GREEN" CHEMISTRY OR INDUSTRIAL BIOTECHNOLOGY COMPANIES

- Is your product better than the competition (not just as good, but better)?
- Do you have sufficient patents and related intellectual property in your space?
- Can you succeed without incentives or government subsidies?
- Does this work in Ohio, Florida, Maine _and_ California (or is this business dependent on geographical conditions)?

WATER COMPANIES

- Does your product remain attractive given increasing pricing and scarcity of water resources?
- Does your product or service model continue to work as end-users are more efficient with their water and energy consumption?
- Will your product scale to industrial, utility or consumer applications?

SOLAR COMPANIES

- Do you have long-term access to the raw materials needed to produce your product?
- Do you have a strategy to reduce costs per watt to compete with other energy sources?
- Do you have the key partners needed across the entire value chain?
- Do you have access to the capital necessary to execute?

WIND COMPANIES

- Do you have access to the necessary technology (turbines) or do you have access to the necessary raw materials for new technology?
- Do you have the rights and permissions necessary for your product and/or projects?
- Can you obtain the necessary funding for your product and/or projects?

Deciding to Raise Money

Raising money will usually be a time-consuming process regardless of whether you are trying to obtain a bank loan, raising money from friends and family, or trying to obtain the backing of an angel group or venture capital firm. The first step in the fundraising process is to actually decide to raise money. Some startups will plan to avoid or postpone the need to find outside investors for as long as possible (perhaps forever, in some cases). This approach is oftentimes termed "Bootstrapping," since the company will "pull itself up by its own bootstraps" without taking any external or minimal outside investments. Startup companies that elect to bootstrap the business will use internal cash flow, loans, and some founder money and be extremely cautious with expenses in order to finance product development and grow the business.

Utilizing bootstrapping allows the founders and other stockholders to avoid diluting their ownership interests and build the company with just the profits of the company. And even if a startup plans to raise funds down the road, using bootstrapping can be a way to minimize the amount of funds needed to get the business off the ground. This approach may be difficult for some green businesses that are capital intensive or require funding in order to get the product to market. However, companies may decide to bootstrap for a period of time in order to extend the rate of cash spending for the company.

HOW TO BOOTSTRAP A BUSINESS

Below are a few examples of approaches companies may take to bootstrap the business until profitability or future funding:

- Providing *consulting services* related to the product you are developing
- *Licensing your technology* for an alternative application
- *Factoring* your accounts receivable
- *Leasing* equipment or buildings

- Utilizing *trade credit* (purchases made on net 30, 60, or 90 days terms, for example)
- Requiring *up-front payment* from customers
- Developing favorable relationships with *key vendors*
- Enter into a *sale-leaseback* arrangement where a third party will purchase corporate assets such as computers, furniture or other company equipment for cash and lease them back to you

Sources of Funds

There are a number of sources where a new startup can obtain the money that it will need to grow. There is no one-size-fits-all strategy for funding a business—it often depends on numerous factors, many of which may be out of your control.

Entrepreneurs often will be required to self-finance the initial capital outlays for their business through personal savings, credit cards, second mortgages, and traditional bank loans. Some startup founders can also bootstrap their startup by using profits from early sales to grow the business or through using consulting as a mechanism to generate cash as the product is being developed or further refined. The bootstrapping approach works especially well in the service industry or in small businesses, where start-up expenses are sometimes low and the need for employees may initially be minimal.

For some businesses, the entrepreneur may not have enough personal savings or assets, and bootstrapping may not be an option. In these cases, the business will need to raise additional money from some of the alternative sources listed here in order to fully capitalize their business.

- Friends and family
- Angel investors
- Bank loans
- Government and public sector
- Seed funding firms
- Joint ventures and strategic alliances
- Venture capital firms

Each of these alternatives has its advantages and disadvantages, and they all require your company to form some sort of "partnership" with an investor or financier (Table 8.1). While sorting through these potential sources for funding, you should consider the amount

TABLE 8.1

Summary Details of Various Funding Sources

	Typical Maturity of Companies	Time to Obtain Funds	Amount of Funds Awarded	Likelihood of Utilizing	Costs of Funds	Comments
Founder	Early	Fast	Small	High	Low	Typical for most startup companies
Friends/ family	Early	Fast	Small	Medium	Medium	May be difficult to manage; may not provide strategic value
Angel investors	Early	Fast/ medium	Small	Low/ medium	Low	May require high degrees of effort to obtain
Boot strapping	Early/mid	None	N/A	Low	Low	Unlikely in early stage companies
Bank loans	Mid/mature	Medium/ slow	Medium	Low	Medium	Unlikely in early stage companies; less favorable terms than equity
Govt. grants/ loans	Early	Slow	Medium	Low/ medium	Low	May be difficult and slow to obtain; not all companies will be eligible
JV/strategic	Mid/mature	Medium/ slow	Medium	Low	Medium	May be difficult to obtain and be slow to close; can affect strategy choices
Venture capital	Mid/mature	Medium	Medium/ high	Low/ medium	High	Typically requires a high growth company; may limit control of founders

of control you want to retain, the amount of equity dilution you and your investors are willing to bear, and the rate of growth you wish to achieve.

Also, entrepreneurs should be mindful that the funding choices they make in one round can have effects in later financing rounds. Without proper planning, due diligence, and careful negotiation of such partnerships, an entrepreneur may inadvertently miss opportunities or relinquish future rights to valuable assets in pursuit of immediate funds. All startup companies exploring these alternative forms of financing should obtain sound advice from an attorney, accountant, and financial advisor before entering into a definitive agreement.

Your "Right" Source of Funds

One of the biggest challenges in fundraising is linking the business to the most appropriate source of funding. For example, a company building an energy efficiency software tool may not have sufficient assets to obtain a business bank loan, a biodiesel pumping station may not have luck with angel funding, and a green consulting business probably isn't going to have a great deal of success raising money from venture capitalists. However, the energy efficiency software company may have luck with an angel group focused on clean technology, the biodiesel station might be able to get an SBA loan and the green consulting business may be able to obtain seed funding from a combination of friends and family funds and a social investment club. In each case, chasing the wrong source of funds takes time away from the business and yields a low chance of success.

And identifying the right funding source isn't only based on the type of your business, but also matching the funding source with the stage of development of the business. At different stages of a company's growth, certain funding sources will be more applicable than others. Early stage companies most often rely on funds from personal savings, private investors, government grants, angel funding, and bootstrapping. While mid-stage and later-stage companies are more likely to pursue funds from bank lines of credit, venture capital sources or even private equity sources. For example, a business with only an idea isn't going to be a good fit for venture capital and probably may require personal savings initially to validate the idea.

Business Stage	Typical Investors
• Experienced founder and business plan	• Founders; family and friends; government grants; angels
• Experienced founder and prototype/beta product	• Government grants; angels; early stage venture capital
• Experienced team and developed product & customers	• Later stage venture capital; strategic investor
• Experienced team and sustained revenue growth and profitable or to be profitable in 6–24 months	• Initial public offering or strategic merger/acquisition

Fundraising Process

Your business may need to raise money to start, build, or even expand the business, and if you are unable to raise funds your business could be put on hold. As a result, fundraising efforts could be a "gating item," or something your business must do before you can move ahead. In other cases, without key funding, a company may be unable to make strategic hires, conduct research and development initiatives, purchase equipment, enter into license agreements, and exploit business opportunities.

This can apply a lot of pressure on an entrepreneur, especially for someone that has not raised money from investors, banks, or the government in the past. Raising money for your business may be a new process—and may involve skills, knowledge, or experiences that a new business owner may not have.

UNDERSTANDING WHY FUNDRAISING EFFORTS FAIL

Perhaps the best way to approach the fundraising process is to consider what causes most fundraising efforts to be unsuccessful? The list below offers the top reasons why fundraising efforts for startup businesses often fail:

- Little or no experience in fundraising
- Failure to identify appropriate funding sources

- Lack of knowledge or understanding about a particular funding source
- Not matching your business to the most appropriate funding source
- Failure to focus on fundraising
- Starting fundraising too late
- Lack of well-defined goals
- Failing to consult with outside assistance

What Are Investors and Partners Looking for?

Depending on the stage of your business, investors will be looking for different things from a potential investment oppotunity. For instance, an investment by a family member into an early stage business may be based largely on the reputation of the founder. A bank may look into credit histories of the founders, the availability of collateral and the cash flow of the business. An investment by an established venture fund may focus on current market penetration and the experience of the management team in further exploiting business opportunities.

The reality is that each investor or partner will analyze the business and its future prospects through its own lens. What one investor sees as a futile idea without a strong management team, another investor may see as a surefire success. Therefore, businesses should spend sufficient time researching potential investors and cast a broad net to identify investors that understand the technology, appreciate the relative experience of the management team, and believe in the market potential.

WHAT DO PROFESSIONAL INVESTORS LOOK FOR?

In general, the key aspects that most investors (from angels to banks to venture capitalists) evaluate in a potential investment are:

1. Team
2. Technology
3. Market potential

Investors will evaluate each of those three key areas to make a determination as to the likelihood that, given the right conditions, the business can be a significant success. More specifically, investors will look at the following:

1. Experience of management team
2. Relative skills of management team (including complementary nature of skills)
3. Strategy of management team
4. Stage of product development
5. Protections of the technology
6. Product market potential
7. Current and likely customers
8. Future financial needs of the business

GreenStartup, Inc.: A Fundraising Case Study

The following is a case study detailing some of the potential fundraising options for a green business startup—beginning from funding by the founders of the business through to raising venture capital. Here is a fictional case study of KinetiBaby:

A quick recap …

While Teddy Muir was a researcher at the University of North Dakota, his friend and college lab partner, Rebecca Carson came to him with an idea to solve the world's energy troubles. Rebecca was working as a financial analyst and had been interested in starting a business since college. Now, five years after Teddy and Rebecca had graduated, they finally had their idea. At first Teddy was skeptical, but after doing some research and analysis he figured it actually had a chance. They sat down over dinner one night and hatched a plan to leave their current employers and start their new green business: KinetiBaby. Teddy and Rebecca, with the help of their attorney Derek, chose to incorporate their business in Delaware as a corporation and then make an S-election with the IRS in order to initially be treated as a pass-through entity. They had a busy first couple of months, incorporating the business as KinetiBaby, Inc., issuing stock to the founders, forming their board of directors with Alan Smith as their third director, making their first hire of Julie Bowen and bringing on several advisors. Rebecca had put $20,000 into the business as a loan which they intended to use to pay Julie's salary and to pay for some of their startup expenses.

RAISING SOME SEED MONEY TO GET STARTED

After about a month, Teddy and Rebecca began to realize that the $20,000 loan just wasn't going to be enough to get them through their first six months—especially since Julie's salary was going to eat all of that up on its own. At their first board meeting, Teddy, Rebecca, and Alan had discussed their options and decided they needed to get enough money to get a demonstration product ready before they could even consider trying to raise money needed to commercialize their product. In the meantime, Alan had suggested that they find out if any family or friends might be willing to make a small investment to get them through building the demonstration product.

Teddy estimated that the business would need about $25,000 more in funding to get their demonstration product ready for launch. Neither of them had any more money saved up, so Rebecca suggested they talk to her sister Dorothy and her husband (who both happened to be accredited investors). It had actually been Dorothy's kids who had inspired the idea, so Rebecca thought she might be the right person to speak with.

Dorothy had already told her sister that she thought the idea was bound to be a success, given Rebecca's ingenuity and personal understanding of the energy of toddlers.

"I'm willing to make a $25,000 investment," said Dorothy. "Given where you guys are in the business development stage and all the research and development to date, how much of the company does that get me?"

"That's wonderful," said Rebecca. "Teddy, Alan and I were thinking that the business is probably worth about $500,000 right now, given what other similar startups have been valued at and all the work already done so far. So we are prepared to give you 5% of the company for your investment."

Dorothy agreed and wrote Rebecca a check for $25,000. The first step in issuing Dorothy her stake in the business required amending the Certificate of Incorporation to increase the authorized stock to 10.5 million shares (since all the other shares had been issued or reserved in the option pool). This change was approved by the board as well as Teddy, Rebecca, and Alan as individual shareholders (but not the optionees since they didn't actually hold stock, just the right to acquire stock). Once the amended Certificate of Incorporation was approved and filed with the state, Derek prepared a common stock purchase agreement in which Dorothy purchased 5% of the common stock of the business and the capitalization table looked as in Table 8.2.

The sale to Dorothy is referred to as a "priced round" since KinetiBaby is selling 5% of the company in exchange for her $25,000. If you are wondering why it looks like Dorothy only has 4.8% of KinetiBaby, remember that she is purchasing 5% of KinetiBaby *before* her investment—which we refer to as "premoney." Therefore, you calculate how many shares Dorothy was to receive by taking 5% of the 10 million shares outstanding immediately *before* her investment, then adding those to the capitalization table.

In addition, this sale determined the current value of the business—which we refer to as the "valuation." The sale of 5% of KinetiBaby for $25,000 means that before Dorothy's investment, the "premoney valuation" is $500,000 (to calculate this simply divide the investment amount of $25,000 by the percentage purchased of 5%). Then, after Dorothy's investment of $25,000, the "postmoney" valuation would be $525,000, meaning you take the premoney valuation of $500,000 and add the additional investment of $25,000.

With its first investment by Dorothy, KinetiBaby now has its $25,000 of seed money (plus the $20,000 of startup capital from Rebecca) and Rebecca and Teddy can start building their first demonstration product: the KinetiBaby One.

A CHANGE OF PLANS

With sufficient capital in the bank to get its demonstration product built, Teddy and Julie focused on the product build while Rebecca was meeting with prospective customers

TABLE 8.2

Capitalization

Shareholder	Number of Shares	Ownership (%)
Theodore Muir	3,920,000	37.3
Rebecca Carson	3,920,000	37.3
Alan Smith	160,000	1.5
Dorothy Carson	500,000	4.8
Reserved option pool	1,600,000	15.2
Julie Bowen	*300,000*	*2.9*
John Muir	*25,000*	*0.2*
Mary Orosco	*25,000*	*0.2*
Tim Gotten	*50,000*	*0.5*
Total	10,500,000	100.0

and sales partners to build some demand for the product. The team remained on track to finish the demonstration product and had lined up a few parents willing to test out their new product on their toddlers.

At their board meeting, Teddy, Rebecca, and Alan discussed the status of the KinetiBaby One and their plans for launch. Dispensing with the normal conversations, Rebecca jumped right into things.

"So I've got some good news and some bad news here," she began. "I think I stumbled upon something we hadn't thought of based on a recent meeting I'd just had. It's going to change our strategy a bit."

Rebecca told the board about the enthusiasm she'd gotten from a meeting with the purchasing department from one of the leading baby good retailers. Mary Orosco, one of their advisors, had helped her prepare a pitch and knew exactly what the baby goods retailer would be looking for. Based on a single meeting and showing the retailer some of the sketches of the products, the head of purchasing was ready to put in an order for the KinetiBaby One immediately. He'd thought the idea could be huge and wanted to be the first distributor of their product. But before he readied a purchase order, he'd asked Rebecca about some product safety issues—in particular if they'd gotten any of the product safety certifications for electronics and baby goods. Rebecca told the head of purchasing that they had not yet received the certifications, but intended to do so. The head of purchasing was disappointed and relayed the news that their stores couldn't carry the product in that case and would not even consider putting it in the hands of consumers.

"One on hand we've got a buyer ready to stock this on the shelves," Rebecca said, "but first, we've got to get a bunch of product safety certifications, which probably puts us at least a year away from launching."

"But we don't have a year's worth of cash," said Teddy. "We knew we needed to raise money, but we were hoping to make a few sales to at least have some cash on hand. This changes everything."

Adam interrupted. "It changes everything, but it also makes everything a bit easier. You just had a commitment from a huge distributor once you get your product certified for safety of children. You were going to need to do that anyways, now you've just got to do it sooner. And we just need to get some investors to help us get there."

The board discussed their funding needs to finish the demonstration product, put the product through the certification process and begin manufacturing the KinetiBaby One for the stores. Based on Adam's experience, the board felt that it would take another six months get the product ready for and through the certification process. Then they could begin manufacturing and start selling their products in earnest.

"But we'll need half a million dollars just to get us through all those hoops," said Rebecca. "And then, if this is as big as the purchaser indicated it was, we're going to need even more money to get this product into the hands of every parent we can."

"I think it's about time we had a discussion with Tim Gotten, our advisor and angel-in-waiting," replied Adam.

AN ANGEL INVESTMENT

Tim had been floored by the demonstration product that Teddy had just shown him. Not only did the outfit generate power to a battery pack, but it also looked like any other set of stylish children's clothing. All a parent had to do was snap off the batteries from the garment after a full day of play, plug them into the power station and the energy would be used throughout the house. It was brilliant, he thought. And after hearing that a major retailer in the baby market was ready to write a purchase order, he was beside himself.

"I just don't get why you guys aren't bouncing off the walls," Tim said.

"It looks like we've still got to change our plans and delay the launch," replied Rebecca. She continued on to tell Tim about the need for certain product safety certifications and the delay this would put in their plans. "We're talking about a six- to twelve-month delay here and a shortfall of about half a million dollars."

Tim sat in silence for a minute thinking about what to say next. He smiled and replied, "We're talking about more than a half a million really. We're talking about several millions once this thing is certified and you need to get selling. But you are also talking about a huge opportunity. So for now we need a half a million. You've shown me the product and the customers—now, I'll get you the money."

Tim sat down and laid out a plan to personally make an investment of $200,000 into the business and to identify another $300,000 from people in the community. He planned to bring the KinetiBaby team to the next angel group meeting the following month to introduce them to some of his investor friends and try and round up some interest. Since they were planning to raise several million dollars after the product got certified, Tim suggested they make this investment in the form of convertible debt.

"You've got a great story and I'm very confident that we can find a venture capitalist willing to write a big check once the product is certified and on the shelves at one major retailer. So I want to structure this investment as a bridge loan. That'll get the money in the door quickly and save us from having to spend a bunch of time and energy on determining a valuation and getting deal documents in place."

Tim discussed structuring his investment as a loan that would automatically convert into equity when KinetiBaby closed an investment from venture capitalists. Tim and each of the investors would be the "bridge" to the venture investment. When their loan converted into equity, it would receive a discount on the share price the venture capitalists paid. So if the share price was $1.00 per share for the venture capital fund, Tim and his other angel investors would get their stock at a discount of 20% or $0.80 per share. This way Tim and the angels would essentially piggyback off of the price the venture capital fund could negotiate and still get a discount.

And if KinetiBaby couldn't get a venture capital deal done, then the debt would continue to earn interest and would become due down the road. Tim would negotiate the deal and would help Rebecca and Teddy line up the other investors. Derek drew up a convertible promissory note purchase agreement and a promissory note for the investors and had the board approve the investment.

True to his word, Tim helped the team get connected with Angels Unlimited, the angel group that met monthly to review deals. They presented at their next meeting and one of the attendees, Mario Anderson immediately approached Tim and asked to invest in the round. The director of Angels Unlimited also approached Tim and Mario about coinvesting in the round from the fund that the angel group had setup. Rebecca agreed to convert her promissory note for her initial $20,000 loan into the note round, and Alan Smith and John Muir both agreed to make investments. Following the closing of the bridge round, KinetiBaby had the noteholders' details shown in Table 8.3.

LOOKING FOR VENTURE CAPITAL

With sufficient funds to get the product tested and certified for safety purposes, Rebecca turned her sights to identifying venture capital investors for the business.

TABLE 8.3

Noteholders and Investments

Noteholder	Amount ($)
Rebecca Carson	20,000
Tim Gotten	200,000
Alan Smith	50,000
John Muir	25,000
Mario Anderson	125,000
Angels Unlimited LLC	100,000
Total	520,000

Their business had been fortunate to receive a bit of good press and she'd turned that press and some introductions from her advisors into a few meetings with some partners at venture capital firms. Their story seemed to resonate with the investors and the investors continued to dive into the business to understand the risks and the opportunities.

Finally, Rebecca found Green Horizon Ventures who was willing to put $3 million into the business to help expand their sales beyond the first retail chain and develop other product lines. Green Horizons asked for a copy of their capitalization table which they provided (Table 8.4).

Green Horizons provided KinetiBaby with a term sheet and offered to make a $3 million investment at a premoney valuation for the business of $4 million dollars in exchange for the purchase of Series A Preferred Stock. In addition, as a condition to the deal, they want KinetiBaby to increase the size of its reserved option pool to 15% before closing the transaction, in order to ensure that the company has sufficient shares reserved for the new hiring it will need to do. Green Horizons would also receive two board seats. The convertible notes that were outstanding would convert into equity at a 20% discount to the price paid by Green Horizons as a part of the deal.

Teddy, Rebecca, and Adam met to consider the offer and discuss whether or not to approve the investment. They invited Derek, their attorney, to join them and help understand the deal.

TABLE 8.4

Capitalization

Shareholder	Number of Shares	Ownership (%)
Theodore Muir	3,920,000	37.3
Rebecca Carson	3,920,000	37.3
Alan Smith	160,000	1.5
Reserved option pool	850,000	8.1
Julie Bowen	*300,000*	*2.9*
Recent employee grants	*750,000*	*7.1*
John Muir	*25,000*	*0.2*
Mary Orosco	*25,000*	*0.2*
Tim Gotten	*50,000*	*0.5*
Dorothy Carson	500,000	4.8
Total	10,500,000	100.0

"I don't even know where to begin," said Teddy. "How do we know if this is a good deal?"

"Let's start with the basics," said Adam. "Let's look at what the company will look like immediately prior to the deal and what it looks like immediately afterwards. That'll help you understand the impact of the investment."

Derek laid out a series of spreadsheets. In order to increase the unissued portion of the option pool to 15% before the investment, the board would need to reserve another 850,000 shares, bringing the total shares issued and reserved to 11,350,000. The angel investors had just over $560,000 outstanding in principal and interest that had accrued.

Derek pointed to a spreadsheet and continued, "In order to determine the per-share price that Green Horizons is willing to pay, you'll divide the $4 million valuation by the 11.35 million shares outstanding. Here, that means a per-share price of $0.352 per share, which means the angel investors who get a 20% discount will receive a share price of $0.282 per share." Derek wrote the calculation on a white board:

$$\$4,000,000/11,350,000 = \$0.352$$

$$\$0.352 \times 80\% = \$0.282$$

"The next step is to determine how many shares they are going to purchase," he continued. "Assuming Green Horizons invests just over $3 million, they'll receive 8,525,000 shares. The angel investors who have just about $563,000 dollars outstanding will receive 2,000,000 shares."

$$\text{Green Horizons: } \$3,000,800/\$0.352 = 8,525,000 \text{ shares}$$

$$\text{Angel Investors: } \$563,200/\$0.282 = 2,000,000 \text{ shares}$$

TABLE 8.5

Premoney and Postmoney Ownership

Shareholder	Premoney		Postmoney	
	Number of Shares	Ownership (%)	Number of Shares	Ownership (%)
Theodore Muir	3,920,000	34.5	3,920,000	17.9
Rebecca Carson	3,920,000	34.5	3,920,000	17.9
Alan Smith	160,000	1.4	160,000	0.7
Reserved option pool	850,000	7.5	850,000	3.9
Increase to option pool	850,000	7.5	850,000	3.9
Julie Bowen	*300,000*	*2.6*	*300,000*	*1.4*
Employee grants	*750,000*	*6.6*	*750,000*	*3.4*
John Muir	*25,000*	*0.2*	*25,000*	*0.1*
Mary Orosco	*25,000*	*0.2*	*25,000*	*0.1*
Tim Gotten	*50,000*	*0.4*	*50,000*	*0.2*
Dorothy Carson	500,000	4.4	500,000	2.3
Green Horizons	—	—	8,525,000	39.0
Angel Investors	—	—	2,000,000	9.1
Total	11,350,000	100.0	21,875,000	100.0

Derek handed the board members a spreadsheet showing the ownership premoney (before the investment) and postmoney (after the investment) (Table 8.5).

"Okay, so basically, we are giving up about 48% of our company for $3.5 million?" asked Rebecca. "I mean it seems like a lot."

"It is," replied Adam. "But that falls about within the median range for Series A investments by venture capitalists. Generally, a company can expect to give up about 40–50% of the company in a Series A investment of under $10 million. Here, you are giving just under 40% to the venture capital fund and about 9% to your angel investors."

The board discussed the investment in more detail including understanding the liquidation preference, dividends, and the addition of two more board members.

"It is a lot of the company," said Teddy, "but it really is the only way for us to build a company that can put the KinetiBaby One in every store around the country. So I vote we do it."

Rebecca and Adam nodded in agreement and approved the term sheet, instructing Derek to get started on documenting the investment.

For more information on some of the issues discussed in this case study, check out Chapter 27 "Raising Money" and Chapter 28 "More about Fundraising." These include topics such as Raising Money from Friends and Family, Raising Money from Angels and Angel Groups, Raising Money from Venture Capital, Obtaining a Business Loan, Developing your Fundraising Strategy and Plan, Convertible Note Financing, Sample Term Sheet for Convertible Note Financing and the Venture Capital "Fit" Test.

9

Green Intellectual Property

Much like computers and pharmaceuticals before it, clean technology is fast becoming a business sector defined by research, development, and intellectual property ("IP"). Studies have found that the number of patents issued in the clean energy space (so called "green patents") have risen dramatically in the first decade of the twenty-first century and continue to ramp up growth rates—increasing by more than 135% between 1998 and 2007. Similar to prior "boom" sectors such as life sciences, software, and computer hardware, green startups share the challenge of adequately protecting their intellectual assets with the limited capital resources of an early stage business. As we all know, protecting your IP costs money (oftentimes more than you have handy), but failing to protect the IP could cost money down the road.

The obvious next question is: what should my company do to protect its intellectual property? The answer, unfortunately, is not as intuitive as the question. While the need to protect the crown jewels of your organization (your key concepts, developments, inventions, and advances) is a given, many of these protections cost money. And, unfortunately, these costs can add up quickly. Filing a patent can cost $25,000 and take five years—money that many early-stage startups just don't have. This conflict between protecting the most valuable assets of the company and finding a way to pay for it in the early stages of your company are a difficult tug-of-war.

For more information on clean energy and other "green" patents visit

http://cepgi.typepad.com/

http://greenpatentblog.com/

The practical answer for high-tech startups is that you must develop a strategy that balances the substantial upfront costs of a broad intellectual property approach with a modest budget. You need to stretch the almighty dollar to the maximum by utilizing the least expensive protections where possible and spending whatever is necessary to protect the key intellectual property for your startup. We've include a handy resource in Part V (Green Business Fundamentals) on *Strategies for Managing Startup Intellectual Property*, which provides information on the various protections for your company's intellectual property and offers some insights into developing a comprehensive strategy to utilize these tools to provide maximum protection on a startup budget.

Green IP

Law firm Heslin Rothenberg Farley & Mesiti P.C. tracks the growth of green trademark applications and has seen a dramatic spike. Specifically, so-called "green" trademark applications were 3.5 times greater in 2007 than in 2006, and had increased tenfold since just 2004. And this isn't just trademarks. The number of patent applications for green technologies has risen so rapidly that it was estimated at the end of 2009 that more than 25,000 green technologies patent applications were pending with the USPTO—nearly half of which had been filed in 2009 alone.

What does this all mean for a new company tackling its own challenging green issue? Well, the first lesson is that intellectual property is an important consideration for green businesses. It is not only important for a company to protect its intellectual assets, but it is important for a new company to spend the time and energy to determine if there is an existing patent already in the space. With a flood of new patents and patent applications issued in the clean technology sector in just the past few years, there represents a much greater likelihood that you could unintentionally infringe on an existing patent. So get familiar with the USPTO.gov Web site.

But while this increase may frighten some entrepreneurs away given the feeling that "everything good has probably already been patented," you should take heart. Sectors like software, hardware, biotechnology, and medical devices continue to see countless patents issued each year despite the billions that have already poured into researching these fields. While there remains a risk of a field becoming cluttered with patents, creative inventors continue to find ways to get patents issued each year.

In addition, green inventors and entrepreneurs should be aware that while the green movement seems like a recent phenomenon, the reality is that many of today's inventions and technologies are built on research and innovation from 30 to 40 years ago. As has been mentioned throughout this book, the rise of the modern green movement began in the 1960s and 1970s. From that movement, we saw a great deal of research, development, and commercialization of various clean and green technologies, including photo-voltaic solar, wind, biomass, and several others. And while those initial technologies were unable to gain mass adoption due to the inability to compete on price with traditional technologies, those renewable technologies have come back this time around as the basis for much of today's innovation.

According to research by Professor John Barton, many of today's green patents are not highly specialized developments—noting that the basic technologies underlying most of today's patents in the areas of photo-voltaic solar, wind, and biomass are off-patent. And since the underlying technologies are off-patent, the recent patents are related to improvements or additional features and not a fundamental technology. This finding suggests that it will be difficult for a single company to hold a monopoly over a key clean technology.

Yes, the rise of green patents, green trademarks and various other developments in the field may have cluttered what was perceived to be a wide open field at the end of the twentieth century (Figure 9.1). However, opportunities remain, and both the public and private sector appear to have embraced creating new ways to develop and utilize green intellectual property.

Basics of Intellectual Property

Intellectual property is a term used to cover a variety of intangible exclusionary rights in inventions, trade secrets, and creative works. These rights may be protected by federal law

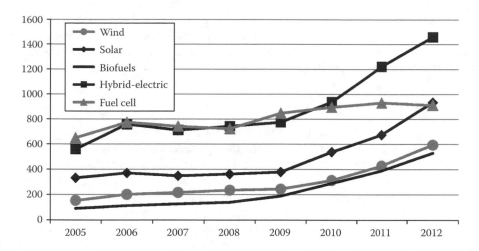

FIGURE 9.1
Patents issued by sector (actual and projected)—Analysis by Woodcock Washburn.

(such as patents) or by state law (such as trade secrets). And for many green businesses, these rights form the foundation of the business' opportunities and growth potential.

For purposes of this section, the four most important forms of intellectual property rights are those associated with patents, copyrights, trademarks, and trade secrets. Certain other systems of rights exist, such as those governing design patents, but they are relevant only in very limited circumstances. Of these four principal intellectual property rights, two are based solely on federal law (patents and copyrights); one is based on both state and federal law (trademarks); and the last is based primarily on state law (trade secrets).

Patents

A patent gives its inventor the right to prevent others from making, using, or selling the patented subject matter described in words in the patent's claims. To be eligible for a patent, the subject matter in question must fall within the statutory subject matter of the United States patent statute and additionally meet the stringent tests of novelty, utility, and non-obviousness set out in the statute. Patents may protect inventions ranging from electronic circuitry to new drugs to living organisms to software. Patents are initially owned by the individual inventors. Although state law implies an assignment for certain employees (employees who are "hired to invent"), the company should ensure that all employees (and appropriate independent contractors) execute appropriate assignments immediately after being hired.

Patents in the United States are obtained by application to the United States Patent and Trademark Office. This process is an adversarial one and takes several years. The applications must be filed by patent lawyers (or patent agents) who are licensed to practice before the Patent and Trademark Office. Patents are granted on a national basis. Consequently, a patent issued in the United States will not provide any rights in other countries. Patent laws in other countries differ significantly from those in the United States and companies should be careful to employ an experienced attorney for these matters.

Obtaining a patent is traditionally a relatively expensive and time-consuming process, generally costing between $10,000 and $30,000 per patent and taking three to five years. But many technology entrepreneurs will tell you that obtaining patent rights to your intellectual property is a crucial step. These rights provide small startups with a protected marketplace and are key factors that investors and potential members of your management team or board will consider. As such, many startups will invest significantly in protecting their inventions by patents.

Copyright

The owner of the copyright in an original work of authorship has the right to prevent others from reproducing, distributing, modifying (creating a derivative work), publicly performing, or publicly displaying the work or one which is substantially similar. Copyright law protects the expression of an idea, but not the idea itself. Copyright protection is available for works ranging from books to computer software to films.

In the United States, the employer is automatically the owner of the copyright in works created by employees "within the scope of their employment." Generally, this rule applies only to full-time employees. However, this rule does not exist in many foreign countries and the company should ensure that all employees (and appropriate independent contractors) execute assignment agreements immediately after being hired. Once again, the existence and scope of protection under copyright law varies from country to country.

Trademarks

A trademark right protects the inherent or acquired symbolic value of a word, name, symbol, or device (or a combination of the foregoing) which the trademark owner uses to identify or distinguish his or her goods (or those which he or she sponsors or endorses) from those of others. Service marks resemble trademarks, but are used to identify services. The owner of a trademark (or service mark) can prevent the use of a "confusingly similar" mark or trade name. Confusing similarity is based on a comparison of both the appearance of the two marks and the goods (services) and channels of distribution for the goods (or services).

For a green startup company, many companies choose to trademark their brand name to ensure a consistent identity of their product and to provide a consumer brand. Genentech, Google, Microsoft, Yahoo!, and Facebook all began as names of the company but have since become synonymous with the products the company produces. Protecting the "next" Yahoo! or Google is why most high tech startups will consider trademark protection for their emerging brand.

THE RISE OF "GREEN" TRADEMARKS

It shouldn't be a big surprise to anyone reading this book that there is likely to be strong growth in companies looking to trademark their green products, phrases, and verbiage. In fact, as of the end of 2009, there were over 8000 live registrations and pending applications that include the word "green" (including translations like "verde") on U.S. Trademark Office search database.

Other popular "green" terms include trademarks that contain words like "eco," "enviro," "clean," "cleantech," "organic," "earth," and several others. Some intellectual property experts have suggested that we could begin to run out of these eco-friendly trademarks—but while that could be a reality, new marks continue to be issued containing various environmentally friendly terminology, colors, and visuals.

However, the fact that there has been a rapid growth of these eco-friendly trademarks does lead to a risk of less protection of these marks down the road. According to trademark attorney Thomas Williams, we should look back just a few years to the late 1990s tech boom and the rapid increase of trademark filings relating to "i," "e," "dot-com," where the USPTO had a restrictive manner to dispense with eco-friendly marks.

Said Williams, the trademark office "is likely to afford a very narrow scope of protection to GREEN marks. As a result, if you're a commercial enterprise 'thinking green,' you're not alone in the marketplace and you won't be alone on the trademark registry. From a trademark perspective, the term 'green' is unlikely to distinguish your goods or services from those of your competitors."

Trademark protection in the United States arises under common law based upon the use of the mark in commerce, but many of the most important means of enforcing trademark rights are granted under the federal Lanham Act and specific state laws. Marks issued since 1989 under federal registration are registered and protected for 10 years, and can be renewed indefinitely. Generally, registrations of your mark under state-specific laws are for 10-year periods.

The Lanham Act permits the registration of a trademark (or service mark) based not on use, but on "intent to use." These applications permit a company to "reserve" a mark for up to four years. The registration of a trademark under the Lanham Act provides nationwide rights instead of rights limited to the geographic area of actual use provided under the common law. Registration under the Lanham Act requires application to the United States Patent and Trademark Office. The process is an adversarial one in which the application is reviewed by trademark examiners. In addition, a number of states have separate trademark statutes patterned on the Lanham Act which create separate causes of action. The protection of trademarks (and service marks) in foreign jurisdictions is very complex and varies from country to country. Many countries do _not_ provide any "common law" rights and all rights are based on registration.

Trade Secrets

Trade secrets are an underappreciated tool in the protection of intellectual assets of high-technology startups. And it is surprisingly easy—just keep it a secret. Of course, there is a bit more to it, but that's the concept in a nutshell. For a new company without the financial resources to build a substantial patent portfolio, trade secrets may be an important tool for your company as it grows.

Green businesses may use trade secrets to protect a key component of their business. For example, perhaps a biofuels company has a unique compound they use in their processing, a unique grinding or milling protocol that increases their yields or a very specific timetable for heating, cooling, and storing their biofuel. Each of these may not be something the company chooses or desires to patent, but they may insist their employees and partners keep these aspects secret in order to gain certain protections under trade secret law.

The commonly understood definition of a trade secret is any information, including a formula, technique, pattern, physical device, program, idea, process, compilation of information, or other information (1) that provides a business with a competitive advantage (that is generally unknown and not readily discoverable), and (2) where the individual or company takes reasonable steps to protect the secret and maintain these protections, absent improper acquisition or theft. A trade secret right permits the owner of the right to act against persons who breach an agreement or a confidential relationship, or who otherwise use improper means to misappropriate secret information. This allows you to retain the right to the secrets that give you a competitive advantage.

Examples of trade secrets

- Customer lists
- Software code
- Supplier lists
- Blueprints
- Maps
- Design drawings
- Business plans
- Company records (e.g., personnel, financial, sales)
- Chemical compounds
- Business processes
- Survey results
- Prototypes
- Research results
- Sales and marketing plans

In most cases, the owner of the trade secret has expended some costs to develop or exploit this trade secret. The scope of what qualifies for a trade secret is quite broad, even including negative information (where your efforts show that something is not possible and shouldn't be researched or exploited further). The key for most startup companies is that the information provides you a distinct competitive advantage—and therefore you take steps to protect this information.

Trade secret rights are, for the most part, governed by state law. California courts have generally considered the following factors in deciding whether information constitutes protectible trade secrets of a company: (1) whether the information has economic value due to its relative anonymity in the industry; (2) the company's efforts to keep the information secret, both outside the company and within the company; (3) the time and money spent by the company in developing the information; (4) the relative commercial value of the information; and (5) the ease or difficulty with which the information could be independently obtained by outsiders. The nature of the "reasonable efforts" necessary to protect a trade secret varies depending on the nature of the trade secret. They include nondisclosure agreements with employees (and other companies to whom the trade secrets are disclosed), marking any lab books and other materials as confidential, and restricting access to trade secrets on a "need-to-know" basis. Trade secrets can range from computer programs to customer lists to the formula for Coca Cola®.

Trade secret law protects owners from the wrongful appropriation of their trade secrets, but, unlike the patent law, not from independent development of the same information by other parties. Trade secrets are protected in most foreign countries, but the statutory protection is generally much weaker than in the United States. This weakness of the statutory scheme makes the use of contracts much more important in foreign countries.

WHY UTILIZE TRADE SECRETS LAWS AND RIGHTS?

- You are considering applying for a patent or have already applied for a patent, but have not yet received the patent.
- You have a trade secret that can be kept confidential over an extended period of time without unusual effort (which can remain a trade secret longer than the information can be protected by a patent).
- You have information that can't be patented.

- You have a unique process, procedure, operating manner, and so on that differentiates the way you produce a product.
- You have valuable information, but it is not the "crown jewel" of the company.

Fast-Tracking Your Green Patent

If you weren't convinced that the U.S. government is doing everything in its power to advance the green agenda, look no further than a program unveiled by the U.S. Patent and Trademark Office (USPTO) in late 2009. The goal of the new expedited review process for green technologies is designed to shave an average of one year off of the time currently required to patent a new technology. Estimates as of the date of the program launch were that there were more than 25,000 pending applications of green technologies that would be eligible for this program, of which approximately 12,000 were filed in 2009.

The program is limited to patent application for inventions which materially enhance the quality of the environment by contributing to the restoration or maintenance of the basic life-sustaining natural elements. In addition, the USPTO will extend the expedited review to applications for inventions that materially contribute to: (1) the discovery or development of renewable energy resources—including inventions relating to hydroelectric, solar, wind, renewable biomass, landfill gas, ocean (including tidal, wave, current, and thermal), geothermal, and municipal solid waste, as well as transmission, distribution, or other services directly used in providing electrical energy from these sources; (2) the more efficient utilization and conservation of energy resources—including inventions relating to the reduction of energy consumption in combustion systems, industrial equipment, and household appliances; and (3) greenhouse gas emission reduction—including inventions that contribute to advances in nuclear power generation technology, fossil fuel power generation, or industrial processes with greenhouse gas-abatement technology (e.g., inventions that significantly improve the safety and reliability of these technologies). The program is limited to only new nonreissue, nonprovisional utility applications filed prior to December 8, 2009.

For more information on the Eco-Patent Commons visit

http://www.wbcsd.org

The normal course of action for any patent application is that they are reviewed in the order they are filed—which may lead to a several year lag before a patent application even makes it onto an examiner's desk. According to the USPTO, prior to the adoption of this program "the average pendency time for applications in green technology areas is approximately 30 months to a first office action and 40 months to a final decision." Any patent applications accepted into the expedited "green" review process will be placed on the patent examiner's special docket and will be given a special status with respect to any appeals and publication processes, with the goal of knocking the time to a final decision to less than 28 months.

The initial program is limited to just 3000 patent applications, although depending on the success of the program and the political support, that number could grow. Applicants wishing to receive this review must submit an application by the announcement date of this program, December 8, 2010. However, if your green technology patent wasn't already in the queue by this date, then sadly you are not eligible for this expedited review request (again, something that could be extended depending on the success and political support).

For more information on the "fast-track" USPTO Program visit

http://www.uspto.gov/patents/init_events/green_tech.jsp

Internationally, the UK Intellectual Property Office introduced a similar program earlier in 2009 designed to move green technology patents through the process more quickly.

Eco-Patent Commons

In order to combat the perception that certain technologies that help the environment are being hoarded by multinational corporations, a number of companies have partnered to create the Eco-Patent Commons. The Eco-Patent Commons was developed in partnership with World Business Council for Sustainable Development (WBCSD), a global association of some 200 companies dealing exclusively with business and sustainable development. Since the launch of the Eco-Patent Commons in January 2008, one hundred eco-friendly patents have been pledged by 11 companies representing a variety of industries worldwide: Bosch, Dow, DuPont, Fuji-Xerox, IBM, Nokia, Pitney Bowes, Ricoh, Sony, Taisei, and Xerox.

The aim of the Eco-Patent Commons is to create a database of patents pledged allowing any company to share those environmental innovations in order to develop further innovations. The companies providing their patents to the Eco-Patent Commons promise not to sue if anyone else uses their patent, as does the user, through a so-called "defensive termination" clause most typically seen in open-source software. The Commons contains patents in the areas of alternative packaging materials, water recycling technologies, plastic recycling technologies, and eco-friendly solvent chemical compounds among others. Some experts suggest the impact of the Eco-Patent Commons will be limited due to the limited patents being offered compared with the thousands of new patents filed each year; however, organizers are hopeful that this can be a resource to help spur new innovations.

Litigation of Green IP

Copenhagen and IP

Submissions by certain developing countries (including Brazil, Saudi Arabia, Pakistan, and the Philippines) have proposed weakening Intellectual Property Rights for technologies needed to abate climate change.

These proposals assert that the industrialized countries bear historical responsibility for climate change, and in turn, must accept a greater share of the cost burden of mitigation.

While there are many people who are building green businesses to help the environment, you should not forget that green still represents big business and big money. Many large companies have staked out their future growth on patents and other intellectual properties for their green strategy.

For example, Toyota Motor Corp. one of the leading automobile manufacturers that has been ahead of the American automakers with its hybrid vehicles was a party to major litigation in 2007 related to its drive train technology for the Toyota hybrid electric vehicles. The courts ruled in favor of the inventor of the drive train technology and ruled that

Toyota's technology infringed—ordering a $25 royalty on each car sold with the infringing technology until the patent expired. According to attorney Eric Lane, the most commonly litigated category of clean technology patents is the light-emitting diode (LED) patents, possibly due to the extensive market for LED products. While the relative youth of the clean technology field has limited the number of lawsuits into matters related to green businesses and, more specifically, green intellectual property, the opportunities and the money being spent make it highly likely that we will see an increase in the litigation of these matters.

STRATEGIES FOR MANAGING STARTUP INTELLECTUAL PROPERTY

For more information on practical strategies and examples of managing intellectual property for a "resource-constrained" startup company, see Chapter 29 in Part V of the book on Green Business Fundamentals entitled *Strategies for Managing Startup Intellectual Property*.

GreenStartup, Inc.: A Case Study on Managing Intellectual Property

The following is a case study detailing the management of intellectual property by a startup company. Here is a fictional case study of KinetiBaby:

A quick recap …

While Teddy Muir was a researcher at the University of North Dakota, his friend and college lab partner, Rebecca Carson came to him with an idea to solve the world's energy troubles. Rebecca was working as a financial analyst and had been interested in starting a business since college. Now, five years after Teddy and Rebecca had graduated, they finally had their idea. At first Teddy was skeptical, but after doing some research and analysis he figured it actually had a chance. They sat down over dinner one night and hatched a plan to leave their current employers and start their new green business: KinetiBaby. Teddy and Rebecca, with the help of their attorney Derek, chose to incorporate their business in Delaware as a corporation and then make an S-election with the IRS in order to initially be treated as a pass-through entity, which following an investment from a venture capital fund had been withdrawn. The business had a busy first couple of months, incorporating the business as KinetiBaby, Inc., issuing stock to the founders, forming their board of directors with Alan Smith as their third director, making their first hire of Julie Bowen and bringing on several advisors.

After their first six months, the board recognized that the business was going to require additional funding and the company closed a convertible note financing from a series of angel investors. These funds helped the company certify the product's safety and begin selling into a major distributor. Green Horizons, a leading venture capital fund investing in clean technology, decided to invest in the business supplying KinetiBaby with funds to further product development and begin selling the product to consumers across the country.

KEEPING IP COSTS LOW

Teddy and Rebecca recognized early on that they did not have sufficient capital to take an aggressive stance on intellectual property protections. Filing patents takes money that KinetiBaby just didn't have. Teddy had initially assigned a patent he owned to the business that served as the underlying technology for their products. He knew they didn't have $20,000 to pay for each patent. However, they'd made some substantial developments and incremental inventions to the initial patent that Rebecca and Teddy knew were keys to the business and were likely to be patentable.

Teddy and Rebecca had asked Derek about ways to keep their technology protected without incurring the full costs of a patent filing.

"Obviously, there are some things that you will want to consider patenting," said Derek. "The core technology is something worth getting protected, especially if you plan to show it to potential customers or sell it. For other technology, you can consider relying on the protections offered under trade secret law, which basically requires you to keep your technology a secret."

Derek explained some of the key aspects of protecting technologies under trade secret law, including the use of nondisclosure agreements (NDAs), putting in place confidentiality agreements among employees, ensuring inventions are assigned from contractors and employees, and putting in place a trade secret policy within the company. By utilizing these protections and limiting disclosure of the core aspects of technology, you could keep the technology protected. The key to this approach is thorough documentation of the invention process—including the date of the conception of the idea and the actual date of invention.

"This strategy won't necessarily protect you internationally however," suggested Derek. "While the United States is a 'first-to-invent' jurisdiction, internationally, it is a 'first-to-file' invention policy."

PROVISIONALS, PATENTS, AND MORE

Teddy had been responsible for the internal IP protection strategy—ensuring all the engineers and researchers were using lab notebooks to document everything, insisting on NDAs and thorough agreements for confidentiality for every consultant and employee and he had ensured that any publicity was first vetted through him. This approach had done the trick as far as keeping costs related to intellectual property low while the company tried to keep itself running lean.

However, Teddy was growing concerned that the KinetaBaby One really needed to be protected before they began showing it to the potential investors, partners, and the team that would be handling their certifications. There was just too much risk that an investor would refuse to sign an NDA or some photos and diagrams could get spread around. So while Rebecca had begun to prepare to talk to investors, Teddy had begun to tackle the formal protection of the KinetiBaby technology.

Teddy knew that KinetiBaby still didn't have enough money to pay for a full patent filing, so he planned to utilize a Provisional Application for Patent. This would give them a one-year window of time to file for a full patent, but would lock in their discoveries in the meantime. Teddy and Rebecca knew that their plan required them to raise money within the year following their provisional applications in order to pay for the full filings—otherwise, it was going to be tough for them to come up with the money they needed to cover all the costs. Teddy had identified four key inventions in the KinetiBaby One and decided that those inventions each probably required a patent to maximize the protections for the complete system they'd developed.

Derek had suggested that they spend a bit of time with a patent attorney to help guide them in preparing their initial provisional applications, but to take responsibility for the bulk of the work on their applications. He'd also suggested they be careful if they choose not to patent something. Little slip ups could be the difference.

"There are three common mistakes startups make that can immediately blow your chance at a patent," said Derek. "First, if the KinetiBaby One has been 'offered for sale' for greater than one year you are not allowed to file a patent. This can be demonstration versions, partial products or even just an advertisement for the product. Second, be sure that your product has not been used publicly for more than a year. For example, if you develop a demonstration product and let your niece or nephew wear this around town in public for a year that could be enough to sink you. And finally, make sure that none of your scientists or researchers decides they want to get some industry recognition by publishing something in an industry magazine or journal. It's all about being cautious."

With a little help from Derek's colleague who was a patent attorney, Teddy prepared his four provisional applications and sent them off to the U.S. PTO. With a bit of luck and hard work, Teddy hoped to convert those provisional applications into full filings a year from now.

KINETIBABY™

"So what's the difference between using ™ and ®?" asked Rebecca.

Teddy had been discussing the reasons why he wanted to file for a trademark of their name with Rebecca.

"It really puts a stake in the ground," replied Teddy. "By registering our mark we can use the ®. We want to register our mark to show other potential companies that we are reserving this mark and to keep anyone else from accidently using the same mark. Plus, by registering our name, we reserve the trademark everywhere across the country."

Rebecca thought for a moment before responding. "So for now we can use ™ on KinetiBaby as long as we've done some diligence, but the registration mark tells people that the U.S. PTO has signed off on our name as a registered mark. Do we want to register anything else while we are at it?"

Teddy and Rebecca discussed trademarking KinetiBaby One and some other product names they'd been considering, but Teddy reminded Rebecca that the goal was to give themselves the maximum protection at the lowest costs. They agreed to look into trademarking their specific products nearer to the launches.

"Okay, what about international patents?" asked Teddy. "Is there anywhere we definitely want to sell in the near term?"

They talked about opportunities in Europe and Asia, particularly with the increased emphasis on environmental technologies in countries such as Germany, the United kingdom, and China.

"I think we are going to want to sell into Europe, but are we ready yet?" asked Rebecca. "Truthfully, we've got to get our product through U.S. certifications and testing before we start thinking about doing the same abroad. For now, I think we hold off."

For more information on some of the issues discussed in this case study, check out Chapter 29 "Strategies for Managing Startup Intellectual Property." This includes topics such as General IP Protections, Patenting for Startups and Trademarking for Startups.

10

Making the Sale

Sales are the lifeblood of any business. The green marketplace is a complicated one (as described in the first part of this book). In addition, selling into some common channels in green business has its own share of unique challenges. This chapter will examine the sales process for a startup business and specifically the unique processes for selling "green" electricity—through measures such as power purchase agreements, net-metering, and renewable energy credits. We will further examine positioning of green products and try to understand more about the issue of greenwashing in order to avoid the risks of misusing environmental claims. In addition, we will review selling green products abroad.

Ultimately, the success of a green business will depend on its ability to sell its products to its end customers, be it consumers, retailers, utilities, or large corporations. The greatest products may not always be the ones that are the highest sellers—so making the sale remains one of the most important factors that will distinguish between winners and losers in the green marketplace.

Selling as a Green Startup

Development of an effective sales organization (if that is just the entrepreneur himself or a sales staff as the business grows) is a crucial aspect of business growth and ultimately success. Green business is no different from any other business in that regard. According to a HR Chally survey of more than 50,000 North American corporate buyers, salesperson effectiveness counted for more than 39% of customer choice in purchasing decisions—impacting the ultimate product choice of a customer more than price, quality or solution. Developing an effective sales organization is crucial to expanding your product and reaching key customer markets.

While it is impossible to compare expenses and costs related to sales across all varieties of green businesses, it is possible to compare the relative costs of sales, marketing, and engineering in technology companies generally. Launch Pad, a marketing consulting organization, surveyed a group of technology companies with average revenues of $7.4 million about their relative expenses related to sales, marketing, and engineering. According to that survey, sales expenses represented approximately 24.2% of the amount spent collectively in these areas. Engineering made up the largest expense at 54.2% of their spending, while marketing was the smallest expenditure at 21.6%.

Startup companies face a unique challenge in their sales strategy: convincing decision makers that it makes sense to partner with a smaller company that may lack a proven track record. According to a survey of chief information officers (CIOs) by Launch Pad, only 48% of the individuals surveyed were willing to purchase products from a startup company. In fact, only 5% of those surveyed said they were very willing to purchase from a startup

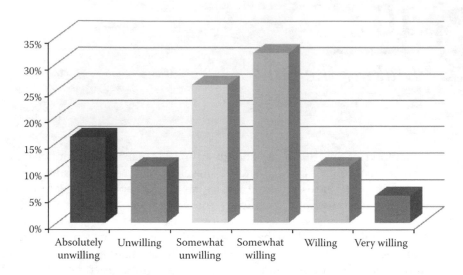

FIGURE 10.1
Willingness of chief information officers to purchase from a startup company.

company. In the survey, several CIOs raised the point that they were less willing to purchase a "mission critical" product from a startup company due to concerns of long-term
sustainability and service (Figure 10.1).

Positioning Green Products

If you are selling electricity from renewable sources to a utility that has renewable portfolio standard to meet, your sales process may be very different from a company attempting
to sell a consumer on an organic product, energy-efficient lighting, or nontoxic, all natural
paints. Each sale is different, but generally there are certain lessons being understood
about the buying process related to green products.

Arnt Meyer of the Institute for Economy and the Environment has studied green buying
habits of consumers and companies, focusing on the psychological process. He found that
with many green products, the sales process tended to focus on or lead with the "green" or
"environmental" benefits (see the case study on lighting in Chapter 4 as an example).
However, he observed that while many consumers had stated they wanted to purchase
green products, when making a buying decision, the green benefits were not typically a
significant part of the thought process while purchasing. He also noted that environmental benefits are more of a "selfless" benefit (a benefit that is not personally beneficial to the
buyer) rather than the typical "selfish" benefits (good feelings associated with the buying
process) received when buying a product.

For example, when a consumer buys a new television, the consumer typically receives
both positive mental and physical emotions, but those benefits are completely selfish in
nature (think of euphoria from buying something the consumer wants). However, this
euphoria was not necessarily the case when buying a product with benefits that do not
accrue *directly* to the purchaser. If you were to purchase an eco-friendly television, those
eco-friendly qualities do not produce the same positive emotions for a normal consumer,

potentially reducing the anticipated positive emotions a consumer expects to receive, potentially resulting in disappointment. A consumer may not feel *as good as* if they'd purchased a product entirely for selfish reasoning. (If this all seems hard to fathom, you can now understand why many experts have struggled to understand the disconnect between green talk and green action.)

Therefore, while it seems obvious that green products should tout those green benefits in the sales process (since all the studies show that consumers say they want to purchase green products), the reality is that the green message does not resonate with many typical consumers. Touting green benefits actually was not compelling as a direct consumer benefit. At the same time, the green benefits were seen as unclear, uncertain, and potentially untrue. Meyer found that consumers were easily confused by green claims and even tended to think cynically about claims from products without a strong underlying eco-friendly reputation. This lack of information further led consumers to question whether that "selfless" benefit was real or at all valuable.

Meyer offered a list of suggestions to companies looking to sell products that had environmentally beneficial qualities:

- Combine green benefits with conventional benefits.
- Provide animated, emotional green information that is easy to learn.
- Build up image and credibility.
- Contribute to the development of green labels.
- Try the unusual.

CASE STUDY ON GREEN BUILDING

SELLING GREEN IN REMODELS

The green building market accounted for 2% of all nonresidential construction starts in 2005; 10–12% in 2008; and is expected to grow to 20–25% by 2013. What are the "benefits" being sold and touted in sales of green building products, methods and practices?

MONEY SAVING

- The cost per square foot for buildings seeking LEED Certification falls into the existing range of costs for buildings not seeking LEED Certification.
- An upfront investment of 2% in green building design, on average, results in life cycle savings of 20% of the total construction costs—more than 10 times the initial investment.
- Building sale prices for energy efficient buildings are as much as 10% higher per square foot than conventional buildings.

CONSUMES LESS ENERGY AND FEWER RESOURCES

- In comparison to the average commercial buildings, green buildings:
  - Consume 26% less energy
  - Have 13% lower maintenance costs
  - Have 27% higher occupant satisfaction
  - Have 33% less greenhouse gas emissions

GOOD INVESTMENT

- Operating costs decrease 8–9%.
- Building value increases 7.5%.
- Return on investment improves 6.6%.
- Occupancy ratio increases 3.5%.
- Rent ratio increases 3%.

HEALTH CONCERNS

- Use of nontoxic paints and other green building products can improve air quality and remove potential toxins from the home.

GOVERNMENT CREDITS

- The Recovery Act included $5 billion for low-income home weatherization projects.

Sources: O'Keefe, C, *Selling Green*, *Upscale Remodeling Magazine*, October 1, 2006 and U.S. Green Building Council.

Greenwashing

SEVEN DEADLY SINS OF GREENWASHING

Terrachoice Environmental Marketing has created a list of the top seven forms of greenwashing on its site http://www.sinsofgreenwashing.org/:

- Sin of the Hidden Trade-Off
- Sin of No Proof
- Sin of Vagueness
- Sin of Worshipping False Labels
- Sin of Irrelevance
- Sin of Fibbing
- Sin of the Lesser of Two Evils

As you have read throughout this chapter, today's consumers are skeptical of environmental marketing claims and are hesitant to believe the claims are true. This lack of credibility coupled with the lack of effective standards for many green products have led many involved in green business to note that marketing green products is a huge challenge.

The term "greenwashing" is used to describe the use of misleading, confusing, exaggerated, or outright incorrect claims about the environmental benefits of a particular product.

Not only are these types of statements causing confusion to consumers, but the Federal Trade Commission (FTC) is filing claims against companies engaging in misleading advertising. In June 2009, the FTC charged three companies, Kmart Corp., Tender Corp. and Dyna-E International, with making false and unsubstantiated claims that their paper products were "biodegradable."

The FTC has developed a guide to green marketing to help companies ensure that any claims of environmental benefits are measurable, clear, verified by a third party, and relevant. Ultimately, many green marketers have been guilty of stretching the benefits of green products in order to pull consumers from traditional products.

The FTC guides offer some specific suggestions for companies to aid in their marketing efforts. Any company making any green claims in their product should review the FTC green guide as well as any guidelines published by the states where you are selling your products.

- *Substantiation.* When it comes to environmental claims, a reasonable basis often may require competent and reliable scientific evidence, which is defined as tests, analyses, research, studies, or other evidence based on the expertise of professionals in the relevant area conducted and evaluated in an objective way by qualified people using procedures generally accepted in the profession to yield accurate and reliable results.

  To download the Federal Trade Commission's guide to environmental marketing claims, visit

  http://www.ftc.gov/bcp/edu/pubs/business/energy/bus42.shtm

- *Specificity.* Claims need to be clear whether the environmental attribute or benefit refers to the product, the product packaging, or something else.

- *Limit claims.* Environmental claims should not exaggerate or overstate attributes or benefits. Comparative environmental claims should be clear to avoid consumer confusion about what is being compared.

The FTC guide also contains information on third-party certifications and claims related to biodegradability, compostability, recyclability, ozone safety, source reduction, and others. In the event your company has questions, you should consider contacting the Division of Enforcement of the Federal Trade Commission via telephone at (202) 326-2996 to receive answers to questions and clarifications.

BIOBASED VS. BIODEGRADABLE VS. BIOERODABLE VS. COMPOSTABLE VS. PHOTODEGRADABLE

When marketing products to consumers (and also complying with the FTC green labeling rules), it is important to note the differences between key "green" product features.

In the case of green plastics, some companies proudly claim to produce biobased products—including food-grade plastics from corn or sugar cane raw materials. However, just because the plastic is from a biobased feedstock doesn't mean the product is biodegradable. Some biobased products will only biodegrade in certain environments, are only partially biodegradable, may only break into smaller pieces of the material (bioerode), may only "melt" due to hydrolytic breakdown (but not

effectively biodegrade, by being digested by microbes), or may only be photodegradable by requiring sunlight. And compostability is a whole additional feature—
whereby a product is deemed to be able to be composted along with food scraps, yard
waste, and other compostable products to be used as a soil or fertilizer.

Understanding the differences between the key "green" product features is important. In an age where increasing requirements will be made with respect to green
product claims, be sure not to mislead which features fit a product.

Electricity Sales

There are a number of ways that a renewable energy startup can sell to an electric utility.
Generating renewable energy such as solar or wind power is expensive, requiring unique
business structures to make it profitable. Project financing will be covered in a later
chapter, but this section will discuss four business models that have emerged to support
renewable energy generation. These business models include (1) power purchase
agreements, (2) net-metering, (3) utility-marketer partnerships, and (4) solar renewable
energy certificates.

Power Purchase Agreement

The power purchase agreement (PPA) is a unique contractual arrangement to encourage
the development of renewable energy. This arrangement provides an alternative to a building (called the host) financing and owning a renewable
energy generation system, while still allowing electricity
customers to benefit from green energy. It has been a typical approach for solar generation, but some companies
have begun using it for small-scale wind, hydro, and other
generation approaches.

How power purchase agreements work

A third party installs and maintains renewable
energy generation (solar, wind, biomass, other)
and sells the electricity generated to a utility
at a preset price on behalf of the party leasing
its space for the generation equipment.

The transaction itself is fairly complicated, but the purpose is to allow a company or facility (think, a Walmart
store or a FedEx facility) to partner with a solar or wind company (think, SunEdison or
SunPower) and a financing company (think, individual investors or a banking entity) to
allow the company to have solar panels or wind turbines on its roof and to sell/use the
electricity (Figure 10.2, Table 10.1). The idea is that everyone in the process makes money
(the company, the solar/wind company, and the financing company) and the utility has
additional renewable energy.

Three main parties are involved in this contract: an electricity customer (host), a renewable energy provider and an entity that finances the project. In the usual arrangement, the
electricity customer provides a location for the renewable energy source (such as its roof or
its land) and benefits by being able to purchase the electricity produced at a preset rate for
the contractual period (usually more than 15 years) without providing up-front capital or
maintaining the system.

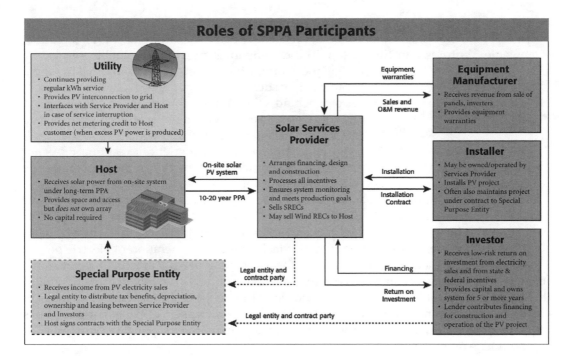

FIGURE 10.2
Roles of SPPA participants. (Data from Solar PPA, Environmental Protection Agency, http://www.epa.gov/grnpower/buygp/sppa.htm)

Financing is provided by a separate entity that obtains the income from the electricity sales, and also receives any tax benefits resulting from the solar installation. The renewable energy provider arranges the financing, chooses the solar equipment, has the responsibility for the maintenance of the system, and facilitates the relationship between the host and the investor. Other parties are also involved in this arrangement, although they are not part of the PPA. For example, the utility provides regular electricity service to the host and provides the interconnection of the solar panels to the grid. The solar equipment manufacturer also benefits from this arrangement, because it receives revenue from the project. Finally, the solar system may be installed by a separate entity that may also be responsible for the maintenance of the system.

The PPA model has been projected to drive as much as 75% of the commercial and industrial solar sales in the coming years. However, this was mostly due to the federal

TABLE 10.1

SPPA Participants

Solar Host	PPA Partner	Electricity Produced (MW)
Nellis Air Force Base (Nevada)	SunPower	14.2
FedEx (New Jersey)	BP Solar	2.42
Walmart (Puerto Rico)[a]	SunEdison	4.7
JCPenney (New Jersey and California)[b]	SunPower	4

[a] Over five Puerto Rico stores.
[b] Over 10 stores.

investment tax credit, which creates a greater incentive for investors to support solar projects. There are some companies that are already active in setting up PPAs, including Renewable Ventures and SunEdison. These companies provide a valuable service by taking the complicated aspects of solar and making them more easily understandable for the consumer. The technology behind solar is often confusing, and figuring out the complicated regulatory regimes in each state as well as the current federal energy and tax policy can be challenging. Companies with experience executing PPAs not only bring financers together with interested electricity customers, but they also make the complicated regulatory system more comprehensible.

More information on PPAs

The U.S. EPA has developed a webinar on Solar Purchase Power Agreements available at: http://www.epa.gov/grnpower/events/july28_webinar.htm

Executing a PPA requires executing two contracts: a site lease agreement and the power purchase agreement. The site lease agreement allows the solar provider to place solar panels on the host's property. The power purchase agreement specifies who will construct, own, operate, and maintain the solar facility for the generation of electric power. This contract also includes a time limit for the agreement, usually at least 15 years in the future. A minimum output of electricity is set, and contracts usually specify that all of that electricity is sold to the utility. In addition, a price is set for the host's purchase of electricity, which is not subject to any regulators' price increases. These major provisions are supplemented by a host of other provisions to ensure the proper functioning of the contractual arrangement.

BENEFITS AND CHALLENGES OF POWER PURCHASE AGREEMENT STRUCTURE

Benefits for Host Customer	Challenges for Host Customer
• No upfront capital cost	• More complex negotiations and potentially higher transaction costs than buying PV system outright
• Predictable energy pricing	
• No system performance or operating risk	
• Projects can be cash flow positive from day one	• Administrative cost of paying two separate electricity bills if system does not meet 100% of site's electric load
• Visibly demonstrable environmental commitment	
• Potential to make claims about being solar powered (if associated Renewable Energy Credits (RECs) are retained)	• Potential increase in property taxes if property value is reassessed
	• Site lease may limit ability to make changes to property that would affect PV system performance or access to the system
• Potential reduction in carbon footprint (if associated RECs are retained)	
• Potential increase in property value	• Understand tradeoffs related to REC ownership/sale
• Support for local economy and job creation	

Source: Environmental Protection Agency (http://www.epa.gov/grnpower/buygp/solarpower.htm).

CASE STUDIES ON PURCHASE POWER AGREEMENTS

SUNPOWER AND NELLIS AIR FORCE BASE

This solar project in Las Vegas, Nevada covers 140 acres and produces 14.2 MW of electricity. SunPower provides the power purchase agreement, Renewable Ventures provided project financing and owns the system, and Nevada Power Company obtains renewable energy credits. The array produces approximately 25% of the total power used at the Air Force base and saves approximately $1 million each year.

BP SOLAR AND FEDEX

BP Solar installed and operates a 2.4 MW solar system on the roof of a FedEx distribution hub in New Jersey. The 12,400 solar panel installation will provide approximately 30% of the site's energy needs. FedEx will purchase the power generated at a specified, consistent price.

SUNEDISON AND WALMART

SunEdison will finance, own, and operate rooftop solar installations over a 15-year contractual period at five Walmart stores in Puerto Rico. In return, Walmart will have a guaranteed long-term source of renewable electricity at a predictable rate and will be closer to attain its goal of using 100% renewable energy in its stores. These solar installations will provide between 25% and 35% of the total electricity consumed by the store. In addition to using renewable energy sources, Walmart has made strides to have greater energy efficiency at its stores, therefore requiring less electricity.

SUNPOWER CORPORATION AND JCPENNEY

SunPower Corporation is installing solar systems on the rooftops of 10 JCPenney stores in California and New Jersey. While a third party will finance, own, and operate the systems, JC Penney will purchase the solar-generated electricity under a power purchase agreement from SunPower. This total project is expected to produce at least 4 MW of power, which is equivalent to about 25% of the electricity needs at each store. JCPenney has also coupled this solar installation with a greater commitment to energy efficiency in its stores. As of 2007, one hundred sixty-seven JCPenney stores have undergone lighting retrofits that will save more than 27 GWh of electricity annually.

Net-Metering

An electricity customer may buy and install solar panels or wind turbines on their own property. Net-metering refers to the mechanism for calculating how much electricity a customer contributes to the utility, and how much electricity the customer uses from the customer's own generation of electricity. The customer pays nothing for electricity if the solar panels or wind turbines generate enough electricity to cover the customer's electricity needs. If the customer produces more electricity than the customer needs, that electricity is bought by the utility.

While net-metering can reduce an electric bill to zero, or even result in a check *from* the electric utility, the up-front costs can be high. For example, a company sells solar equipment or, in some cases, consumer wind turbines to the customer and links the customer into the utility.

How net-metering works

End user (home, office, etc.) purchases solar panels, wind turbines, and so on and pays for ongoing operations and maintenance costs. For any electricity produced above amounts used, the utility credits the end user.

Buying the equipment outright results in high capital expenses, which may be impossible for most electricity customers to afford. However, some cities are developing special loan programs to help alleviate these high costs. Some companies may also offer lease-to-own programs that make the high cost more manageable.

Advantages of net-metering

- Increased building value with the addition of the renewable energy system
- Set system costs (protection from escalating electricity costs)
- Free fuel from the sun/wind
- Less contracting complexity (especially when compared to power purchase agreements)
- If adequate additional electricity is produced, that electricity can be sold in the form of renewable energy certificates (RECs), resulting in additional income

Challenges of net-metering

- Maintenance of the system and replacing system components
- Monitoring the system performance and ensuring proper performance
- Dealing with the sale of RECs

CASE STUDY ON NET-METERING

BERKELEY SOLAR FINANCING PLAN

In 2007, Berkeley, California began a solar financing plan that set up a special tax district allowing property owners (both residential and commercial) to install solar panels and make other energy improvements, by repaying those improvements over a 20-year period through a property tax assessment. A typical solar installation costs about $20,000, and the property tax assessment costs about $180 a month. However, that monthly payment is offset by the electricity cost saved from generating solar power.

Many cities have adopted similar models and 11 states have passed legislation making it easier for municipalities to create their own financing plans. Vice President Joe Biden recently announced a national program called Recovery Through Retrofit, which is based on Berkeley's solar financing plan.

Third-Party Marketing Agreements with Utilities

A renewable energy company may also sell electricity to a utility by creating a special relationship with it to sell renewable energy at a premium to its customers. In these situations, a utility will pair with an independent renewable energy marketer and a renewable energy provider. The customer will in turn contract with the renewable energy marketer and provider to pay a special premium for purchasing electricity produced by renewable sources. The premium is usually between one and three cents above the regular rate for electricity. However, it is important to note that the actual electricity used by the customers paying a premium rate is not necessarily produced by renewable sources. Their premium payments ensure that more renewable energy is part of the grid, not that they are using only electricity produced by the renewable sources.

Partnerships between utilities and renewable energy marketers have proven more successful than attempts by utilities to sell renewable energy to its customers. Utilities often have limited experience in procuring renewable energy supplies and marketing renewable energy. Renewable energy marketers can bring their expertise to the arrangement, along with any special relationships they may have with renewable energy providers. Sometimes these partnerships are mandated by state law after a restructuring agreement or as part of a merger agreement, but a number of utilities have entered into these agreements voluntarily.

How third-party marketing agreements works

Customer pays a premium to an electric utility to purchase "green electricity" that has been generated by renewable sources.

Benefits of third-party marketing agreements

- Building on utility strengths (access to customers, reputation, name recognition)
- Sharing the startup costs with the third party
- Supplementing in-house marketing capabilities and access to renewable energy technologies
- Reducing utility's risks associated in offering the program
- Access to larger marketing budgets

Challenges

- Requiring greater coordination and collaboration with an external entity
- Reducing the profit potential for the utility
- Potential customer confusion
- Possible lengthy or delayed regulatory approval process

CASE STUDY ON THIRD-PARTY MARKETING AGREEMENTS

PALOALTOGREEN

City of Palo Alto Utilities voluntarily partnered with 3 Phases Energy in 2003 to establish a green pricing program. Residential customers pay a premium of 1.5 cents/kWh, while industrial customers can buy renewable energy in blocks of 1000 kWh for $15 per block. This renewable electricity comes from a blend of 97.5% wind (Oregon, Washington, Wyoming) and 2.5% solar (northern California). Within two years of the beginning of the program, over 14% of the customer base was enrolled in the program and collectively purchased more than 30 GWh of renewable energy annually. This was a huge improvement over the prior utility-run program where less than 1% of the customer base was enrolled in the renewable energy program over three years.

Renewable Energy Certificates

Besides being good for the environment, renewable energy generation has a number of other valuable aspects that make it an attractive investment opportunity. For example, a wind project creates tax credits for every watt of energy the wind turbines generate. Some investors may wish to acquire those tax credits to offset tax gains they have elsewhere. Other renewable generation projects such as solar panels or a biomass facility will also generate a valuable resource—renewable energy certificates (RECs). And these RECs can be sold or traded as another asset tied to renewable energy generation.

Generally, RECs are tradeable, nontangible energy commodities that represent one megawatt-hour (MWh) of electricity generated from a renewable energy resource. Their value is most often tied to certain state laws mandating that utilities in the state generate a certain portion of the total energy consumed in the state from renewable sources. And if a utility is unable to put in enough wind turbines or hydroelectric turbines themselves, they can look to buy RECs to subsidize their shortfalls.

Nearly 30 states have renewable portfolio standards (RPSs) that provide for escalating renewable energy generation requirements for the state's utilities. If a utility can only get 5% of its electricity generation through installation of its own renewable sources, the utility may not need to put up a wind firm itself, instead meeting the 7% RPS through the use of RECs. What makes this approach unique is that the utility and its customers are not necessarily using the green electricity produced by the renewable energy source. For example, a solar farm in Arizona may produce electricity that is used locally, while a utility in Northern California purchases RECs from that solar farm representing the electricity consumed locally. The REC symbolizes the purchase of renewable energy, even if the utility's customers are using electricity that was actually produced by fossil fuel sources.

How solar renewable energy certificates work

Renewable energy generated in one location can sell its SRECs to another utility, organization or other state as a nontangible energy commodity.

A startup company may be uniquely placed to help utilities meet their RPS requirements and can build the sale of RECs into their financial models. An RPS is the percentage of renewable energy required in the electricity mix for a utility in a particular state. Twenty-nine states and DC have developed RPS requirements, and meeting these standards can often be expensive for utilities. For example, California requires that 33% of electricity be obtained from renewable sources by 2020, while Colorado requires 20% by 2020. In addition to this compliance market created through state legislation, there is also a voluntary market made up of companies wanting to support renewable electricity production.

CASE STUDY ON RENEWABLE ENERGY CERTIFICATES

SUN CHIPS

Sun Chips, a product of Frito-Lay, installed solar panels to generate power to be used in its manufacturing facilities. However, the company wanted to do more to show it was working to make its chips entire from the power of the sun, so Frito-Lay and Sun Chips purchased RECs from solar generation.

Selling Your Products Abroad

While the United States remains the largest single market in the world, international markets may be significant and more willing to purchase products with environmental benefits. Within much of the European Union, Japan, and even China, consumers are closely monitoring environmentally friendly products. Today, sales into global markets represent an increasing part of any company's long-term growth strategy. But such a strategy is not without its share of challenges and obstacles to overcome. Determining where to sell the products, how to reformulate or reconfigure the product, how to ship the product, and understanding the rules, domestically and internationally, represent a few questions you'll face.

EXPORT ASSISTANCE CENTER

For companies that are new to exporting and are considering exporting into certain target markets, consider scheduling an appointment with your local Export Assistance Center, a service of the U.S. Department of Commerce. These local offices are located in over 100 U.S. cities and 80 international cities.

To find a local office in your city or state, visit: http://www.export.gov/eac/

For general questions about exporting such as tariff rates or U.S. Federal Government export assistance programs, call the Trade Information Center at 1-800-USATRAD(E).

Understanding International Issues

The sale of products abroad is not without its own share of challenges. Labeling products for international markets is itself a challenge and coupling that with the requirements for making green certifications and claims makes the process even more difficult. The European Union has a voluntary eco-labeling scheme known as European Ecolabel. More information is available at http://ec.europa.eu/environment/ecolabel/.

What are a few of the key issues to watch for when considering selling your products to green consumers around the globe?

- *Packaging and labeling.* Language will make labeling a product abroad different than the packaging utilized within the United States. Even countries that speak English may not associate certain phrases, markings, or graphics in the same manner as Americans. Be aware of local restrictions on the use of certain international trademarks or confusion that may result from the use of a name associated with another product. Some countries have rules requiring packaging to be readable in numerous languages as well. And, as discussed above, green claims must comply with rules within the geography where the product is being sold. Greenwashing has not only been an issue within the United States, but in other countries as well. Ensure that any claims and validation meets the standard of the jurisdiction in question.

- *Shipping and distribution issues.* Shipping internationally involves a number of issues from protecting the products as they are shipped, international tariffs, taxes, and fees, distribution routes, documentation, insurance, and language barriers. To overcome these challenges, many companies will hire a freight forwarder to assist in the process. A freight forwarder is generally aware of the majority of potential problems, can provide you information about the process before you ship the product, will be able to give you packaging and distribution guidance, and can assist with distribution once the product passes through the local port. The costs of a freight forwarder can usually be included in the overall costs to ship your products to an international destination.

- *Legal issues.* Many countries have very specific rules and regulations governing imports from another country. In addition, you may find that sales will trigger certain licensing or filing requirements. Some products such as technology, medicines or biotechnology, and agricultural products have stringent restrictions on international transactions. Likewise, you should be cautious with items that could be used for potential terrorism acts which are subject to both export and import restrictions. In most cases, it is helpful to retain local counsel to address some of the likely issues that will arise entering into a new country.

- *Product modifications.* Entering into some regions or countries may require substantial changes to your products. In particular, companies will face obstacles with respect to the exclusive use of the metric system, differing electrical components, varied mobile and wireless infrastructure, and requirements for use of different languages for certain products. For example, software companies have been required to change the language displayed for users. Before you consider entering a new country, be certain to research the product modifications necessary to determine if these changes represent a major overhaul or simply a minor modification and the associated costs.

- *Instructions and warranties.* As with packaging or labeling, product instructions or warranties will require translation into the local language. With technical products, it is even more crucial to find a local service that can provide translation of technical terminology to avoid confusion (remember the time you tried to read instructions to a product manufactured abroad with poorly translated English). Likewise, be certain to provide clear and understandable instructions related to any product warranties as well as contact information.

Identifying International Markets

For most new market introductions, you will need to consider a feasibility study looking at the market demand and the opportunities for your product to fill market needs. The feasibility study will be based on an understanding of individual country or regional markets, including local demand, pricing structures, distribution mechanisms, and delivery means. Experts will recommend that a company focus its efforts on the top two or three best prospect markets, rather than attempt a broad platform of international growth up front. Determine which markets should be based on a larger review of a number of markets, perhaps beginning research to understand a particular region or regions such as Europe, Southeast Asia, or South America, for example, before selecting individual countries or localities.

U.S. DEPARTMENT OF COMMERCE RESOURCES

One of the first places to begin your study is from information compiled by the United States Department of Commerce, specifically on their Export.gov Web site (found at http://export.gov). Visit the U.S. Commercial Service Market Research Library which houses more than 100,000 industry and country-specific market reports that contain local information useful in understanding numerous markets.

More specifically, the U.S. Commercial Service Market Research Library contains:

- Country commercial guides
- Industry overviews
- Market updates
- Multilateral development bank reports
- Best markets
- Industry/regional reports

One of the first steps in identifying potential global markets is to research trading information to identify countries currently purchasing products in your market. One of the mistakes some companies make in identifying a country for growth is selecting a market that is doing little importation in a particular market of interest. Therefore, it is helpful to research legal considerations, taxes, tariffs, market openness, common practices, and distribution channels to identify targets.

Once you have narrowed the search down to a few key markets of interest, you will want to further research macro trends in the country, including economic and demographic trends. At this stage, you should also examine the price and cost implications of a marketplace. In addition, be cognizant of barriers to entry in any new marketplace, including tariffs, or the potential existence of U.S. or foreign incentives to exportation of your product or service. If you are planning to build a facility to generate electricity or refine biofuels in a foreign market, be sure that you are aware of intellectual property rules in that region as well.

Finally, once you've selected a market or markets, you may wish to perform actual testing of demand within an international market. The U.S. Commercial Service offers the following services to assist with these efforts including catalog exhibitions, access to the magazine *Commercial News USA*, and services such as foreign partner matching and trade lead services.

LESSONS LEARNED ABOUT SELLING ABROAD

Selling your products internationally is significant. Here are a few of the key items you should consider when selling internationally:

- *Failing to secure a local partner (or to secure a good one).* Much the same way as selling domestically can be driven by the talent or lack of talent you employ in your sales and marketing forces, so too can you struggle in your selection of local partners. Selling products in international locations involves a series of challenges from import–export rules and tariffs, local distribution, finding local retailers, and transactions sometimes without face-to-face interactions. In some cases, international operations can involve different ethical considerations which may not come to light until after the transaction has taken place. Be sure to find a partner to work with in the local jurisdiction that is trustworthy and capable of assisting with your business efforts.
- *Failing to adequately commit to international operations.* International markets aren't the place to enter quickly and hope for the best. International companies may be viewed with a suspicious eye at first and will require a sustained commitment to secure significant penetration. Before you consider entering a market, be certain that you are prepared to commit for the time necessary.
- *Failing to understand how to sell and market your product.* If you invested a great deal of time and energy into determining how to sell your product domestically and, after a bit of trial and error, found success, you may be tempted to apply those same lessons to international markets. Remember that international markets may have some similarities, but will also require customizations. Starbucks found that it needed to market to younger women in Japan to get a foothold. So remember that just getting your product onto shelves won't be enough. Be sure to develop a specific marketing and sales plan for the locality.
- *Failing to modify products to meet local needs or demands.* Selling a product into a foreign market will oftentimes involve a number of modifications from sizes (oftentimes metric) to languages to electrical conversions to packaging. Have you ever traveled to Europe or Asia and tasted the differences in the flavor of sodas such as Coke or Pepsi? If you did, you'd know that the beverages were reformulated to match the tastes of the local consumers. For some products, the packaging will need to be completely changed to make your product appealing to a consumer from another country. Likewise, you may need to modify the product or packaging to comply with local laws, rules, or even customs. Be certain to research effectively before making a single sale into an international jurisdiction.

GreenStartup, Inc.: A Case Study about Selling Green Products

The following is a case study detailing the process a green business goes through in selling its products. Here is a fictional case study of KinetiBaby:

A quick recap …

While Teddy Muir was a researcher at the University of North Dakota, his friend and college lab partner, Rebecca Carson came to him with an idea to solve the world's energy troubles. Rebecca was working as a financial analyst and had been interested in starting a business since college. Now, five years after Teddy and Rebecca had graduated, they finally had their idea. At first Teddy was skeptical, but after doing some research and analysis he figured it actually had a chance. They sat down over dinner one night and hatched a plan to leave their current employers and start their new green business: KinetiBaby. After their first six months, the board recognized that the business was going to require additional funding and the company closed a convertible note financing from a series of angel investors. These funds helped the company certify the product's safety and begin preparing the product for sale to a major distributor. Green Horizons, a leading venture capital fund investing in clean technology, decided to invest in the business supplying KinetiBaby with funds to further product development and begin selling the product to consumers across the country.

POSITIONING KINETIBABY

The KinetiBaby One was a technological marvel. Teddy and the entire development team had created a product that was lightweight, easily integrated into children's clothing, had sensor and battery storage systems that could be moved from one outfit to another in a snap, and had a stylish energy converter pack that plugged right into an outlet. The clothing choices were hip and parents were pleased with both the look and feel of the clothing. Plus, the product was even more effective at converting kinetic energy into stored electricity—20 to 25% effective according to testing. Take a normal five-year old playing for a day and the KinetiBaby One would charge batteries able to supply 30% of a typical home's power.

"People are going to be fighting to get their hands on this," said Teddy as he finished up his reporting on the product at a Board meeting. "It's going to fly off the shelves."

Tony DePetrillo, one of the board members from Green Horizons, nodded in approval. "That's truly fantastic," he said. "The KinetiBaby One is an outstanding product. But, frankly I'm worried it won't sell. Maybe it'll sell a few, but I don't think we can sell to the mass market."

Rebecca looked at Tony and stared quizzically. "What do you mean 'it won't sell'? The product is fantastic—that's what you just said. I don't see why it wouldn't sell.

The product is eco-friendly baby clothing that cuts your energy costs by a third. Parents are going to be totally enthused to get their hands on this."

"Listen, I get it and I agree that the product is top notch," he replied, "but there are bigger problems here than just making a technically superior product." Tony explained his concerns about what made this product a "must-have" for a parent. There were some outstanding benefits—energy generation, eco-friendly manufacturing, and an ease of use. But at the same time, they were asking parents to plunk down at nearly $500 for a couple sets of sensors, battery packs and the energy converter system that made it all run. Then, parents were being asked to buy clothing that was specially manufactured for these sensors—clothing that was 5–10% more expensive than most baby clothing and 50% more expensive than some of the discount brands.

"And then we had to convince these same parents to spend all this money just to wrap their precious child in a bunch of wires and batteries for some electricity savings," Tony finished. "All that is why I think we've got a potential sales problem here."

MOMS AND GREEN CONSUMERS

Adam Smith, their initial outside director also chimed in, "Tony really has a good point here, guys. I've been wondering all along how we convince someone to buy this. The product is neat and has some great green benefits, but help us out here: describe the person who is going to buy this system and why they'll part with their hard-earned money to get their hands on it?"

"We thought that any middle-class or upper-middle class parent would buy it," Teddy replied. "These parents spend tons on their kids anyways. With this, sure, there is an upfront cost, but people pay more for green, right? This helps them cut their carbon footprint. People want to buy green—and we know that is a fact."

"A fact, huh? See, I think that's where the problem lies," replied Adam. "Companies in the past thought that consumers would pay more for their hybrid car or for their energy efficient lightbulb, but the truth is they *won't* actually pay more just to help the environment. We've got to give them more than that."

Rebecca chimed in, "But why isn't a reduced dependence on fossil fuels enough? They don't have to do anything—just dress up their kids. Won't people pay for greener energy that comes from their kid running around?"

"I don't think they will either," replied Tony. "But, parents may pay something in order to save some money down the road. If this pays for itself, maybe that's enough. What's the payback time on the system?"

Teddy walked them through the calculation to show how he'd arrived at the savings of about $30 per month in an electricity bill with an average toddler using the device. "Assuming the system costs a total of $500–$600," said Teddy, "you can pay it off before your walking one-year old turns three."

"That's just too long, I think. That makes me concerned that we are thinking about this all wrong," said Tony. "Why is a mom going to pay $600 to buy a set of wires and sensors to strap on her kid? It requires too many steps in your brain to become convinced it is a good idea. The reality is, is it worth saving $30 per month for that small chance my kid gets electrocuted wearing this stuff?"

Teddy looked horrified and began to protest citing all the safety tests they'd passed. Adam stopped him. "Teddy. Teddy, whoa. We all know this thing is safe. What we are talking about is *what* are you selling here? You and Rebecca keep talking about green energy and selling parents on green energy. But that's not it. When a parent goes into the store and sees this on the shelf, they aren't buying energy savings; they are buying baby clothing."

The meeting got silent for a minute as that thought sunk in. Perhaps they had been thinking about things all wrong. The energy thing was important, but was it enough?

"But we are selling clothing. These kids' clothes are stylish," said Rebecca. "We've had lots of parents tell us that. They also seem to like the tracking aspect of the clothing—you can see if your child is getting their exercise and how much energy they generate, if your daughter is being active and even if they have any hard falls. Teddy has some great tracking mechanisms in the system. Plus, the clothing is made from natural fibers and use natural, chemical-free dyes. And the other added benefit we've seen is that the sensors come with just a bit of padding that parents love—it protects their kids when they are young and may fall. Who knew, but all the moms and dads we talk to think that is one of the greatest ..."

"That's it!" interrupted Tony and Adam shook his head in agreement. "I know it isn't at all what you guys set out to do, but you've created outfits that are stylish, all-natural, allow a parent to worry less and give your kids a bit of extra protection in the world. That's it. Now *that's* our angle."

The board discussed positioning the product as baby clothing that was priced above other clothing because it protected your child from the elements. It was a way for parents to better protect and care for their child. All the benefits of electricity generation were great and something that parents would certainly appreciate, but didn't seem to be enough to get someone to hand over their hard-earned cash.

"Then what do we do with all these electricity converter boxes," asked Teddy.

PARTNERING

"Oh, I'm still with you there. But I'm still concerned about the upfront costs," said Tony. "Listen, we are talking about a $500 starting price on top of diapers, day care, strollers, and everything else. I think we need to figure out a way to cut the startup costs way down to use this product. Teddy, let me ask you something, how does a parent know how much electricity this system is generating?"

"The whole thing is wireless," he said. "Plug it in and it sends a signal to your home computer which tracks all the energy you are saving. We've got the whole thing integrated so that all you need to do is click on our monitoring software or login through the web and you can figure out exactly how much electricity you are saving."

"I see where you are going," replied Adam. "You may be onto something, Tony."

Rebecca and Teddy looked from Adam back to Tony waiting for an explanation.

"We need to get parents a way to fund their purchase of the KinetiBaby One without having to eat $500," said Tony. "I've got a couple ideas for you guys to consider: rebates and leasing. We've got to think about the value of what we are producing here."

Tony and Adam explained that KinetiBaby needed to look into finding creative ways to have someone other than the parents fund the $500 of initial costs. On one hand, they should see if any government agencies might offer tax credits for the purchase of these energy generation devices, much like purchasing solar panels or an electric car. If the credit was sufficient, parents might be able to only pay a couple of hundred bucks out of pocket or better yet, nothing. "We're helping utilities and the state governments become greener," said Adam, "so they should want to help you gain adoption." The problem was that this process could take a while unless they were able to get a quick ruling for their product. But everyone agreed they needed to see if there were opportunities to find rebates, credits or other programs to spur adoption and a lower price point.

"Our other option may be skipping the sales approach altogether," said Tony. "What about utilizing a model much like a purchase power agreement—but on a much smaller scale?"

Tony went on to suggest that the team look into leasing the equipment to parents and utilizing the energy being generated as a revenue stream and a source of renewable energy credits. By having a financing partner actually own the energy conversion system and taking a cut of the green energy being generated may be a way to get a third party to buy the product and get it into the hands of parents.

"We're talking about a complicated strategy here, we understand that," suggested Tony. "But we need to look to partners at the local utilities, the state and local government, and potentially some financial partners as well. This gets the biggest hurdle out of the way and may get our device into the hands of every parent."

Rebecca and Teddy were surprised at the turn things had taken in KinetiBaby. What started out as a product designed to help the world's energy problems, turned around entirely. Sure, they'd be generating electricity and helping the utilities meet their growing renewable energy generation demands, but they were in the business of safe and healthy baby clothing. KinetiBaby still became a green business—just not how Rebecca and Teddy ever imagined…

Part III

The Green Playing Field

11

Understanding Utilities

Electricity. It remains one of the most vital resources in our economy. And, as a result, technologies for creating clean electricity, transmitting it more effectively, and using it more efficiently represent three of the largest and the most significant markets within the clean tech sector. In fact, well over half of the investment dollars into the broader category of "clean technology" has been into technologies associated with electricity generation, energy efficiency, and related technologies.

And while not all clean technologies will directly interface with an electric utility, one of reasons for much of the optimism for clean technologies comes from the huge opportunities of revolutionizing today's electric utilities. From solar, wind, tidal, biomass, or countless other energy generation technologies from renewable sources to new means of transmissions and storage of electricity to smart grid technologies to energy efficiency—all of these emerging sectors interface with electric utilities. For this reason, we've devoted an entire chapter to help a greentrepreneur understand utilities.

THE BASICS OF ELECTRICITY PRODUCTION

Electricity is the flow of electrons—tiny, charged atomic particles. The methods for producing these electrons are different for different fuel sources; however, all use the same basic mechanism. The process for creating electricity is based on discoveries in the first few decades of the nineteenth century.

In the case of some traditional electricity generators, fossil fuels (such as coal, natural gas, or oil) are used to heat water until it becomes steam. The steam then turns a turbine blade attached to a magnet surrounded by coils of wire. In the case of hydropower, falling water turns a waterwheel that has its shaft attached to a magnet. For wind power, the wind moves the blades of the wind turbine, which causes a magnet inside the generator to move past copper wires, which converts mechanical energy into electricity. In summary, the rotation of the turbine or waterwheel moves the magnet so that its lines of force move through the wire, which creates an electric current in the wire.

The electricity generated operates under a pressure (called voltage) which can be varied depending on whether the electricity is being transported over long distances or being used in residences and industrial or commercial centers. The current may be direct current (DC) or alternating current (AC). The alternating current (AC) reverses its polarity at a standard frequency.

The Electricity Business

The electricity market makes up 7% of the gross domestic product (GDP) in the United States and is vital to the functioning of our economy and our society. To understand the

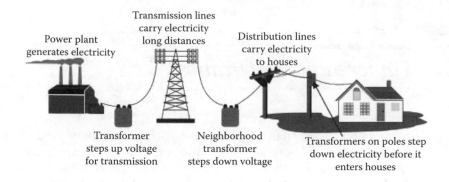

FIGURE 11.1

Generation and distribution of electricity. (Data from U.S. Energy Information Administration, http://tonto.eia. doe.gov/energyexplained/)

impact of electricity on our society, one only has to look back to the recent rolling blackouts in California and large power outages on the East Coast in 2003 to demonstrate the disruption caused by the loss of power. Experts have estimated that this Northeast blackout resulted in a six billion dollar economic loss to the region. Sun Microsystems estimates that any electricity blackout costs the company one million dollars every minute.

And while the electricity market itself represents such a large and vital piece of the world's economic stability, most utilities continue to rely on technology for generation and distribution that was developed in the early twentieth century, in some cases, is over a hundred years old (Figure 11.1). The electricity industry is capital intensive, with startup costs for renewable energy installations reaching into the hundreds of millions of dollars, and costs for traditional fossil fuel plants being tied directly to the fluctuating price of fuel.

Electric utilities play an obvious and important role in our daily lives. A utility must generate enough power each day to meet the forecasted peak demand for electricity plus an adequate reserve margin. If the power needed in a given period exceeds the amount generated, we experience blackouts. There are numerous options for energy generation, although fossil fuels, such as coal, natural gas, and oil, generate most of the energy for electric utilities. Nuclear power and renewable power sources, including solar, wind, and hydropower, provide the balance of the energy generation, and these types of sources will likely increase in the future. Once produced, electricity is distributed through a system of power lines to residential, industrial, and commercial consumers.

BALANCING DEMAND FOR POWER

Utilities often have multiple generating stations operating to meet the need of customers serviced by the utility. Electricity is not stored once it is produced by a generator and is delivered directly to customers across the electric lines. Therefore, a utility must have an idea of what the *peak demand* for electricity will be on any given day. The utility must not only meet that peak demand, but must also have a *reserve margin* in case a power station breaks down or there are any other unforeseen disruptions in service.

Most utilities have a generator that operates continuously at a fixed level each day (*base-loaded generator*). This base-loaded generator will produce most of the required power, but will be supplemented by a peaking generator when customer demand peaks. Peaks can occur at different times of the day during different seasons of the year. For example, during the summer, afternoons are often peak times because customers use their airconditioners. Similarly, cold snaps during the winter can cause a strain on the power system because more people are using their heaters (although many people use alternate forms of heating, such as natural gas).

Utilities must be aware of the weather conditions and the usage patterns of their customers in making decisions about the peak use and their reserve for any given day. If a utility does not produce enough electricity, and there is no supplemental electricity from another part of the electric grid (to be discussed in more detail later), then customers will be without power, which can disrupt the economy (loss of production) and public safety (traffic lights failing to work). Utilities also often have spinning reserves, which are power plants that are in operation, but not producing electricity, just waiting for the moment when they need to be put online quickly.

Structure of Electric Utilities

Electric utilities are natural monopolies. Rather than stringing up several duplicate sets of electric power lines in each neighborhood, there is one set of power lines to transmit electricity to the customers. Not only in the transmission of electricity, but also in the generation of electricity, having one utility provide electricity for a single geographic area utilizes economy of scale and ensures minimum levels of service (some have even likened the electricity grid to the local police or fire department—a basic right of each citizen). However, this natural monopoly also has the potential to lead to abuses by the utility companies because a lack of competition may lead to unfair pricing by the utilities. To avoid this, the federal government has put in place systems and bodies to regulate the activities of utility companies and the pricing of electricity. The federal government has also become more involved in regulating the environmental effects of electricity production, and in encouraging increased use of renewable energy. In addition, states also regulate electric utilities in varying degrees.

Types of Utilities: POUs, IOUs, and Rural Co-Ops

The structure of these types of utilities varies, but all feed into the interconnected power system. There are three primary types of utilities: (1) an investor-owned utility (IOU), (2) a public-owned utility (POU), or (3) a rural electric co-op (Table 11.1).

- *Investor-owned utilities* dominate the generating market, but have recently begun increasing their sales of generation capacity to nonutility generating companies. As much as 25% of electricity output may now be by nonutility generating companies.

- *Public-owned utilities* include over 2000 electric utilities owned by city, state and federal governments, and agencies. For example, the federal government owns the

TABLE 11.1

Electricity Production by Generator Type

Generator Type	Capacity[a]		Example
	%	GW	
Investor-owned utilities (IOUs)	67.0	556.9	Pacific Gas & Electric (CA); Puget Sound Energy (WA)
Nonutility sources (primarily investor-owned)	11.9	99.2	Combined heat and power plants that produce process heat (e.g., steam) for business activities and use the surplus heat to generate electricity for sale to utilities
Federally owned POUs	8.3	68.9	Tennessee Valley Authority (TN)
Municipality-owned POUs	5.3	43.8	Seattle Department of Lighting, Seattle City Light (WA)
State agencies and public power districts	4.4	36.7	New York Power Authority (NY)
Rural electric co-ops	3.1	26.1	Old Dominion Electric Co-operative (VA)

[a] Capacity figures as of 1998.

Tennessee Valley Authority in Tennessee and hydroelectric facilities on the Columbia River Valley in the Northwest. Public-owned utilities are exempt from many of the restrictions put in place by the regulatory agencies and are not required to pay income tax on any profits earned.

- *Rural electric co-ops:* There are approximately 900 electric cooperatives owned by members serving rural areas of the country.

Electricity Generation

In the United States, most electric power is generated by fossil fuels: particularly coal, natural gas, and some oil (Figure 11.2). There is also a good portion of electricity produced by nuclear sources (Table 11.2). Renewable sources of energy started with hydroelectric power from the construction of dams, particularly in the West. Renewable sources of energy are becoming more important, as concerns about the environment have led to more innovation, stricter regulations on the use of fossil fuels, and the development of alternative fuel sources. Geothermal, solar, wind, and biomass still make up relatively small sources of electric power generation, but these are likely to grow under sympathetic governmental regulations and policies. Each of these sources of electricity generation has a number of benefits and also drawbacks (Table 11.3).

Fossil Fuels

Coal was the first fuel source used to produce electricity on a massive scale in the United States. It was readily available and relatively inexpensive. However, today we recognize the potential environmental impacts from the burning of coal. As the government has passed more regulations requiring environmental protections, the relative use of coal for electricity production has decreased. Yet, it still accounts for nearly half of the electricity production in this country (nearly two-third of all current production comes from fossil fuels, coal, and natural gas).

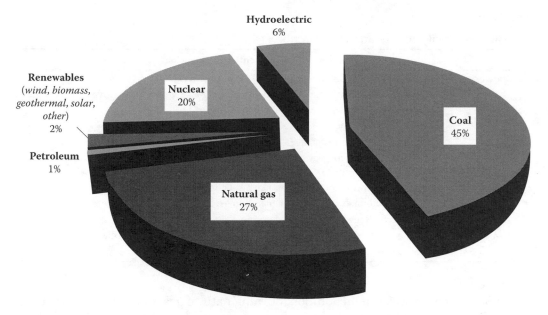

FIGURE 11.2
U.S. electricity generation in 2009 (excluding that by commercial and industrial sources).

In the 1960s, politicians began to take seriously the concerns raised by scientists about the negative environmental impact of burning coal. As a result, certain utilities began increasing their use of petroleum (oil) for electricity production. While burning petroleum was relatively cleaner than coal, concerns over the U.S. dependence on foreign petroleum sources led utilities to scale back their use of petroleum as a source of electricity in the early 1980s. Today, only approximately 1% of U.S. electricity generation comes from petroleum.

Natural gas has now become the fossil fuel of choice for new power plants. Natural gas is a cleaner burning fuel than both coal and petroleum and is more readily available in the

TABLE 11.2

U.S. Electricity Generation in 2009 (Excluding Generation by Commercial and Industrial Sources)

Source of Fuel	Capacity per Day (GW)
Coal	5.073
Natural gas	3.055
Petroleum	0.102
Biomass	0.032
Wind	0.147
Geothermal	0.041
Solar	0.003
Other renewables	0.042
Nuclear	2.278
Hydroelectric	0.657

TABLE 11.3

Renewable Electricity Sources and Their Drawbacks

Source	Potential Drawbacks
Hydropower	Damming a river disrupts the natural flow of the river
	Damming a river interrupts the ecosystem and negatively affects native fish populations
Solar	Electricity generation limited to daylight hours
	Limited number of locations to make most effective use of solar energy
	Solar technology is expensive
	Limited solutions for nighttime electricity generation or storage from day-time electricity production
Wind	Limited number of locations to make most effective use of wind energy
	Variable amounts of electricity generation—dependent on the weather
	Wind technology is relatively expensive
Geothermal	Limited number of locations to make most effective use of geothermal energy
	Geothermal technology is expensive

United States than petroleum. In addition, burning natural gas for electricity has about a 60% efficiency rate versus a 35% efficiency rate for coal. Despite these positive features, natural gas remains a limited resource. In order to continue relying on this fossil fuel source there must be an improvement in exploration and drilling to find new sources.

Why do fossil fuels continue to be the dominant fuel source of the United States electrical grid? Cost. Simply put, the use of fossil fuels for electricity generation remains the most cost-effective approach to electricity generation (taking into account both fuel and capital costs). Yet, environmental issues and limitations of supply of these fuel sources provide opportunities for alternatives.

Sunlight is transformed into electricity through different methods

Photovoltaics (PV) produce electricity when sunlight hits the surface of the solar panel and excites electrons in the panel. The particular construction of the panel directs those excited electrons in one direction, and that flow of electrons is electricity.

Concentrating solar power (CSP) utilizes solar panels to reflect light to a point on a central tower. This generates a great deal of heat, which is used in the same way heat is used in conventional power plants, that is, to heat water to steam which turns a turbine or waterwheel.

Nuclear

Nuclear power was initially considered to be an economical and safe way to generate electricity. In fact, a number of European countries still rely heavily on nuclear power. However, a number of accidents related to nuclear power plants (e.g., Chernobyl and Three Mile Island) made the American public wary of nuclear power. In addition, there are unresolved environmental concerns as there is no permanent storage facility for nuclear wastes produced by nuclear reactors. The nuclear industry remains heavily regulated and lacks strong public support. As a result, there have been no new nuclear reactors since the mid-1990s.

Renewable Sources

Renewable energy sources for electricity generation remain an important part of the future of utility-scale power. Currently, hydroelectric power supplies nearly 6% of the nation's electricity. Other sources including solar, wind, and geothermal generation have seen dramatic growth in recent decades, but still only account for a small percentage of the U.S.

electricity generation. Although the mechanisms for energy production from renewable sources are different for each of these sources, the underlying benefit of using renewable resources lies in the decreased environmental impacts and a "fuel" source that is essentially free.

OPPORTUNITIES FOR INNOVATION

ELECTRICITY GENERATION

- Greater use of *renewable sources* for electricity generation
  - Solar installations (both utility-scale and residential- or commercial-scale)
  - Wind farms (both utility-scale and residential- or commercial-scale)
  - Biomass
  - Geothermal
- Creating *cleaner fuels* to use in traditional fossil-fuel generators (i.e., lower sulfur coal)
- Better technology to *remove pollutants from water and air waste* from traditional fossil-fuel generators
- *Increasing the efficiency* of traditional fossil-fuel generators (i.e., natural gas being significantly more efficient than coal)

Electricity Transmission and Distribution

Once electricity is produced in a generating station it is placed into the high-voltage transmission system. Most of these transmission lines are owned by the electric utilities generating the electricity. The transmission system delivers electricity from the power plants to distribution substations and from these substations to consumers. This process of transmission results in electricity loss along the journey from generating station to the consumer. As a result, it is often impracticable to have generating stations located far from the end destination. To limit electricity loss and promote efficient distribution, electricity is distributed at a very high voltage. Transformers convert electricity at one voltage to a higher or lower voltage, depending on their location in the power line.

Electricity typically leaves a generating station at a voltage of about 25,000 volts (V). A volt is the measurement of how much electric force is pushing electrons around a circuit. As it leaves the generating station, a transformer boosts the electricity up to 400,000 V. This higher voltage helps the electricity travel more efficiently to long distances. Once the electricity reaches a neighborhood for distribution to the end-user, smaller transformers in utility boxes lower the voltage of the electricity. Most home uses including lights, televisions, and other smaller appliances operate at 120 V, while larger ones including cloth dryers and electric stoves operate at 220 V.

A complicated system of power lines and transformers serves to connect generators with urban and rural consumers alike. Overhead primary lines usually serve rural and suburban areas, with transformers and other transmission and distribution equipment mounted on poles. Although these lines are susceptible to weather, they are quicker to repair. In urban areas, a high-density load is served by underground cable systems with transformers and other transmission and distribution equipment in underground vaults or ground-level cabinets. These lines are protected from weather or similar disruptions, but this method of delivery is more expensive and can take longer to repair.

OPPORTUNITIES FOR INNOVATION

ELECTRICITY TRANSMISSION AND DISTRIBUTION

- *Constructing renewable energy generators closer to consumers.* By effectively reducing the need to construct more transmission and distribution lines, it is possible to decrease loss of electricity that comes from the transportation of electricity over long distances
- *Using different technologies for transmission of electricity.* Use of new high voltage DC transmission acts more effectively as a long-distance carrier when compared to the traditional use of AC
- *Increase distribution efficiency*

Electricity Consumption

Residential, commercial, and industrial customers consume electricity in a myriad ways. Most consumers do not have a choice of an electricity provider and must utilize their local utility. However, customers may affect the cost of their electricity by using off-peak or implementing energy-efficient strategies.

In order for increased efficiencies to come from better usage and monitoring by consumer and business end-users of electricity, many expect technologies to be developed making the electricity grid more efficient, more reliable, and generally "smarter," known as Smart Grid. These technologies include a variety of monitoring, reporting, tracking, real-time adjustments, and storage technologies. The National Institute of Standards and Technology (NIST) has announced plans to establish 77 smart grid standards over the next few years and finalize 14 priority standards in 2010 alone to help provide standardized rules to advance the Smart Grid concept.

OPPORTUNITIES FOR INNOVATION

ELECTRICITY CONSUMPTION

- *Increasing energy efficiency.* Utilities have begun encouraging customers to improve their efficiency of energy usage rather than increasing the load capacity of the utility's generators. Examples include energy-efficient appliances, and more efficient home and commercial lighting.

- *Smart grid.* Consumers are connected to the power grid through electric meters. These meters currently record the amount of electricity used by each consumer, which the electric company physically checks to accurately charge for the electricity consumed (that's the guy known as the "meter reader"). The concept of the "Smart Grid" describes a scenario where a utility integrates modern technologies (similar to technologies utilized in the distribution of the Internet) to give utilities and consumers a better understanding of the demands for electricity at any given time. Utilities would use this information to find out the times it was required to generate more electricity, and consumers could understand and make choices to refrain from using additional electricity during peak times when electricity is in greater demand (assuming that utilities would make this electricity more expensive). In addition, the Smart Grid could allow the electric utility to know if there was a disruption in service, whereas today most consumers must call in to the utility to inform them that the power is out.
- *Increasing education to consumers about green energy options.* Educating consumers about available green energy sources may encourage greater support for green energy development.

Organization and Regulation of Electric Utilities

One of the critiques of utilities is that they are slow to change—relying on outdated operating methods and hesitant to adopt newer technologies. And while these criticisms may or may not have truth in them, utilities receive little reward for increasing efficiencies, but experience the wrath of angry consumers with power outages or blackouts.

In short, utilities are required to deliver electricity to customers uniformly and consistently. This means that utilities need to provide electricity at a uniform voltage and frequency, since low voltage will dim light bulbs or cause devices to malfunction and high voltage can destroy certain appliances (hence your surge protector attached to your PC). Additionally, a steady and consistent supply of electricity must be made available to meet the needs of the customers dependent on the delivery of electricity. In order to meet these two goals of uniformity and consistency, the electric power system has developed into a large, complex network requiring coordination from numerous parties.

Summary

North America is a large geographical area of nearly 9.45 million square miles—over 16.5% of the Earth's landmass. More than 500 million people live in North America. And to serve that population across such a large landmass, a large, interconnected power system has emerged. Two key entities, Federal Energy Regulatory Commission (FERC) and North American Electric Reliability Council (NERC), supervise this entire international power system, while regional entities will often act to oversee certain reliability of the system and resolve disputes between owners and operators of the grid. At a local level, the control areas provide most of the day-to-day oversight and management of the power system, with approximately 150 control areas in the United States managing these activities within a set boundary of the power system. And not to be outdone, the states get into the act through Public Utility Commissions (PUCs) which regulate the retail rates of electricity for consumers in their respective states, approve sites for new generation facilities and transmission lines, and issue state environmental regulations.

Are you still confused? It is okay if you are—regulations of electricity and utilities keep numerous government employees, corporate executives, accountants, lawyers, and countless others gainfully employed to help oversee and manage electricity. Below is a further summary of the key regulatory bodies and entities that oversee the complex utility system.

FERC: Federal Regulatory Authority over the Entire Transmission System

The FERC was established under the Federal Power Act of 1935 and exercises principal regulatory authority over the transmission system. FERC regulates wholesale electricity rates, approves sale or leasing of transmission facilities, approves mergers and acquisitions between IOUs, and exercises jurisdiction over the interstate commerce of electricity. FERC also has regulation power over the transmission of natural gas and oil, and plays a vital role in the licensing and inspection of nonfederal hydroelectric projects.

In the early 1990s, FERC established the Open Access Same-Time Information System (OASIS), which is an Internet-based system for obtaining services related to the transmission of electricity. Commercial entities involved in contractual arrangements for the sale or purchase of electricity outside of their control area, must report these arrangements to their control area to make provisions for transmission losses. These reports are entered into OASIS, which allows energy marketers, utilities, and other wholesale energy customers real-time access to information regarding the availability of transmission line capacity.

Case study: Western Electricity Coordinating Council (WECC)

In the West, the Western Electricity Coordinating Council (WECC) is a regional entity responsible for coordinating and promoting bulk electric system reliability in the Western Interconnection. In addition, the WECC provides a forum for resolving disputes about transmission access, and facilitates coordination of operating and planning activities of its members.

The WECC covers the widest and most geographically diverse area of any of the eight regional organizations responsible for overseeing the power system. Its territory includes the Canadian provinces of Alberta and British Columbia, the northern portion of Baja California, Mexico, and the 14 Western states in between.

NERC: Overseeing the Reliability of the Power System

The NERC was established in 1968 (renamed in 1981 to reflect the inclusion of Canada into the territory covered by the Council) to ensure the reliability of the bulk power system in North America. NERC develops and enforces reliability standards, assesses reliability, monitors the bulk power system, and oversees the education and training of industry personnel. While NERC is a self-regulatory organization, it is subject to oversight by FERC and Canadian authorities. NERC defines the reliability of the power system according to two considerations: adequacy and security. Adequacy is defined as the ability of the power system to supply the electricity demands of all customers at all times, which requires an examination of scheduled and reasonably expected unscheduled outages of certain parts of the system. Security is defined as the ability of the power system to respond to sudden disturbances, such as an unanticipated loss of part of the system (Figure 11.3).

The Western and Eastern Interconnections: An Interconnected Power System

North America has two large interconnected power systems that operate independently of one another, but are linked in certain places. These interconnections allow the two

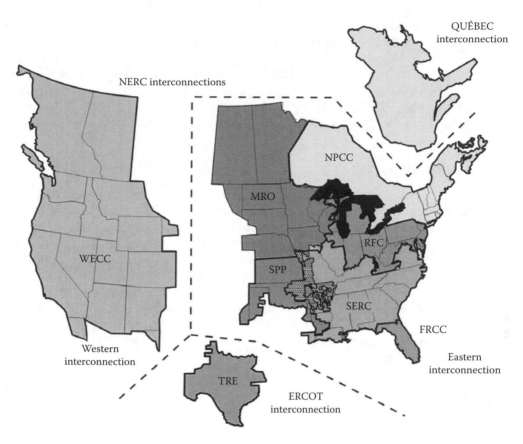

FIGURE 11.3
NERC interconnections. (Data from North American Electric Reliability Council, http://www.nerc.com/
fileUploads/File/AboutNERC/maps/)

power systems to share reserves and sell power not being utilized by the utility customers
in one area to another utility that has customers in need of additional electricity. The
Eastern interconnection comprises most of the land east of the Rockies, including Canada,
while the Western interconnection controls most of the land west of the Rockies. There
are a few smaller interconnections, such as the Electric Reliability Council of Texas that
covers most of Texas, but these smaller interconnections are fed into the larger Eastern
and Western interconnections. These interconnections are then broken up into smaller
areas, called control areas, which are responsible for the day-to-day management of that
area's electricity system.

Regional Entities: Supervising the Power System in the Region

The aim of Eastern and Western interconnections is the same—to ensure consistent and
uniform transmission of electricity. These power systems are supervised by regional organizations, which while providing for regulation and supervision on a scale smaller than
FERC or NERC, are not without their own challenges. Each regional organization may be
made up of dozens of utilities, states, provinces, cities, and other interested entities, which
demonstrates the difficulty in planning and organizing power generation, not only on a

day-to-day basis, but also for long-range planning to provide reliable electric service in the future as the population grows and different needs for electricity arise.

Regional entities like the Western Electric Coordinating Council (WECC) are also assisted by advisory boards such as the Western Interconnection Regional Advisory Board that was created under Section 215 of the Federal Power Act and advises the WECC and other entities on whether the proposed reliability standards and budget are in the public interest. FERC may also request additional advice on other related topics.

Control Areas: Coordinating the Day-to-Day Operation of the Power System

Approximately 150 control areas exist within the United States, providing most of the day-to-day management of the power system. While FERC and NERC supervise the international power system, the control areas are responsible for generation and operations within their own boundaries. A control area operator sets operating policies for their specific area and designates the various power plants to generate at certain levels on each given day. In the event there are any problems or disturbances, the control area operator must react quickly to utilize other resources inside or outside of the control area. In some cases, this requires cutting off the affected area from the rest of the power system, which would lead to a power outage for some customers but would prevent disruption for the entire system. Much of this work is computerized to balance the supply and demand of electricity safely and reliably.

Other important bodies for local regulation and oversight are the Independent System Operator (ISO) and the Regional Transmission Operator (RTO). Both ISO and RTO are formed at the direction or recommendation of FERC. These bodies each coordinate, control, and monitor the operation of the electrical power system. An ISO typically has jurisdiction over a single U.S. state, while an RTO exercises its power over a wider area that crosses state borders.

PUCs: Regulation on the State Level

The states are also involved in this process—managing activities of utilities and transmission within its borders. A PUC regulates the retail rates of electricity for consumers, approves sites for generation facilities and transmission lines, and issues state environmental regulations. Most PUCs are organized by state and help execute state's laws and rules regarding the provision of electricity to consumers. PUCs are often responsible for all utilities in the state and have an important role in regulating privately owned water, natural gas, telecommunications, and even railroad or rail transit companies. For information on rate-setting and the role of PUCs, see the section on Regulation of Pricing of Electric Utilities.

Electricity Industry Restructuring and Deregulation

Until the early 1990s, nearly all electricity systems in developed economies around the world were operated as public or quasipublic agencies (as within France and the United Kingdom), or as highly regulated, privately owned companies (as within the United States). Because of the importance of providing reliable, low-cost electricity to consumers, electricity was treated as a public right and a requirement that was too vital to be left to a competitive marketplace. And while these systems did promote investment in the substantial infrastructure necessary for utility-scale generation and encouraged the spread of electricity to regions outside of urban areas, by the early 1990s more politicians and industry experts believed it was necessary to introduce a greater competitive nature to electrical generation. One of the reasons for this movement was that this noncompetitive system did not reward utilities for taking risks on new technologies or for creating operating efficiencies (an example is the current use of analog manual-read electricity meters in many homes and businesses).

The Public Utility Regulatory Policies Act (PURPA) opened wholesale power markets to nonutility producers of electricity. In 1992, the passage of the Energy Policy Act served to increase competition in the bulk power market, furthered by FERC with Orders 888 and 889, intended to "remove impediments to competition in wholesale trade and to bring more efficient, lower cost power to the nation's electricity customers." These FERC orders fundamentally mandated equal and open access for all producers of electricity to jurisdictional utilities' transmission lines. These regulatory changes permitted states to effect legislation allowing the electric power industry to transition from highly regulated, local monopolies toward competitive companies that provide the electricity with utilities providing transmission or distribution services. And, as a result, nearly half of the states now have enacted legislation removing its rate-setting responsibilities for electricity and creating more of an oversight role within the resulting deregulated industry where prices are determined by competitive markets.

CALIFORNIA ROTATING BLACKOUTS DURING THE SUMMER OF 2000

What happened in the summer of 2000 in California that led to calls for new regulations and challenged the results of deregulation efforts in California utilities?

A growing demand for power and limited increases in generation and transmission capacity set the stage for the blackouts that plagued California during the summer of 2000. These problems were exacerbated by a number of electricity generation plants being off-line for various reasons and a hot summer that resulted in more demands for electricity from customers to power their air conditioners. In addition, there was inadequate market design, where the price of electricity to consumers did not correspond to the demand for electricity at any given time. Therefore, price signals were not available to moderate demand and the power system was overloaded.

When the supply was insufficient to meet the demand, a "rotating blackout" was utilized to maintain a balance of available electricity supplies. A rotating blackout is a controlled interruption of customer demand, so the overall reliability of the interconnected system may be sustained to maintain a balance with available supplies. Although this was a workable stop-gap measure in an emergency, the disruption to customer service resulted in lost profits and dissatisfaction with the power system.

Regulation of Pricing of Electric Utilities

Electric utilities are natural monopolies, which provide utility customers without a choice between electricity providers (unless a consumer wanted to take matters into one's own hands and install wind turbines and solar panels on the roof of one's house … and have a backup generator!) To prevent abuses in pricing by utilities, the price of electricity is regulated by FERC, a federal agency, and agencies located in each individual state. These state organizations, PUCs, will often be given control over all utilities within a particular state, including water and natural gas.

While these protections offer consumers greater transparency into the pricing process for electricity, this regulatory control over pricing is not without its share of issues—which have been brought to light for startups attempting to break into this marketplace. For example, electricity pricing may be set to meet a certain rate of return, rather than promoting economic efficiency, thereby leading to hesitancy by utilities to adopt new technologies for a smarter grid or cleaner generation. And while the pricing method may provide a certain rate of return to utilities and their investors, some have argued that this approach fails to send the right price signals to consumers in order to encourage greater energy efficiency.

FERC: Regulating Wholesale Electricity Rates

FERC regulates wholesale electricity rates and exercises jurisdiction over interstate commerce in electricity. FERC sets the price of service for transmission facilities of investor-owned utilities that fall within its jurisdiction, although most transmission assets reside within the jurisdiction of the state regulatory agencies. FERC has no jurisdiction over prices set by transmission facilities owned by public power agencies.

FERC also has authority to grant market-based rate authorization for wholesale sales of electric energy, capacity, and other services by public utility sellers. This allows for greater competition, but only if the utility and its affiliates lack or have adequately mitigated horizontal and vertical market power, which will decrease the likelihood of pricing abuses as the utility acts as a monopoly. In this case, ongoing restrictions are put in place for any entity authorized to set market-based rates. For instance, the market-regulated power sales entity and its relationship with a franchised public utility entity with captive customers is restricted by FERC to prevent any pricing abuses.

PUCs: Setting Retail Rates for Electricity

While FERC has jurisdiction over the trade of electricity in interstate commerce, the state PUCs regulate intrastate trade of electricity and set retail rates for customers. These are in charge of the retail rates for electricity in that state sold to customers of IOUs. Under the traditional regulatory system, the PUC sets the retail rates for electricity, as described below. In setting these rates, each state agency must consider a number of different factors, including the cost to the utility for generated and purchased power, the capital costs of power, transmission, and distribution plants, all operations and maintenance expenses, and the costs to provide programs often mandated by the PUC for consumer protections and energy efficiency, as well as taxes.

The National Association of Regulatory Utility Commissioners (NARUC) is an organization representing public utility commissioners from each state. These commissioners regulate utility services, including electricity, as well as natural gas, telecommunications, water, and transportation. NARUC ensures that rates charged by regulated utilities are fair, just, and reasonable. The commissioners are either appointed by their governor or legislature or are elected. Additionally, NARUC acts to represent these commissioners in cases of litigation and also serves as a resource for them.

EXAMPLES OF PUCs

While each state has a different PUC, Massachusetts and California each provide a basic overview of the work accomplished by these PUCs:

Massachusetts: Department of Public Utilities

In Massachusetts, the Department of Public Utilities regulates investor-owned electric power, natural gas, and water industries. The stated mission of the department is to "ensure that utility consumers are provided with the most reliable service at the lowest possible cost; to protect the public safety from transportation and gas pipeline related accidents; to oversee the energy facilities siting process; and to ensure that residential ratepayers' rights are protected."

Beginning March 1998, Massachusetts's customers were able to choose to purchase electricity from a competitive retail supplier or to continue to receive electricity from their existing distribution company. As a result, a retail electricity market was created in Massachusetts. The department still retains some level of oversight on the pricing for this "competitive" retail pricing. Today, mostly large electricity consumers, like large commercial or industrial customers, are utilizing the new retail market for electricity.

California: California Public Utilities Commission

The California Public Utilities Commission (CPUC) is responsible for oversight of electricity rates, as well as overseeing water, telecommunications, and transportation in the state. CPUC sets rates based on the utility's cost of service, which "includes the utility's cost of owning and operating its transmission, distribution, generation facilities, its fuel, and purchased power expenses including the cost of paying the California Department of Water Resources, revenue requirements, and its cost for implementing public purpose programs such as energy efficiency programs, low-income discounts and energy efficiency assistance, renewable programs, and research and development programs." This complicated list of factors informs the rate set on an annual forecast basis.

In California, the bundled rates are divided into approximately four categories: (1) electricity generation, which accounts for 60% of the price of the electricity; (2) electricity distribution, which accounts for approximately 23%; (3) electricity transmission, which accounts for 3%; and (4) other costs, such as under-collection amounts, nuclear decommissioning and energy efficiency, which accounts for 14% of the price of electricity. CPUC has also made efforts to encourage renewable energy sources for energy generation, particularly the use of solar. Each of the three major electric utilities in California, Southern California Edison, Pacific Gas & Electric, and San Diego Gas & Electric, have taken part in the Solar Incentive Program, which provides cash incentives for energy consumers who "Go Solar."

Process for Rate Setting: Allowing an Appropriate Rate of Return

FERC and state agencies play different roles in setting the price of electricity. State agencies also have different considerations in setting prices from other states. The general process of rate-setting is similar, however. These agencies use cost of service ratemaking, which includes the rate of return on capital invested in the business. Most regulators base the rates on a projected year, rather than a historical year altered to meet the current year's projected needs. Here are the basic steps a regulator undertakes to set pricing:

- First, the regulator considers revenue, adjusted for abnormal weather conditions.
- Second, the regulator considers operating expenses (also adjusted for abnormal weather conditions) and taxes.
- Third, the regulator subtracts operating expenses and taxes from revenue, which results in operating income, or income available to provide a return on invested capital.
- Fourth, the regulator will divide operating income by the rate base, which produces a rate of return.
- Finally, the regulating agency will examine the utility's rate of return to determine if it is adequate, and may make adjustments to the price to meet the adequate rate of return (which increases or decreases revenue to effectively increase or decrease the rate of return).

The ultimate rate of return will be based on the cost of each segment of the capital structure for the utility, which will include the debt, preferred stock, common equity, and deferred credits (usually deferred taxes).

When a utility sets a budget for a new expenditure or project, they will set certain expectations and pricing assumptions, but ultimately the ability of a utility to undertake a new program, expenditure, or project is based on whether or not the regulatory agencies accept the utility's figures. Different types of accounting specific to the utility industry can also affect the resulting figures. The rate setting process is extremely complicated and involves a great deal of estimation due to the external factors outside of the control of the utility affecting electricity production in a given year.

PURPA: Encouraging Renewable Energy

The PURPA was passed in 1978 to promote the greater use of renewable energy. This Act encouraged the development of alternative generation sources designated as "qualifying facilities," which included small power producers that used renewable resources as their primary energy source. Electric utilities were required to buy power from these nonutility electric power producers at the "avoided cost" rate, which was the cost for the electric utility if it were to generate that power itself, or purchase it from another source. These prices were fixed for the length of the contract, which often resulted with inaccurate pricing, with the utility purchaser paying too much. Implementation of PURPA was left to the states, so there was not much done in many states. In addition, many of the contracts formed under PURPA are now about to expire.

PURPA was amended in 2005 by the Energy Policy Act (EPAct), which shows a greater commitment to the creation of renewable energy. The Energy Policy Act of 2005 goes further in encouraging renewable energy by authorizing loan guarantees for technologies that avoid greenhouse gases, such as renewable energy and cleaner coal. In addition, the Act authorized subsidies and other incentives for renewable energy sources, including a number of tax breaks that have been used to help finance renewable energy projects.

Environmental Regulations: Encouraging Renewable Energy

The Environmental Protection Agency was established in 1970 to protect human health and safeguard the national environment. Since its inception it has established a number of regulations and policies intended to protect the environment. While few of these regulations directly affect the electricity sector, they have an important indirect effect. As emissions standards become more stringent, many traditional forms of fossil fuel energy production must modify their production processes, making energy production more expensive. This makes alternative energy production processes that are often very expensive more competitive. State and local environmental laws can also affect the siting of generation facilities, and thereby encourage nontraditional forms of electricity generation by state-specific regulations.

OPPORTUNITIES FOR INNOVATION

SELLING "GREEN" ELECTRICITY

While it is impossible for a consumer to tell the difference between "green" electricity (generated through renewable sources) and "brown" electricity (generated through nonrenewable sources such as fossil fuels), that hasn't stopped utilities from selling consumers "green" electricity.

Sales of the first green power programs began in the United States in 1993, and today some programs have signed up 3–6% of their residential customers. Green Mountain Energy has signed up over half a million customers to purchase "green" electricity in seven states; however, more typical sign-up levels have been less than 1% of residential customers.

These programs have largely been driven by marketing campaigns encouraging customers to purchase green power (for an additional 1–3 cents per kWh).

CASE STUDY

ELECTRICITY REGULATION AFFECTING A CONSUMER IN SEATTLE, WASHINGTON

A resident of the suburbs of Seattle, Washington may receive power from Puget Sound Energy, which serves more than 1 million electric customers and 750,000 natural gas customers.

Puget Sound Energy is regulated by a PUC, the Washington Utilities and Transportation Commission. In addition to regulating private utilities, the Commission also regulates wireline telecommunications, private water, solid waste collection, and commercial ferries.

The Washington Utilities and Transportation Commission is part of the Northwest Power Pool Area (NWPP) subregion of the WECC. The map at left illustrates this subregion, and the stars on the

map signify electricity trading points, or interconnections. The WECC is a region controlled by NERC, while all NERC regions are regulated by FERC.

Summary: These varying regulating bodies are organized as follows:

FERC
↓
NERC
↓
NERC Regions (e.g., WECC)
↓
WECC sub-regions (e.g., NWPP)
↓
PUCs
(e.g., Washington Utilities and Transportation Commission)
↓
Electric Utility
(e.g., Puget Sound Energy)

Puget Sound Energy Service Map:
Source: http://www.pse.com/insidePSE/corporateinfo/Pages/CorporateInfo_serviceArea.aspx

Western Electric Coordinating Council (WECC) Services Map:
Source: http://www.ferc.gov/market-oversight/mkt-electric/northwest.asp

12

How Project Finance Works

The term "project finance" is commonly used while talking about green businesses, but for many people little is understood about how it actually works and its importance in the large scale, capital-intensive clean technology projects. The reality is that project finance has helped move solar, wind, geothermal, and other projects from the labs and small facilities into utility-scale projects. Need to build a power plant for generation from renewable sources? Project finance may be the way to fund it. Need to build a biofuel production facility? Again, this may be the right fit for project finance. Trying to fund a renewable project in a developing country? Traditional finance vehicles may not be an option. What about creating an alternative vehicle assembly line, a biomass facility, or a landfill gas capture facility? You guessed it; project finance could well be the trick. This wealth of appropriate opportunities is the reason why industry leaders all note the importance of project finance in growth of the sector.

What you may not know is that project finance has its roots in the ancient Greeks and Romans who turned to project finance schemes to finance shipping voyages. The Panama Canal became one of the first uses of project finance in a large infrastructure project, but then it continued to become common practice in oil and gas infrastructure projects. Historically, project finance has been implemented in the public utility, mining, transportation, and telecommunication industries. Today, project finance has become an important way to finance large projects including solar, wind, biofuels, alternative vehicles, and others.

The Basics

Project finance is different from other traditional forms of financing in that the financing is based on the risk and future cash flows of the project rather than the inherent risk of a specific company. Rather than giving a loan to a company, the loan is tied to a particular project—perhaps a specific wind farm in Massachusetts, a solar installation in Arizona, or a biofuel facility in Iowa. Project finance is a mechanism for financing these types of projects where the lenders will have no or limited rights of recourse to the parent company that sponsors or develops the project.

Why would a lender give up recourse against the parent company? They usually won't unless the project can show a future revenue stream that is reasonably certain that can be used to secure the repayment of the loans. In the case of a solar, geothermal, or wind project, this means having a contract in place with a local utility from a power purchase agreement; for a biofuel plant, it means having a contract for the purchase and use of the fuel by a corporate fleet or for integration into another nonrenewable fuel source; for a waste-to-energy facility, it means a contract for the supply of waste from a city or municipality and a power purchase agreement (PPA) with a utility for the electricity generated. Based on the lender's understanding of those future revenues, determination can be made that a project is within the risk tolerances of the lender. Wind farms have developed to the point where project finance is a very common approach, while solar power projects are still only common when a PPA is in place for the purchase of the bulk of the output from the project. Due to the relative immaturity of the solar market, project financing for other solar plans still remains somewhat unusual unless the project has the ability to provide a greater level of certainty and risk reduction through obtaining a PPA.

The actual research and analysis of a project requires a review of historical pricing and purchase needs as well as forecasting the future price curves and demand. Additionally, in the case of renewable energy projects, the importance of tax benefits into the project financing analysis cannot be overlooked. The tax assets can provide another mechanism to make the project attractive to potential investors.

Why Is Project Finance Important in Green Business?

How much does it cost to build a full-scale biofuel plant? Wind farm? Solar cells across commercial rooftops? Solar thermal plant? Waste-to-energy facility? Bioplastic manufacturing plan? Geothermal power plant? Green automobile production facility? In each case, the answer usually starts at the $50 million dollar mark and only goes up from there. Money that large isn't easy to come by. Venture capitalists very infrequently will fund amounts that large. Public markets are also unlikely to be the right source of funding (although sometimes it can be). This means that project financing has to step in. And the examples of institutional investors from banks, governmental entities, hedge funds, and even in some cases groups of venture capital funds making project finance investments of hundreds of millions of dollars in "green" projects is becoming more common.

A quick flip through some of the recent financing headlines in the news (many of which show dollar values in hundreds of millions) give a glimpse into the importance of project financing in many of the clean technology commercial development projects:

- Raser Technologies signed an agreement with Merrill Lynch for the structuring and financing of up to 155 MW of geothermal power plants.
- WaterHealth International using project financing for water purification and disinfection systems for 600 communities across India from International Finance Corp., a division of the World Bank.
- Ormat Technologies has used project finance for geothermal projects in Kenya, Turkey, New Zealand, Costa Rica, and three other Central American countries.
- VeraSun acquired three ethanol plants for a combination of cash, equity, and project finance.

- Finavera Renewables is using project financing for wind and wave power projects.
- Wind project developer enXco used project financing to build the largest wind farm in Minnesota.
- A proposed hazardous-waste-to-energy facility in Indiana was to be funded by project finance.
- Numerous solar companies have used project financing, including First Solar, SolFocus, eSolar, NRG Energy, and Acciona (just to name a few).
- In 2009, both SunRun and Solar City raised project financing from U.S. Bancorp. SunRun will use the funds for residential solar PPAs and SolarCity will use funds for residential installation.
- Tioga Energy, Solar Power Partners, SunBorne Energy, and Recurrent Energy have each raised funds for project financing for solar power purchase agreements from venture capital investors—an alternative model for project financing from banking institutions or governments.

Just why are all of these entities above so willing to invest this amount of money into these projects? Largely because these projects have sufficiently limited risk profiles to look like a "sure thing" to the investors. Of course there is risk, but the projects are able to put together a compelling financial model incorporating the cash the project generates and potential tax savings. For example, the bank puts in $100 million. The company installs wind turbines (or other technology). The utility agrees to buy all the electricity that wind farm can produce. The utility gets green energy while the bank and the company share in the proceeds of the project. It's a win-win-win.

Sounds simple, right? But there's a bit more to it than a quick way to raise hundreds of millions of dollars. While project finance can provide a viable strategy for funding clean technology, the approach is confined to projects that meet a fairly narrow set of criteria and is subject to risks in the broader credit markets.

Impact of Project Finance on Clean Technology

One of the catalysts for the explosive growth in clean technology investment in the past decade has come from project finance, and the opportunities it offers to certain investors. To date, the bulk of renewable project finance has been provided by tax equity investors (usually large investment banks and insurance companies) who partner with project developers through highly specialized financing structures structured to capitalize on federal support for renewable power technologies, which has historically come in the form of tax credits and accelerated depreciation deductions. Solar has benefited greatly with total global investment increasing from $66 million in 2000 to $12.4 billion in 2007, investment amounts accelerated by the opportunities tied to project financing.

Yet, with the increase of project finance investments in clean technologies, comes a reliance on the public debt markets. In late 2008, as the credit markets tightened due to the global credit crisis, project financing sources for clean technologies tightened as well. Congress responded by including provisions in The American Recovery and Reinvestment Act of 2009 (ARRA 2009) to make federal incentives for renewable power technologies more useful, including permitting projects eligible to receive the production tax credits (PTC) elect the investment tax credits (ITC), and enabling ITC-eligible projects to receive (for a limited time) a cash grant of equivalent value in tax credits.

The states have also helped spur the rapid commercialization of renewable energy generation with the enactment of renewable portfolio standards (RPS). These standards require the use of renewable energy to meet certain percentages of overall electricity consumption by certain dates. This has driven utilities to be aggressive in partnering with projects generating renewable electricity, and therefore can help give certainty to lenders in the case of project financing. By having long-term purchase power agreements in place with the utilities, the lenders can mitigate risks and ensure that long-term cash flows will be available until these large loans are repaid.

When Project Finance Makes Sense

Companies from FirstSolar to enXco to Finavera to Raser all want to be able to utilize their knowledge and intellectual property to build numerous large-scale production facilities. This allows a project sponsor to learn from experience and scale the business. Likewise early investors in the technology development of those sponsors want the sponsors to leverage the early investment in technology to deploy their technologies to maximize the return for the initial investors. The sponsors will receive a part of the cash flows from each facility, so deploying wind farm technology in 20 locations can rapidly increase financial rewards for both the sponsor and its investors.

The real problem project finance helps solve is leverage. If a bank were to evaluate a company like FirstSolar for a $100 million dollar loan, the bank may be able to approve a loan for the first project to be built, but what about the second or the third? This is where project finance comes into play. By reducing or limiting the debt load on the sponsor developing the projects, the sponsor is able to undertake a number of concurrent or closely timed projects. In the example of a company like Raser building a geothermal electricity generation facility that were to "go under" (from bankruptcy, bank default, foreclosure, or other reasons), the intellectual property, key employees, and other projects of Raser (the corporate sponsor) will not necessarily be put at risk. At the same time, utilizing project finance allows for the efficient allocation of taxes and credits, keeps the investment decision focused on the project and its risks, and brings together parties that may have not invested together in alternative financing vehicles.

All that said, project finance is by no means simple. The creation of the corporate structure of a project finance vehicle is time-consuming, requires expertise, and document preparation that can stretch for years putting significant limitations on each project, particularly over how cash generated by the project is able to be spent. For executives that have built a successful technology, the transition to project development involves a very different skill set and change in personnel.

IS PROJECT FINANCE A REALISTIC OPPORTUNITY?

- Is the project or group of projects of a sufficient size to make project finance worthwhile? Lenders may be reluctant to utilize project financing for a project under $50 million, with a preference for $100 million projects.
- Does a revenue stream exist to support such a highly leveraged project of that size?

- Does the project have sufficient physical assets to ensure repayment to the lender in the event of foreclosure? This allows the lender to sell the assets or continue to operate the project until repaid.
- Has the technology been proven on a commercial scale? Lenders typically do not want to be the first to finance a cutting edge technology without a demonstrated ability to show a project operating at scale.
- Do relationships or contracts exist with key suppliers of raw materials, technology, or other key needs of the project?
- Are reputable partners participating to provide key services for the business in the event the sponsor is unable to manage certain aspects of the project?
- Does the sponsoring company want a quick exit? Lenders are hesitant to partner with a company looking to divest the project quickly as such divestitures are complicated and time-consuming.
- Is the sponsoring company willing to give some amount of high-level operational control to the lenders and other parties?

Source: Jefferies & Company, Inc.

Structuring a Project Finance Transaction

The details behind a project finance transaction can be complicated and confusing. However, there are some key things that are important to understand regarding how the transaction works and the parties that are involved. For many transactions, the documentation linking all the parties and ensuring everyone correctly shares in the revenues, the assets, and the operational control could fill a small apartment. Structuring the financial flow of funds and tax assets requires the expertise of financial analysts and tax practioners since the smallest change in assumptions can change whether the transaction makes sense for any one of the dozens of participants, including a slew of equity and debt investors.

The key entity in a project financing transaction is the project company which serves as the home for all of the project's assets as well as all contractual rights and obligations for the project. The project company will likely enter into contracts with each of the parties engaged directly in the projects from lenders, services providers, technology providers, utilities as well as, in some cases, equity investors into the project (Figure 12.1).

In many cases, the lenders will require that the project company obtain a certain portion of the total project funding from equity investments. Oftentimes, between 10% and 25% of the total project costs will come from equity investors. In the case of many of the clean technology projects, these equity holders will not only receive a portion of the cash flows from the business (after the operating costs are paid and the debt is serviced) but will also receive all or a portion of the tax benefits. In particular, wind projects have utilized the production tax credits from the generation of renewable electricity to allocate those tax benefits to equity investors. The discussion of allocating tax benefits in the context of project finance is discussed in more detail below.

The project company will usually enter into a series of intercompany agreements with the sponsor related to operations, administrative services, technology licenses, and others. Oftentimes, the sponsor will create a series of affiliated entities that contract directly with

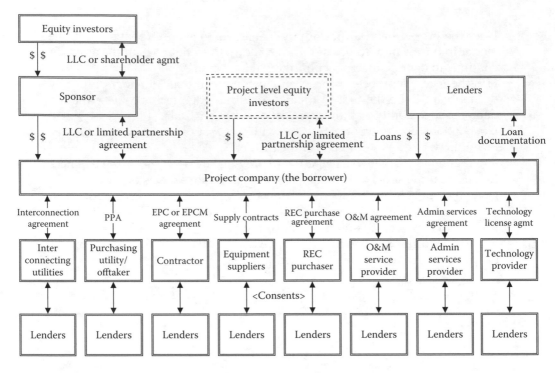

FIGURE 12.1
Example of project financing structure. (Data from Baker & McKenzie LLP).

the various project companies to manage these relationships and services. For example, an intellectual property holding company owned by the sponsor will enter into a technology license agreement with the project allowing for the development of the project based on the intellectual property developed or commercialized by the sponsor. Depending on the scope, size, and sophistication of the sponsor, more of the management of the project can be handled by the sponsor and its affiliates. In other cases, the sponsor will primarily manage the technology deployment and certain services of the project.

Depending on the scope of the sponsor's involvement, numerous other parties will play roles in the project from engineering, purchasing, construction, and management. The project company will typically enter into power purchase agreements with a local utility, lease agreements for the property, renewable energy credit agreements, interconnection agreements for electricity generation projects, feedstock supply agreements, and equity flip agreements for the sale and assignment of federal tax incentives.

Limitations on the Project Company

As discussed previously, one of the challenges of operating a project in a project financing structure is the limitations on what the project can do. In particular, there are limitations on the use of funds for the project to ensure the lenders and investors are able to maximize their cash return.

Typically, these limitations require the project company to comply with regulations and obtain necessary permits, follow industry standards in construction and operations, maintain agreed-upon insurance coverage, adopt budgets, and comply with financing limitations. Additionally, the project company will be limited in its ability to incur additional debt, make investments, issue equity, dispose of assets, abandon the project, and make changes to the budget above certain preset limits.

Following the Money

The next important question is how the parties involved in the project have the potential to make money from the transaction. Ultimately, the project is built on the assumption that the project will be able to generate fairly consistent revenues and profits after it has been constructed and is operating at capacity. To ensure that this is the case, project financiers will require that PPAs, supply agreements, and partnership agreements with customers are in place. This gives a greater certainty that an investor will be able to get their piece of the project revenues after construction.

For most projects, the loan documentation will contain a "project waterfall." This is meant to describe how any funds received by the project company will be distributed—first to the parties at the top of the waterfall until they ultimately reach the parties at the bottom of the funds flow waterfall (Figure 12.2). Ordering of parties and the size of each allocation ensures that parties can forecast their payouts depending on the best and worst case scenarios.

The sponsor will typically have several areas where they receive payment in the project, beginning with payment of costs associated with the construction, operation, maintenance, and related expenses of developing and operating the project. In the case of the payments to the sponsor for certain services, many of those will be predetermined in the contractual intercompany agreements with the project company. Following the operating payments, the lenders will be paid fees, interest, and principal payments. Thereafter, a series of reserves and payment amounts for other debt holders are made. Finally, after payment of these amounts, the remaining cash left in the project revenues account allows for distribution to equity holders of the project company, assuming that there are no defaults and the project company meets the financial tests required in the project financing agreements.

What If It All Goes "South"?

While banks and lenders try to do their best to limit the potential risk of a project failing, there are numerous examples of projects financed by project finance going bankrupt, being foreclosed, or shutting their doors. For that reason, one of the most important parts of the negotiation (perhaps second behind the actual allocation of revenues in the "waterfall") is to determine the ownership of the assets of the business in the event of a default.

Most sponsors will want there to be no recourse against them in the event of default or failure of the project. Depending on the negotiating leverage of the sponsor, there may be limited recourse, perhaps for any amounts on the loan unpaid after the sale of the project or its assets. Otherwise, the lender only has the ability to recover from the project company and its assets. After the closing of the loan, all assets of the project and those assets acquired thereafter will be pledged to the lender—much like a typical mortgage on a home. In order to ensure no other party is able to move ahead of the lender in the case of a default, most other lenders and other parties will be required to consent to the security ownership in the collateral assets. Involving dozens of lenders and parties makes this process one of the most tedious and challenging in the negotiations (Table 12.1).

Taxes and Project Finance

Project finance of renewable energy projects also creates opportunity to leverage the tax benefits created from these projects. As you will note in Chapter 17 of this book, there

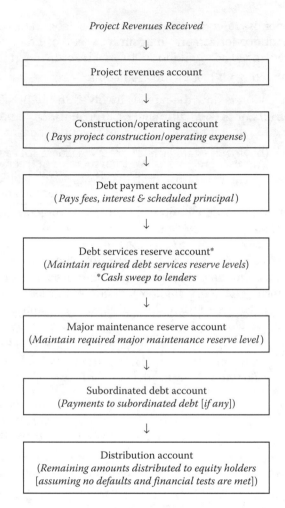

Project Revenues Received
↓

Project revenues account

↓

Construction/operating account
(*Pays project construction/operating expense*)

↓

Debt payment account
(*Pays fees, interest & scheduled principal*)

↓

Debt services reserve account*
(*Maintain required debt services reserve levels*)
*Cash sweep to lenders

↓

Major maintenance reserve account
(*Maintain required major maintenance reserve level*)

↓

Subordinated debt account
(*Payments to subordinated debt* [*if any*])

↓

Distribution account
(*Remaining amounts distributed to equity holders*
[*assuming no defaults and financial tests are met*])

FIGURE 12.2
Example of project finance waterfall.

are a substantial number of tax credits, incentive programs, tax depreciation schemes, and similar programs that are beneficial to green businesses. In certain cases, these credits may not be very useful to the project company or the sponsor, which has made the allocation of these tax benefits a crucial piece of project financing structures. The tax equity investors can purchase the tax benefits as part of their investment in order to maximize the return on their investment (through a combination of cash and tax benefits to reduce their overall tax burden). These structures have helped to drive investment in areas such as wind, solar, biomass, geothermal, landfill gas, hydropower, and others.

One of the most advanced sectors in structuring financing to maximize the benefit of the renewable tax benefits is the wind industry (Tables 12.2 and 12.3). For wind projects, they are able to take advantage of both the production tax credit and accelerated depreciation deductions from the federal government. These credits can provide substantial value, but for most developers of wind projects, they do not have sufficient income tax liability in order to sufficiently utilize the PTC or the accelerated depreciation deductions. That is

TABLE 12.1

Comparison of "Good Deal" vs. "Kill the Deal" Characteristics

	"Good Deal"	**"Kill the Deal"**
Lender Perspective		
Size of the project	More than $50 million	Less than $25 million
Contracted revenue	Yes	No
Liquidation value	Covers debt	None
Technology risk	None	First proof of concept
Contracted suppliers	Blue chip	No contracts
Control of uses of cash flow	Creditor	Sponsor
Quick exit for sponsor	No	Yes
Oversight	Creditor veto	None
Equity Investor Perspective		
Risk of a "cram down"	None	100%
Strategic partners	In place	Uninterested
Path to profitability	Certain	Doubtful
Control	Secured	None
Commercialization timeline	Matches exit strategy	None
Liquidity event	Near-term	Never
Returns on capital	500 basis points above weighted average cost of capital	Below weighted average cost of capital

Source: Adapted from Jefferies & Company, Inc.; Baker & McKenzie LLP.

TABLE 12.2

Common Financing Structures for Wind Projects

Financing Structure Name	Project Capital Structure	Likely Equity Investors	Brief Description of Structure Mechanics
Corporate	All equity	Developer (corporate entity)	Corporate entity develops project and finances all costs. No other investor or lender capital is involved. Corporate entity is able to utilize tax benefits (no flip).
Strategic investor flip	All equity	Developer and strategic investor	Strategic Investor contributes almost all of the equity and receives a *pro rata* percentage of the cash and tax benefits prior to a return-based flip in the allocations.
Institutional investor flip	All equity	Developer and institutional investor	Institutional investor contributes most of the equity and receives *all* of the tax benefits and, after the developer has recouped its investment, *all* of the cash benefits, until a return-based flip in the allocations.
Pay-as-you-go ("PAYGO")	All equity	Developer and institutional investor	Institutional investor finances much of the project, injecting some equity up-front and additional equity over time as the PTCs are generated. Includes a return-based flip in the allocations.

TABLE 12.2   (continued)

Common Financing Structures for Wind Projects

Financing Structure Name	Project Capital Structure	Likely Equity Investors	Brief Description of Structure Mechanics
Cash leveraged	Equity and debt	Developer and institutional investor	Based on the strategic investor flip structure, but adds debt financing. Likely involves institutional investors, rather than strategic investors. Loan size/amortization based on the amount of cash flow from power sales.
Cash and PTC leveraged	Equity and debt	Developer and institutional investor	Similar to the cash leveraged structure, but the loan size and amortization profile are based on the cash flow from power sales *plus* a monetization of the projected PTCs from the project.
Back leveraged	All equity (but developer uses debt outside of the project)	Developer and institutional investor	Virtually identical to the institutional investor flip, but with the developer leveraging its equity stake in the project using debt financing.

Source: Adapted from Harper, J. P., Karcher, M. D., and Bolinger, M., Wind Project Financing Structures: A Review & Comparative Analysis, Lawrence Berkeley National Laboratory technical report LBNL-63434, September 2007.

TABLE 12.3

Wind Developer Financing Structure Decision Matrix

Scenario	Developer Can Use Tax Benefits	Developer Can Fund Project Costs	Developer Wants to Retain Stake in Project Ownership/ Ongoing Cash Flows	Developer Wants Early Cash Distributions	Project Has Low Projected IRR	Project Already Exists (Refinancing/ Acquisition)	Most Suitable Financing Strategy or Structure
1	No	No	No	Yes	N/A	No	Sell project to a strategic investor
2	Yes	Yes	Yes	No	No	No	Corporate
3	No	Limited	Yes	No	No	No	Strategic investor flip
4	No	Limited	Yes	Yes	No	No	Institutional investor flip
5	No	Limited	Yes	No	Yes	No	Cash leveraged *or* Cash and PTC leveraged
6	No	Limited	Yes	Yes	No	Yes	Institutional investor flip
7	No	Yes	Yes	Yes	N/A	Yes	Pay-as-you go
8	No	Limited	Yes	Yes	Yes	No	Back leveraged

Source: Adapted from Harper, J. P., Karcher, M. D., and Bolinger, M., Wind Project Financing Structures: A Review & Comparative Analysis, Lawrence Berkeley National Laboratory technical report LBNL-63434, September 2007.

where certain investors can step in. Oftentimes lenders will require a project company obtain 10–25% of the project costs in equity. In these cases, an investor including banks or insurance companies can make an equity investment in exchange for the allocation of tax benefits.

The choice of which structure to utilize in a project below depends on factors such as the cash available to invest by the sponsor company, the ability of the sponsor to utilize the tax benefits and the role the sponsor wants to play in the project long term. Ultimately, the structure will likely align with the needs of the various parties to try and create the most efficient structure.

Financing a Clean Technology Project

One of the most common technologies utilizing project finance is wind. Wind is a fairly mature renewable energy generation technique (as compared to solar which remains somewhat less mature in the commercialization lifecycle). As a result, the growth in wind production has grown dramatically and the forecasts for growth continue to remain strong. Like solar and geothermal projects, wind farms require huge sums of money. However, smaller-scale wind farms and even single turbines can represent viable businesses.

In the case of the financing example below, we'll examine a large 150 MW wind project and the requirements to put in place the financing and related agreements to build that facility.

WHY ARE PEOPLE SO ENTHUSED ABOUT WIND?

Growth, growth and more growth. According to BTM Consult (one of the leading industry analysts for the Wind Sector), expect strong growth to continue in the near term for newly installed wind capacity. That growth has even outpaced estimates from just a few years ago. In fact, in BTM's 2006 estimates, they forecasted approximately 27,500 MW of annual installations for the year 2011. Those numbers were revised in 2007 upward to 35,000 MW for the year 2011 (an increase of approximately 21%) and in 2008 were again revised upwards to 45,000 MW for the year 2011 (an increase of approximately 64% over the 2006 estimates).

Growth in wind energy is surpassing expectations ...

The 150 MW Wind Farm

WindStartup, Inc. has successfully built and managed several smaller scale wind projects—wind farms that generated 10 MW of electricity at the site. As a result, the company now wants to expand and develop a larger wind farm with the capacity to generate 150 MW of electricity, enough to power more than 50,000 homes. To accomplish this, it believes it will need to raise approximately $150 million to finance the purchase of the wind turbines and the construction. The company had raised approximately $5 million in equity for its company to help fund the project siting and development. It expects the project to be ultimately structured as "Cash Leveraged" financing utilizing debt finance with institutional investors acquiring the tax benefits.

The company identified a promising site and secured a lease of the property from the landowners, with the option to extend the lease terms upon satisfactory test results from

wind speed and duration testing. The location is promising due to initial wind speed tests, a series of construction companies familiar with building wind farms located in the area, a location near to a major metropolitan area and sufficient transportation and access options. In addition, two area utilities have partnered to expand transmission lines to the area in order to handle future growth of wind, solar, and hydroelectric power expected to be generated from this area.

WindStartup decided to proceed ahead with the project and established the project entity Valley Wind LLC. It decided to purchase 100 General Electric 1.5-MW SLE wind turbines for its wind farm. It made a small cash investment into Valley Wind in order to place a deposit on the turbines and entered into a contract with General Electric to purchase the turbines with 25 turbines delivered 24 months from the contract date, and 25 turbines to be delivered each month thereafter.

Before beginning testing, WindStartup had held preliminary discussions with an area utility company. Due to the state's renewable portfolio standards which are due to require at least 11% of the state's electricity consumption to come from renewable sources in the next five years, the utility is more than happy to partner with a company developing a wind farm, and agrees to discuss a purchase power agreement in connection with the financing of the project. Now that the full testing of the site was complete confirming that the wind in the area was satisfactory, the company began work with the local government to obtain the necessary permits for the site. At the same time, they began to lay out the preliminary terms of the PPA between the utility and the company.

WindStartup then began work to assemble the parties for the financing of Valley Wind. In its prior deal, WindStartup had worked with a middle-sized bank to finance its projects. However, this time it would need the assistance of a much larger lender. After a series of discussions, WindStartup identified a lead lender for the project. The loan would have an 8% interest rate and would have limited recourse against WindStartup. The lender also required that Valley Wind also raise 40% of the project costs ($60 million) in the form of an equity investment. Valley Wind identified certain institutional investors interested in funding the project primarily for the tax benefits. The parties agreed to utilize a traditional "flip" structure for the equity investment since WindStartup was unable to utilize the tax benefits. The investors would invest $60 million in exchange for 99% of the PTC, accelerated depreciation deductions, the tax gains and losses, and the distributable cash. The sponsor would retain 1% of these credits and assets initially. After 10 years, the "flip" will occur with the sponsor retaining 90% of the assets credits and distributions while the investors would receive 10%.

Valley Wind (and a bevy of lawyers) went to work on the documentation for the transaction, including the loan agreement with its lender, the equity agreement documentation, and a series of intercompany agreements between Valley Wind, WindStartup, and certain affiliates of WindStartup. Valley Wind also finalized a power purchase agreement with its utility partner for the purchase of up to 1550-MW of electricity generated by the project. With the project finance in place, the parties broke ground and began construction of their 150-MW facility.

WHAT DOES IT TAKE TO DEVELOP A WIND FARM?

The following are the key steps to develop a wind farm beginning with an empty plot of land to a full functioning wind farm.

1. *Secure the site.* Once a location is identified, the developer will enter into a contract with the land owner, typically for a (very) small percentage of the sales.
2. *Test the site.* Testing is done over a series of months (usually more than 12 months) in order to determine that sufficient wind is available and to determine the time the wind blows.
3. *Project planning.* Key steps in the project planning are to determine grid interconnection, site access, and determination of potential local area partners.
4. *Secure planning permission.* Once the planning stage is complete, approval from local government officials is required (which in some locations is more difficult than anticipated).
5. *Secure a power purchase agreement (PPA).* The developer will enter into a PPA with the local utility or some developers may choose to operate as a merchant generator and sell power into the open market.
6. *Finance the project.* Most wind farms are financed with nonrecourse project financing that utilizes an equity ratio of 10–20%. The interest rate levels a project can utilize will play a key role to determine whether or not a project is feasible. Many developers will sell certain components of the project including the tax credits, emissions credits, or depreciation streams in order to defray the upfront project costs.
7. *Secure the turbines.* The lead time to obtain the wind turbines has been as long as 24 months in advance, following a 5–10% deposit.
8. *Construct the project.* This stage of the project will typically be outsourced to local operators.
9. *Operate or sell the project.* Depending on the rate of return on the project, the developers will operate the project or sell it to another party.

Source: *Clean Technology Primer*, Jefferies & Company, Inc. June 2008.

13

Working with the Government

The influence of government on green business cannot be understated. From grants, environmental small business loan programs, loan guarantees, tax credits, government contracts, regulations, and general purchasing patterns, the federal and state governments of the United States have made support of green businesses a key priority. Outside of the United States, governments throughout Europe and Asia have made environmental support key components of their own regulatory and spending programs.

Unlike technology sectors such as software, computers, biotechnology, and medical devices, green business is heavily reliant on the involvement of the government to continue the growth of green businesses. For example, the combination of production tax credits (PTC) and renewable portfolio standards (RPS) has been responsible for much of the growth of the wind industry. The threat of the expiration of the PTC led to changes in investment and development behavior. And the wind industry is not alone. Sectors from biofuels, solar, hybrid, and alternative fuel vehicles have each seen their growth hinge on various aspects of government involvement in these industries.

Government Commitment to Green Business

If there were any doubt that green business is a key priority for the new administration of the U.S. government, including both the executive and legislative branches, that question was answered in the first months of 2009. The American Recovery and Reinvestment Act (Recovery Act) was signed into law by President Obama in February of 2009. The Recovery Act has three main goals: (1) create new jobs and save existing jobs; (2) spur economic activity and invest in long-term economic growth; and (3) foster unprecedented levels of accountability and transparency in government spending. At the same time, the thread of support of green businesses and clean technology was evident throughout the Recovery Act.

For more information on the Recovery Act, visit

http:// www.recovery.gov

In striving to meet these goals, the Recovery Act allocated $787 billion in recovery funds to 28 different federal agencies. Each of those agencies then developed specific plans for how to spend recovery funds. Included in the $787 billion is more than $80 billion in clean energy investments meant to help jump-start the U.S. economy and build clean energy jobs. Several of those investments are summarized as follows:

- $11 billion for a better and smarter energy grid and smart meters to be deployed in American homes
- $5 billion for low-income home weatherization projects

- $4.5 billion to make federal building more energy efficient
- $13 billion to extend tax credits for renewable energy production
- $6.3 billion for state and local renewable energy and energy efficiency efforts
- $600 million in green job training programs
- $2 billion in competitive grants to develop the next generation of batteries to store energy
- $300 million for reducing diesel fuel emissions
- $400 million for electric vehicle technologies

Although the nature of the Recovery Act led to quick dispersal of funding in 2009 and 2010, many Recovery Act funds are being distributed to state governments, who then have further options for creating loan, grant, and incentive programs for residents and businesses located within that state. The nature and amount of funding reserved for clean energy efforts also strongly indicates the federal government's perceived importance of these efforts. The Recovery Act is likely to help pave the way for new and ongoing programs meant to encourage and support energy efficiency and renewable energy projects throughout the country.

The Cast of Government Players in Green Business

Nearly all of the responsibility for environmental efforts of the federal government had originated in the Environmental Protection Agency (EPA). However, the first decade of the twenty-first century has begun to spread the responsibility for developing green businesses and clean technologies to other agencies and organizations throughout the federal government.

Environmental Protection Agency

Headquartered in Washington, DC and operating out of 10 regional offices throughout the country, the EPA's central mission is to protect human health and the environment. In pursuing that mission, the EPA works with federal, state, tribal, and local government entities to regulate and reduce environmental damage. The EPA also offers grants and administers partnership programs in support of companies, organizations, communities, and individuals working to address a wide variety of environmental issues.

The EPA has existed for over 40 years, and it plays an increasingly prominent role in shaping and enforcing U.S. environmental policy. Soon after its creation in 1970, the EPA was given the responsibility of enforcing major pieces of environmental legislation, such as the Clean Air Act and Clean Water Act. Later, in the 1980s, the EPA also became responsible for cleaning up old waste sites under the Superfund program and for initiating emergency response preparations for environmental accidents. In the 1990s and early twenty-first century, the EPA has continued to play a major role in reducing pollution, regulating emissions, and exploring innovative approaches to environmental protection.

For more information on the EPA and some of its programs, visit

http://www.epa.gov

http://www.epa.gov/partners/programs/

The EPA's grants and partnership programs help create a variety of opportunities for green entrepreneurs. For example, through the ENERGY STAR program (a joint program of the EPA and the U.S. Department of Energy) the EPA partners with businesses in the residential, commercial, and industrial sectors to increase sales of energy efficient products, raise energy efficiency standards for new home construction, and improve energy efficiency in commercial and industrial facilities.

Department of Energy

The U.S. Department of Energy (DOE) is a department within the United States' Cabinet that works to advance the energy security of the United States. The DOE aims to promote the nation's energy security by helping to develop and support reliable, clean, and affordable energy. Historically, the DOE allocated the vast majority of its funds and resources to the management of nuclear power facilities. That has begun to change and new emphasis is being put on renewable energy and improving the electricity transmission and distribution systems.

As will be discussed later in greater detail, the DOE has recently played a prominent role in dispersing federal funding meant to support the advancement of both new and conventional renewable energy technologies. The DOE also operates an Office of Energy Efficiency & Renewable Energy, which provides information on saving energy at home, finding financing and tax credits for green projects and initiatives, finding clean energy jobs, and state projects.

The DOE funds several national laboratories and technology centers that carry out cutting-edge research related to energy generation and storage, environmental management, renewable energy, and other topics central to DOE's mission. Often universities or private corporations contract with the DOE to administer, manage, operate, and staff the centers. More information on these facilities (including links to facility Web sites) is available on the DOE Web site at: www.energy.gov/organization/labs-techcenters.htm.

Finally, like the EPA, the DOE is involved in a number of joint initiatives, working with domestic and international partners in both the public and private sectors. Several of these initiatives may be of interest to green entrepreneurs. For example, the Solar America Cities program is designed to accelerate the adoption of solar energy technologies in community infrastructures found in several U.S. cities. Further, the Climate V.I.S.I.O.N. Program (*Voluntary Innovative Sector Initiatives: Opportunities Now*) is a public–private partnership initiative designed to reduce greenhouse gas intensity. More information on these programs, as well as other DOE joint initiatives, can be found at: www.energy.gov/about/jointinitiatives.htm.

For more information on the DOE and some of its programs, visit

http://www.energy.gov

Office of Energy Efficiency & Renewable Energy http://www.eere.energy.gov

United States Department of Agriculture

The United State's Department of Agriculture (USDA) is the U.S. federal executive department responsible for U.S. federal policy on farming, agriculture, and food. With historic roots reaching back to the mid-nineteenth century, the USDA has expanded into a large agency composed of a number of smaller offices and subagencies, including the USDA Office of Rural Development and the United States Forest Service. Pursuant to its mission to "provide leadership on food, agriculture, natural resources, and related issues based on sound public policy, the best available science, and efficient management," the USDA has also become very involved in responding to emerging energy-related issues and opportunities.

The USDA supports both research and production of biodiesel fuels, ethanol fuels, and other sources of biomass energy. In support of such initiatives, the USDA administers a number of loans, grants, and other incentive programs meant to encourage the development of biofuel technology and production.

Like the EPA and the DOE, the USDA is also involved in a number of partnership programs that may be of interest to green entrepreneurs. More information on the USDA's Sustainability and Stewards Programs and Activities is available at: www.usda.gov/oce/ sustainable/Files/SustainabilityWithinUSDA_3.pdf.

Small Business Administration

The U.S. Small Business Administration (SBA) is an independent agency with the responsibility to aid, counsel, and assist entrepreneurs and small businesses. The SBA has an extensive network of field offices and partnerships with public and private organizations. With more than 29 million small businesses in the United States, the SBA offers a variety of services including training and counseling, financial assistance, and contracting assistance.

The SBA does offer numerous programs that are beneficial to entrepreneurs of all kinds, including greentrepreneurs. Some regional offices have begun to offer educational programs designed for green businesses and continue to provide loan guarantees that may be applicable. The SCORE program offers advisors and counselors with backgrounds in green business in many of its regional offices. The SBA has expanded its 504 Loan Program to specific green lending programs, discussed in greater detail in Chapter 15.

For more information on the Small Business Administration, visit

http://www.sba.gov

Internal Revenue Service

The Internal Revenue Service (IRS) is the U.S. federal agency within the Department of the Treasury that is responsible for collecting taxes. With the passage of the American Recovery and Reinvestment Act of 2009, the IRS now helps to encourage clean energy development by offering tax credits to individuals that invest in green products and services.

When filing taxes with the IRS, individuals also may earn tax credits for improving energy efficiency. For example, tax credits are available for such things as: (1) the purchase of hybrid vehicles; (2) the use of approved energy-efficient boilers, furnaces, refrigerators, dishwashers, and other appliances; (3) improvements made to existing structures to improve energy efficiency; and (4) the use of on-site renewable energy sources, such as solar water heaters and small wind turbines. These tax credits, along with other tax incentives, will be discussed in greater detail in Chapter 17. More information is also available at http://energytaxincentives.org/.

Selling to the Government

With purchases totaling more than $425 billion per year, the U.S. federal government is the world's largest buyer of goods and services, and the federal government has made it a clear priority to green its acquisition process. As a result, it is no secret that selling to the federal government can provide significant revenues for your business. While the prospect of securing a government contract may sound intimidating, the process isn't as painful as you'd think if you do the requisite research and allow for adequate time for the process to complete.

There are more than 20 million small businesses in the United States. Only 500,000 or 2.5% of the 20 million small businesses are currently in a position to do business with the federal government—not because they are the only ones able to do the work, but because those 500,000 small businesses know how to get the work. Moreover, the government especially encourages small businesses to bid on contracts for some of its needs. In fact, federal agencies are required to establish contracting goals, with at least 23% of all government expenditures going to small businesses. Believe it or not, the government is required to reserve all federal purchases between $3000 and $100,000 for small businesses, with some exceptions.

Winning Federal Contracts

The U.S. Small Business Administration (SBA) offers a free online course titled "Business Opportunities: A Guide to Winning Federal Contracts". It is a 30-minute program that focuses on the government contracting process.

Available at: http://www.sba.gov/training/governmentcontracting/index.html

---

ARE YOU A GOOD CANDIDATE FOR GOVERNMENT CONTRACTING?

Contracting with the government is not necessarily the right decision for every business. Think about your company's products and services, and what makes your company unique. Consider what the government will look for when considering you for a contract award, such as financial status, staff capabilities, and track record. Before going forward, ask yourself some basic questions:

- Are you willing to learn and follow the rules relating to federal acquisitions?
- Are you willing to put in the legwork to find procurement opportunities and take the time to prepare and present offers (including bids and quotes)?
- Are you confident that your business can finance the performance of a government contract that may involve significant startup costs?
- Are you willing to be a subcontractor to companies that are prime contractors?

If the answer is "yes" to all of these questions, then government contracting may be a viable business strategy for your business.

Identify Contracting Opportunities

FedBizOpps (Federal Business Opportunities) is the Federal Government's one-stop virtual marketplace. This Web site allows commercial vendors to identify opportunities posted by the entire federal contracting community. Agencies can communicate buying requirements for all contracts expected to exceed $25,000. Through FedBizOpps you can receive notices of contract solicitations and requests for information. You can customize the data you want to receive so that you will be notified only about contracts that your business would be eligible to bid on (https://www.fbo.gov/).

Once you are registered with Central Contractor Registration (CCR), you can locate opportunities for contracting.

**HOW TO REGISTER WITH THE CENTRAL
CONTRACTOR REGISTRATION (CCR)**

The CCR is the federal clearinghouse for vendors, including small businesses. Contracting officers go to the database to look for small businesses to fulfill contracting requirements. If your company is not listed in CCR, contracting officers won't know you exist.

1. *Pick your North American Industry Classification System (NAICS) code.* NAICS codes are the standard used by federal agencies to classify businesses for the purposes of collecting and analyzing data pertaining to the nation's business economy. When you register, you should include all NAICS codes that identify the type of contracts you can perform. Please note that you can add or change NAICS codes at any time.

2. *Select your small business size standard based on your NAICS code.* SBA establishes definitions of "small business" for all industries, called size standards. These standards represent the largest size that a business (including its subsidiaries and affiliates) may be to remain classified as a small business. These apply to SBA's financial assistance and to its other programs, as well as to federal government procurement programs. SBA's Table of Small Business Size Standards lists size standards by six-digit NAICS industry codes.

3. *Register for a Dun & Bradstreet DUNS Code.* A DUNS Code is a unique nine digit identification number, for each physical location of your business. A DUNS Code assignment is free for all businesses required to register with the federal government for contracts or grants.

4. *Register for a Federal Tax identification Number (TIN or EIN).* You can apply for this online, by phone, or by fax.

5. *Determine your Standard Industrial Classification (SIC) code.* To determine your code you must examine the SIC manual and find the classification number that applies to your business. You may list up to 20 classification codes that apply to your products and services. SIC codes carry either four or eight numbers.

6. *(Optional) Determine your Federal Supply Classification codes and Product Service codes.* These may be useful when registering with CCR.

7. *Register with Central Contractor Registration (CCR).* You can apply online by going to http://www.ccr.gov. As long as you have an EIN number, the registration process takes one to three business days to complete; otherwise you need to obtain an EIN number prior to starting the process. If your company is already registered with CCR, you must renew your registration annually which may take up to five business days. Additional information by downloading the CCR handbook at https://www.bpn.gov/ccr/doc/UserAccount.pdf.

8. *Complete an Online Representation and Certification Application (ORCA).* ORCA was developed as an Integrated Acquisition Environment (IAE) E-Government initiative designed to reduce the administrative burden on vendors to submit the same paper-based representations and certifications repeatedly for various solicitations.

9. *Enter your small business profile information on the Dynamic Small Business Search (DSBS) page.* Entering your small business profile, which includes your business information and key word descriptions, allows contracting officers, prime contractors, and buyers to learn about your company.
10. *Update and continue to develop your CCR/DSBS profile (ongoing).* Treat your profile as your business' resume; as such, you need to regularly review, update, and strengthen it. Start by performing a search as if you were looking for your business. Look at the profiles of other firms in your area of expertise and use the effective parts of their profiles as a guide as you continue to develop your own profile—they will likely be your competitors. Also, ask contracting officers for their opinion of your profile.

Understanding the Federal Government Buying Process

Unlike private sector buyers, the federal government has special rules and regulations governing its purchasing system. All of the rules and regulations—known as the Federal Acquisition Regulation (FAR)—are set out for everyone from the beginning and are designed to create a highly transparent system. This transparency is prized as a way to promote open and fair competition, prevent corruption, and protect public monies.

In recent years, government acquisition has changed significantly by placing an emphasis on "best value." In seeking "best value," the government doesn't necessarily need to award a contract to the lowest bidder; instead, the government can take other factors into account (i.e., considering what the true "best value" is to the government). However, in its effort to promote transparency, if the government is going to make an award based on best value, the intent to do so must be stated in the solicitation, which must include a description of the evaluation criteria, award factors, and factors other than price that will be considered in making the award.

When the government wants to purchase a certain product or service, it can use a variety of contracting methods.

Simplified Purchases

Federal agencies can solicit and evaluate bids on government purchases of less than $100,000 using simplified procedures that require fewer administrative details, lower approval levels, and less documentation. All purchases of up to $3000 in individual items, or multiple items whose aggregate amount does not exceed $3000, are classified as "micro-purchases" and can be made without obtaining competitive quotes. These purchases are not reserved for small businesses. Agencies can make micro-purchases using a government credit card.

Sealed Bidding

Sealed bidding is a method of contracting that solicits the submission of competitive bids, followed by a public opening of bids. This process involves an invitation for bids (IFB), which typically includes a description of the product or service to be acquired; instructions for preparing a bid; the conditions for purchase, packaging, delivery, shipping, and payment; contract clauses to be included; and the deadline for submitting bids. Each sealed bid is opened in public at the purchasing office at the time designated in the invitation.

All bids are read aloud and recorded. A contract is awarded to the responsive and responsible bidder, whose bid is most advantageous to the government, considering price and price-related factors. In other words, the winning bidder must follow the IFB directions in submitting its bid and must be offering exactly what the government wants.

Contract Negotiation

In some cases, when the value of a government contract exceeds $100,000 or when the product or service being acquired is highly technical, the government may issue a request for proposal (RFP). A typical RFP solicits proposals from prospective contractors on how they intend to deliver the requested product or service, and at what price. Proposals can be subject to negotiation after they have been submitted. When the government is merely checking into the possibility of acquiring a product or service, it may issue a request for quotation (RFQ). A response to an RFQ by a prospective contractor is not considered an offer, and cannot be accepted by the government to form a binding contract.

Consolidated Purchasing Programs

Most government agencies have common purchasing needs for certain types of products or services, such as carpeting, furniture, office machine maintenance, and perishable food supplies. Sometimes the government can realize economies of scale by centralizing the purchasing of such items. The government may use "acquisition vehicles," such as multi-agency contracts and government-wide acquisition contracts (GWACs), to encourage long-term vendor agreements with fewer suppliers. Under the General Services Administration (GSA) Schedules Program (also known as "Multiple Award Schedules" and "Federal Supply Schedules"), GSA establishes long-term, government-wide contracts with commercial firms to provide access to over 11 million commercial supplies and services that can be ordered directly from the GSA contractors on the "GSA Advantage" online shopping and ordering system. State and local governments also use the GSA Schedules. The use of these acquisition vehicles (also called multiple award contracts) has increased significantly during the last few years, since they allow federal buyers to quickly fill requirements by issuing orders against existing contracts without starting a new procurement action from scratch. Agencies can also award multiple task order contracts (which involve indefinite delivery and indefinite quantity) to allow more than one firm to deliver a particular product or service.

For more tips and training on marketing your business to the federal government, visit

"Marketing Your Business" http://www. sba.gov/ContractingOpportunities/owners/ resources/GC_RESC_MARKETING.html

"Small Business Planner" http://www. sba.gov/smallbusinessplanner/manage/ marketandprice/

Contracting Goals and Special Small Business Designations

To ensure that small businesses get their fair share of the government contracting pie, the following goals have been established by law for federal executive agencies:

- 23% of prime contracts for small businesses
- 5% of prime and subcontracts for women-owned small businesses
- 3% of prime contracts for HUBZone small businesses
- 3% of prime and subcontracts for service-disabled veteran-owned small businesses

While these government-wide goals are not always achieved, they are important because federal agencies have a statutory obligation to reach out to and consider small businesses for procurement opportunities. For example, an agency may be looking for a woman- or veteran-owned business to fulfill specific contract requirements and help it achieve government-wide contracting goals. Self-certification and SBA certification are the two special types of programs available for the following classifications:

> **For more information on SBA-Administered Programs, visit**
>
> **HUBZone** https://eweb1sp.sba.gov/hubzone/internet/index.cfm
>
> **8(a) Business Development Program** http://www.sba.gov/aboutsba/sbaprograms/8abd/
>
> **Small Disadvantaged Business certification** http://www.sba.gov/sdb

Self-Certification Programs

These types of businesses are identified solely through a self-certification process: One may simply self-certify by checking the appropriate box when submitting a proposal. Note that there are criminal penalties for false certifications.

> **For more information on Self-Certification Programs, visit**
>
> **SBA's Office of Women's Business Ownership (OWBO) at**
> http://www.sba.gov/aboutsba/sbaprograms/onlinewbc/index.html
>
> **SBA's Office of Veterans Business Development (OVBD) at**
> http://www.sba.gov/aboutsba/sbaprograms/ovbd/index.html
>
> **U.S. Department of Veterans Affairs at**
> http://www.va.gov/ and http://www.sba.gov/vets

- *A woman-owned business* is defined as a business that is owned and controlled 51% or more by a woman or women.
- *A veteran-owned business* is defined as a business that is owned 51% or more by a veteran or veterans.
- *A service-disabled business* is defined as a business that is owned or controlled 51% or more by one or more service-disabled veterans. The Veterans Administration confirms service-related disabilities.

SBA-Administered Programs

The SBA administers three programs to assist specific groups in receiving federal contracts.

- *A HUBZone business.* SBA's HUBZone ("Historically Underutilized Business Zone") Program is designed to promote economic development and employment growth in distressed areas by providing access to more federal contracting opportunities. Certified small business firms have the opportunity to negotiate contracts and participate in restricted competition limited to HUBZone firms. To be eligible for the program, a business must meet the following requirements to obtain certification:
  - Be a small business by SBA standards
  - Be located in a HUBZone
  - Be owned and controlled 51% by person(s) who are U.S. citizens
  - Have at least 35% of its employees residing in a HUBZone
- *8(a) Business Development Program.* The 8(a) Program offers a broad scope of assistance to socially and economically disadvantaged firms, with a special 9-year program that provides special training and counseling services, as well as a Mentor-Protégé Program (http://www.sba.gov/aboutsba/sbaprograms/8abd/

mentorprogram/). The goal is to help these entrepreneurs to compete in the federal contracting arena and to take advantage of greater subcontracting opportunities available from large firms as the result of public/private partnerships.

To qualify for the program, a business must meet the following criteria:

- Qualify as a "small business"
- Be owned and controlled by a socially and economically disadvantaged individual
- Be in business for at least two years (this may be waived)
- Display reasonable success potential
- Display good character

Certain groups of people are presumed to be socially and economically disadvantaged, including African Americans, Hispanic Americans, Asian Pacific Americans, Subcontinent Asian Americans, and Native Americans. Nevertheless, anyone may qualify if they can show by a "preponderance of the evidence" that they are disadvantaged because of race, ethnicity, gender, physical handicap, or residence in an environment isolated from the mainstream of American society.

- *Small Disadvantaged Business certification.* To qualify, a business must be at least 51% owned and controlled by someone who is socially and economically disadvantaged, as defined by the same standards of the 8(a) program above. Note that this is different from the 8(a) program because—although they both serve the same group of people—this is simply a certification, while 8(a) is a development and training program.

Getting the Government to "Buy Green"

The federal government gives preference to the purchase of environmentally friendly products and services. Pursuant to Executive Order (E.O. 12873 on Federal Acquisition, Recycling and Waste Prevention) all agencies of the federal government are to identify and give preference to the purchase of products and services that pose fewer burdens on the environment. Several key criteria are to be considered when a federal agency makes a purchasing decision based on environmental preferability:

To view the EPA's guide to "Selling Environmental Products to the Federal Government," visit

http://www.epa.gov/epp/pubs/selling.htm

- *Pollution prevention*: Consideration of environmental preferability should begin early in the acquisition process and be rooted in pollution prevention that strives to eliminate or reduce, up front, potential risks to human health and the environment.
- *Multiple attributes*: A product or service's environmental preferability is a function of multiple environmental attributes.
- *Life-cycle perspective*: Environmental preferability should reflect life-cycle consideration of products and services to the extent feasible.
- *Magnitude of impact*: Environmental preferability should consider the scale (global versus local) and temporal aspects (reversibility) of the impacts.
- *Local conditions*: Environmental preferability should be tailored to local conditions where appropriate.

- *Competition*: Environmental attributes of products or services should be an important factor or "subfactor" in competition among vendors, where appropriate.
- *Product attribute claims*: Agencies need to examine these carefully.

GOVERNMENT CONTRACTING RESOURCES

- *GSA's Center for Acquisition Excellence* offers an online training course, "How to Become a Contractor—GSA Schedules Program," which provides valuable information for all prospective Schedule contractors (http://www.gsa.gov/Portal/gsa/ep/contentView.do?contentType=GSA_BASIC&contentId=25926).
- *Acquisition Central* is a Web site dedicated to the government acquisition process that allows you to access shared systems and tools to help you conduct business efficiently. This site offers information about regulations, systems, resources, opportunities, and training (https://www.acquisition.gov/).
- *The Office of Small and Disadvantaged Business Utilization Directors*, known as OSDBU, is an office within each federal agency whose mission is to help give contracting dollars to small businesses. The list of offices will help you obtain contact information for the federal agency or agencies that you'd like to work with (http://www.osdbu.gov/).
- *SBA Procurement Resources:* (1) Procurement Center Representatives (PCRs) work with federal agencies to identify prime contracting opportunities, reserve procurements for competitions among small businesses, and provide small business sources to federal buying agencies. (2) Commercial Marketing Representatives (CMRs) conduct compliance reviews of prime contractors, counsel small businesses on how to obtain subcontracts, and conduct matchmaking to facilitate subcontracting to small businesses (http://www.sba.gov/aboutsba/sbaprograms/gcbd/GC_PCRD1.html).
- *Procurement Technical Assistance Centers* (PTACs) provide technical assistance to small businesses seeking government contracts. The centers can help outline the requirements of government contracting and provide further insight into the contracts (http://www.aptac-us.org/new/Govt_Contracting/find.php).
- *SBA's Resource Partners*, including SCORE, Small Business Development Centers (SBDBs), Women's Business Centers and Business Matchmaking can serve as helpful developmental resources for your business (www.sba.gov).
- *Regional Federal Executive Boards* are located in cities with active federal activities. These boards generally look out for the well-being of contractors and conduct community relations activities (http://www.feb.gov/).

14

Laws, Regulations, Initiatives, and More

Where do you start with the analysis of laws and regulations affecting green businesses? Much of the key regulatory framework that led to the green business re-revolution in the first decade of the twenty-first century had its roots in a series of key laws passed in the 1970s. Those laws put in new framework for environmental and energy regulations. And that's where this chapter will start—looking broadly at the laws that govern the environment and energy.

It is simply impossible to give these regulations any level of detailed coverage in a single chapter of a book. The goal here isn't to provide you with a detailed understanding of the Energy Policy Act or National Environmental Policy Act, but to offer a high-level familiarity to some of the key regulations and a general framework on how all these laws fit together. We will also spend a bit of time looking into a crystal ball to see what may lay ahead for the regulation of carbon through a variety of climate change regulatory proposals.

Summary of Green Regulatory Landscape

There are countless regulations and laws that impact green businesses. This chapter will focus primarily on the specific environmental and energy regulations (some of the energy regulations are also covered in Chapter 11 on Understanding Utilities). Other key regulations that affect green businesses include intellectual property, taxation, general corporate regulations, and others.

Much of the foundation for today's environmental and energy regulatory scheme was enacted in the 1970s. Specifically, the National Environmental Policy Act (NEPA), the Environmental Quality Improvement Act, the National Environmental Education Act, and the Environmental Protection Agency (EPA) were all enacted or created in that decade. Energy regulation began in earnest in the early twentieth century, but was accomplished through a number of different agencies. But the energy crisis of the 1970s forced the federal government to centralize its energy industry oversight.

The series of environmental regulations was intended to protect the environment against both public and private actions. Specifically, NEPA was intended to require all governmental agencies to consider the effects of their actions and decisions on the environment, and has been described as one of the country's most far-reaching environmental regulations ever passed. The EPA is tasked with monitoring and analyzing the environment, conducting research, and working in conjunction with state and local governments to devise pollution control policies.

Prior to the twentieth century, there was little federal oversight of energy, largely because many people believed that the supply of energy was unlimited. During the great depression, energy regulation was part of the expansion of the federal government, but was not organized under a single agency. In addition, the Manhattan Project led to the regulation

The Clean Air Act and CO_2

In its 2007 decision, Massachusetts v. Environmental Protection Agency, the U.S. Supreme Court held that carbon dioxide is an "air pollutant" for purposes of the Clean Air Act. Following the decision several states have argued that the EPA now has an obligation to regulate CO_2 and other greenhouse gasses as endangerments to public health and the environment. Such an action by the EPA could have a profound effect on transportation and other industries throughout the country.

More information on this debate is available at: http://www.ag.ca.gov/globalwarming/cleanairact.php

of nuclear energy. Following the energy crisis, a national energy plan was established and the Department of Energy (DOE) was created in 1977 to oversee the complex energy regulatory schema. The federal energy laws and regulations were designed to provide affordable energy by sustaining competitive markets, while protecting the economic, environmental, and security interests of the United States.

The Obama Administration has promised extensive changes to energy and environmental regulations, including a proposed revision to the energy regulations and the implementation of climate change legislation. As of the writing of this book, these proposals have not yet been enacted.

Key Environmental Regulations

National Environmental Policy Act

The National Environmental Policy Act (NEPA) is the basic national charter for protecting the environment. President Nixon, who signed NEPA into law on January 1, 1970, stated that the act was intended to lead the country to "regain a productive harmony between man and nature." NEPA requires all branches of the federal government to consider the environmental impact of any proposed federal action. Should any federal action have the potential to negatively impact the environment, federal actors must also consider whether any reasonable alternatives or mitigation measures may help reduce that impact.

For more information about the National Environmental Policy Act

http://www.epa.gov/Compliance/nepa/

The NEPA procedure for making these considerations requires federal agencies to submit Environmental Impact Statements (EISs) to the Environmental Protection Agency for review and comment. If the EPA finds a particular action to be unsatisfactory, they refer the action to the Council on Environmental Quality, a division of the Executive Office of the President for further review. The EPA maintains a database of all environmental impact statements, and their accompanying EPA comments. This database is available to the public at www.epa.gov/Compliance/nepa/.

Clean Air Act

The Clean Air Act (CAA), passed in 1970 and significantly amended in 1977 and 1990, is a comprehensive law that regulates air emissions from a variety of different sources. Compliance with the CAA is enforced by the EPA. Pursuant to the CAA, the EPA sets acceptable levels of emissions through the creation of National Ambient Air Quality Standards (NAAQS) for pollutants considered harmful to public health and the environment. Currently, the EPA has set NAAQS for six principal pollutants: carbon monoxide, lead, nitrogen dioxide, particulate matter (PM-10), particulate matter (PM-2.5) ozone, and sulfur dioxide. Information on the specific standards associated with these pollutants is available here: www.epa.gov/air/criteria.html.

State, local and tribal governments also have responsibility for enforcing the Clean Air Act. States must develop State Implementation Plans outlining their plan of action for carrying out the CAA. If a state fails to adequately regulate air pollution in compliance with the CAA, the CAA then grants the EPA the authority to issue sanctions against the state, or even take over enforcement of the CAA in that area. More information on state and local enforcement of the Clean Air Act can be found by visiting the EPA Web site: www.epa.gov/air/caa/peg/working.html.

Clean Water Act

The Clean Water Act (CWA) is a comprehensive federal law meant to regulate water pollution and to maintain the chemical, physical, and biological integrity of the nation's waters. Since being passed in 1972, the CWA has led to a reduction of direct pollutant discharge, increased financing of municipal wastewater treatment facilities, and the management of harmful runoff. These achievements have had a profound positive effect on the cleanliness of water used by humans, as well as on the protection and propagation of fish, shellfish, and other wildlife found in U.S. waters.

For more information about the Clean Water Act

http://www.epa.gov/compliance/assistance/bystatute/cwa/

A key provision of the Clean Water Act establishes a permit program called the National Pollutant Discharge Elimination System (NPDES). Under the NPDES, municipal, industrial, and some agricultural facilities are required to obtain a permit from the EPA or an authorized state agency before discharging any potential pollutants into the navigable waters of the United States.

Federal Insecticide, Fungicide, and Rodenticide Act

The Federal Insecticide, Fungicide, and Rodenticide Act (FIFRA) outlines a system for regulating the sale and use of pesticides within the United States. Initially passed in 1947 and significantly amended in 1964, 1970, and 1972, FIFRA requires a user or seller of a pesticide to register the pesticide with the EPA in order to apply for a license. The EPA permits the use or sale of the pesticide only if the applicant can demonstrate that using the pesticide "will not generally cause unreasonable adverse effects on the environment." In making an assessment of what constitutes an "unreasonable adverse effect," the EPA weighs the potential risk of harm posed by the pesticide against the potential benefit of using the pesticide. The EPA also considers whether the residues left by the pesticide may lead to a human dietary risk.

If use of a pesticide is permitted, the EPA then classifies that pesticide as a "general use" or "restricted use" pesticide. Although "general use" pesticides may be applied by anyone, "restricted use" pesticides can only be applied by certified applicators or those supervised by certified applicators. Most new pesticides are initially classified as

For more information about FIFRA

http://www.epa.gov/compliance/civil/fifra/fifraenfstatreq.html

"restricted use" until more information is available on any harmful effect caused by that pesticide. If the EPA finds that a person or company is using a pesticide in violation of FIFRA, EPA has the authority to impose a civil penalty and require that person or company to correct the violation. The EPA may also issue an order prohibiting the person who owns, controls, or has custody of a certain pesticide from selling or using that product, except in accordance with the order.

Toxic Substances Control Act

The Toxic Substances Control Act (TSCA) outlines the EPA's authority to regulate certain chemical substances or mixtures other than food, drugs, cosmetics, and pesticides. For example, TSCA allows for regulation of such chemicals as polychlorinated biphenyls, asbestos, radon, and lead-base paint.

For more information about the Toxic Substances Control Act

http://www.epa.gov/lawsregs/laws/tsca.html

The EPA requires manufacturers, processors, and distributors of such chemicals to take a number of actions to ensure compliance with TCSA. For example, the EPA may (1) require notification for any "new chemical substance" before that chemical is manufactured; (2) require testing of chemicals by manufacturers, importers, and processors where concerns about the safety of that chemical are found; (3) issue "Significant New Use Rules" when the EPA identifies a new use that could result in dangerous exposure to, or dangerous release of, a chemical; (4) require those importing or exporting chemicals to comply with certification reporting and other requirements; (5) require record keeping by those who manufacture, import, export, process, and/or distribute chemical substances in commerce; and (6) require those involved in the processing, manufacture, or distribution of chemical substances report any information obtained regarding potential dangers posed to human health or the environment.

Resource Conservation and Recovery Act

The Resource Conservation and Recovery Act was passed in 1976 to regulate and manage solid waste, hazardous waste, and underground storage tanks holding petroleum products or other certain chemicals. The RCRA's definition of "solid waste" is broad, including such things as any garbage or refuse, sludge from a wastewater treatment plant, and other discarded solid, liquid, semisolid, or contained gaseous materials. "Hazardous waste" is defined as "solid waste" that is ignitable, corrosive, or reactive.

For more information about the Resource Conservation and Recovery Act

http://www/epa.gov/agriculture/lrca.html

RCRA outlines procedures for tracking and regulating hazardous waste from the point of generation to the point of disposal in order to protect human health and the environment. Entities that generate hazardous waste are regulated and subject to strict standards for how waste is manifested, accumulated, and recorded. Similarly, RCRA requires facilities that treat, store, or dispose of hazardous waste to apply for a permit from the EPA or an EPA-authorized state agency. Finally, RCRA contains provisions outlining corrective actions that govern the cleanup of any release of hazardous waste. More information on state-specific waste management can be obtained from a state's Department of Environmental Quality or Department of Environmental Protection.

Comprehensive Environmental Response, Compensation, and Liability Act (a.k.a. "Superfund")

Also known as "Superfund," the Comprehensive Environmental Response, Compensation and Liability Act (CERCLA) was enacted by Congress in 1980 in order to establish broad federal authority to respond to releases or threatened releases of hazardous substances. The Act has three primary components: it (1) enables the revision of the National Contingency Plan, which provides guidelines and procedures for responding to

For more information about Superfund

http://www.epa.gov/superfund/policy/cercla.htm

threats from hazardous substances; (2) provides for the liability of persons responsible for releases of hazardous waste, where such persons can be identified; and (3) earned its Superfund nickname by creating a tax on chemical and petroleum industries that generated $1.6 billion over five years. The tax revenue went into a trust fund to be used for cleaning up abandoned or uncontrolled hazardous waste sites that pose a particular threat to human health and the environment. Superfund sites are included on the National Priorities List, which continues to be updated periodically by the EPA.

In the EPA's efforts to carry out CERCLA objectives, the agency has created opportunities for local communities to get involved in Superfund activities. For example, the Superfund Job Training Initiative was created under the premise that the EPA can help to connect communities, government, businesses, and other organizations through partnerships in order to collaboratively solve a hazardous waste problem. Through this program, the EPA engages local communities and local contractors in order to employ qualified workers to assist in cleanup efforts. The EPA also provides community programs, technical assistance and training, and monetary aid, where available, to encourage site cleanup.

Key Energy Regulations

The early regulation of energy was under the Federal Power Commission which was created with the Federal Power Act of 1920. This Act was later amended in 1935 and 1986, serving as the basis for the energy regulatory framework. The 1973 energy crisis was the motivation for developing a more comprehensive and organized energy regulation plan. Federal Energy Regulatory Commission (FERC) was created under the Department of Energy (DOE) to manage the efforts of various agencies, including the Federal Power Commission. FERC acted as an independent regulatory agency overseeing the natural gas, oil, and electricity markets in the United States.

FERC continues to regulate the transmission and sale of these energies (except for the sale of oil), provides licenses for hydroelectric plants, and reacts to environmental matters that arise. FERC utilizes an internal dispute resolution system, reducing the number of disputes that reach the federal courts. The U.S. Nuclear Regulatory Commission (NRC) continues to oversee nuclear energy and to protect the public health and safety from nuclear radiation and waste. NRC also promotes the common defense through a regime of rulemaking, inspection, and licensing. More information on FERC and utility regulation can be found in Chapter 11.

In the 1990s and 2000s, energy deregulation was a common trend in order to increase market competition to provide cheap, reliable energy. While various levels of deregulation have occurred throughout the energy sector, electricity markets now allow consumers to choose suppliers in many states. The government still plays a significant role in the oversight of energy markets. Deregulation has broken apart historically vertically integrated power companies to create competition throughout the energy chain from production to consumption.

Energy regulation is contained in numerous portions of the federal regulatory scheme. Title 42 of the U.S. Code entitled "The Public Health and Welfare" has many chapters devoted to energy issues, as does Title 16, and Title 30 of the U.S. Code and Title 10 of the Code of Federal Regulations.

Energy Policy Act of 2005

The Energy Policy Act of 2005 (EPAct) addresses energy production in the United States. The EPAct was the first major energy legislation passed by Congress in 13 years, when a previous version of the EPAct (The Energy Policy Act of 1992) was enacted. The Act establishes energy goals reducing energy consumption in federal buildings and facilities and encourages the use of renewable energy through tax incentives, grants and funding programs, and accelerated research and development.

Of particular interest to greentrepreneurs, the Act establishes a renewable electricity production credit, clean renewable energy bonds, renewable energy production incentives, rural and remote community electrification grants, and loan guarantees for projects that both reduce greenhouse gas emissions and employ significantly improved technologies.

Energy Independence and Security Act of 2007

The Energy Independence and Security Act (EISA) is a major piece of legislation that contains a number of provisions designed to increase energy efficiency and the availability of renewable energy. A number of these provisions may be of interest to greentrepreneurs.

One major objective of the EISA was to revise various standards related to energy use and efficiency. For example, EISA establishes a Corporate Average Fuel Economy (CAFE) standard for vehicle manufacturers. In order to meet the CAFE target goal of 35 miles per gallon by the model year 2020, vehicle manufacturers must ensure that their entire fleet of 2020 models obtains an overall average of 35 miles per gallon. EISA also establishes a Renewable Fuel Standard meant to increase the volume of renewable fuel required to be blended into gasoline from 9 billion gallons in 2008 to 36 billion gallons by 2022. Finally, EISA includes a variety of new standards for lighting and for residential and commercial appliance equipment, including residential refrigerators, freezers, refrigerator-freezers, metal halide lamps, and commercial walk-in coolers and freezers.

Additional provisions of EISA are intended to accelerate research and development of alternate energy sources and to provide funding to eligible green businesses. For example, Title VI of EISA outlines a number of strategies and procedures for accelerating research and development in solar energy, geothermal energy, marine and hydrokinetic renewable energy technology, energy storage, and transportation of electric power. Further, Title XII empowers the Small Business Administration to offer energy conservation loans, grants, and debentures to help small businesses develop, invest in, and purchase energy efficient buildings, fixtures, equipment, and technology. The SBA also offers pollution control loans to help eligible small businesses assist in the planning, design, or installation of a pollution control facility.

Fuels Regulations

Natural Gas Regulation

The regulation of natural gas has changed considerably as the industry evolved over the past 150 years. In the mid-1800s, natural gas was generally delivered within the same municipality in which it was produced. Local governments, deeming it a public interest to control potential monopolies, set down regulations that prevented natural gas manufacturers from abusing their market power. As the natural gas industry expanded, however, maintaining local regulations became more difficult. In the early 1900s, natural gas began to be shipped between municipalities, which led to the creation of state regulations and public utility commissions. Later, when natural gas became a part of interstate commerce through interstate pipeline distribution, the U.S. Supreme Court held that state governments no longer had jurisdiction over interstate sales.

The need for federal regulation eventually led to the 1938 passage of the Natural Gas Act (NGA), which gave the Federal Power Commission (FPC) jurisdiction over regulation of interstate natural gas sales. In an effort to effectively regulate wellhead price (the rate at which producers sold natural gas into the interstate market) the FPC adopted national price ceilings for the sale of natural gas into interstate pipelines. Because the set rates for natural gas were below the market value of that gas, demand surged. Due to high local demand, there was little incentive for natural gas producers to either explore for new natural gas reserves or to ship their gas across state lines. This led to a shortage of natural gas in those states that did not produce it. This shortage, in turn, led Congress to enact the Natural Gas Policy Act (NGPA) in 1978.

For more information about natural gas regulations

http:// www.naturalgas.org

Essentially, the NGPA has three main goals: (1) to create a single national natural gas market; (2) to balance supply and demand; and (3) to allow market forces to establish the wellhead price of natural gas. The NGPA initiated the gradual removal of price ceilings at the wellhead, thereby taking steps toward deregulating the natural gas market. However, under the NGA and the NGPA, pipeline customers did not have the option to purchase natural gas and arrange for its transportation separately.

Also in the late 1970s the Department of Energy Organization Act abolished the Federal Power Commission and replaced it with FERC. The latter encouraged the separation of sales and transportation through FERC Order No. 436, issued in 1985. This order allowed interstate pipelines to act solely as transporters of natural gas. Due to the new freedom this afforded natural gas customers, Order No. 436 became generally known as the "Open Access Order." Complete deregulation of wellhead prices soon followed when, in 1989, Congress passed the Natural Gas Wellhead Decontrol Act (NGWDA). Under the NGWDA, "first sales" of natural gas were to be free of any federal price regulations. The Act defined "first sales" as the sale of gas: (1) to a pipeline; (2) to a local distribution company; (3) to an end user; (4) preceding the sale to any of the above; or (5) determined by FERC to be a first sale.

Finally, in 1992, FERC Order No. 636 stated that pipelines *must* separate their transportation and sales services. FERC Order No. 636 is the culmination of deregulating the interstate natural gas industry by unbundling transportation, storage, and marketing of natural gas. The customer now has control over choosing the most efficient method of selecting their gas sales. As a result, today the market structure and operation of the natural gas industry is based heavily on competitive forces, and less on federal, state, or local regulations.

Oil Regulations

The United States oil industry is divided into three sectors: upstream (exploration and production); midstream (processing, storage, and transportation) and downstream (refining, distribution, and marketing). Regulation of the industry is generally differentiated based on these three segments, and each sector is subject to both federal and state regulation.

Upstream

The upstream sector is regulated by both the federal and state governments, depending on the location and ownership of the land containing the oil. The federal Department of the Interior regulates upstream activities through the Bureau of Land Management (BLM), the Minerals Management Service (MMS), and the Bureau of Indian Affairs (BIA). Each of these entities is responsible for the granting and enforcement of leases that outline terms for oil exploration and development. Pursuant to the Mineral Leasing Act of 1920 and the Mineral Leasing Act for Acquired Lands of 1947, the BLM issues leases for exploration and production of oil on federal land. Pursuant to the Outer Continental Lands Act, the MMS issues leases on the Outer Continental Shelf. Finally, the BIA issues leases (subject to certain BLM approvals) on Indian land. State agencies control the exploration and production of oil from state-owned public lands. Oil activity conducted offshore is also subject to state regulation where the oil is located within specified distances of state coastlines.

Oil exploration firms also must pay royalties to state and federal governments in order to explore for oil on public lands. Royalties for exploration on federal lands is generally regulated pursuant to the Federal Oil and Gas Royalty Management Act of 1982. Other pieces of federal legislation, such as the Deep Water Royalty Relief Act of 1995, and the Energy Policy Act of 2005 have provided royalty relief to oil and gas companies. Most royalties paid for oil exploration on federal lands go directly into the general U.S. Treasury. Royalties paid for exploration of state-owned lands are often used to support public programs, such as public education. Alaska's well-known Permanent Fund was established in 1976 in order to share a percentage of oil royalties generated in that state with all permanent residents of Alaska.

Midstream

Regulation of the midstream sector depends in large part on whether there is intrastate or interstate activity involved. Intrastate activity is regulated under state law, while interstate activity is regulated under federal law. For example, rates and terms of service for transportation of oil by interstate pipeline are regulated by the FERC pursuant to the Energy Policy Act of 1992. The Department of Transportation, through the Office of Pipeline Safety, establishes and enforces safety standards for transportation by pipeline. Each state has one or more agencies that also have authority over pipeline regulation within state borders.

For more information about oil regulations

Upstream:
http://www.blm.gov/wo/st/en/prog/energy/oil_and_gas.html
Midstream:
http:// www.gwpc.org http://www.epa.gov/region5oil/plan/spcc.html
Downstream:
http://www.epa.gov/air/caa/peg/working.html

Both the federal and state governments also regulate oil storage. Pursuant to the Clean Water Act, the owner or operator of certain oil storage facilities must prepare a Spill, Prevention, Control, and Countermeasure Plan (SPCC) in order to prepare for potential oil spills. A facility is subject to this requirement if it meets three criteria: (1) it is non-transportation-related; (2) it has an aggregate aboveground storage capacity greater than 1320 gallons, or a completely buried storage capacity greater than 42,000; and (3) there is a reasonable expectation of a discharge into or upon navigable waters of the United States or adjoining shorelines. Each state is also likely to have its own regulations pertaining to oil storage facilities located within that state's borders. Detailed information on many state regulations is available on the Groundwater Protection Counsel's Web site at www.gwpc.org.

Downstream

Finally, state and federal governments are involved in regulating the refining of oil. The Clean Air Act grants the EPA authority to issues permits that place pollution limits on emissions coming from oil refineries (and other potential sources of air pollution). Under the Obama administration, the EPA has grown increasingly strict on the enforcement of the Clean Air Act, which has recently led to tighter regulation of the downstream sector of the oil industry. Because state governments also have responsibility for enforcing the Clean Air Act, states also regulate refinery emissions through permits and other forms of regulation.

Biofuels Regulations

Pursuant to the Clean Air Act, the EPA has the authority to regulate fuels and fuel additives in order to obtain information about emissions and their health effects. Often the EPA also requires health-effects testing before a product is registered or for a product to maintain its registration over time. Should the EPA find a potential risk to public health due to exposure to a certain product, the EPA has the authority to enforce restrictions on that product. The EPA exercises this authority with respect to biofuel production in part through registration requirements.

For more information about biofuels regulations

http://www.epa.gov/OMS/renewablefuels/#compnd

http://www.access.gpo.gov/nara/cfr/waisidx_01/40cfr80_01.html

SAMPLE OF KEY GOVERNMENT REGULATIONS AND INCENTIVES FOR BIOFUELS

Country	Regulation/incentive
United States	Mandate of 15 billion gallons per year; Series of subsidies and credits (between $0.45 and $1.01 per gallon) for producers of biofuels
Brazil	Mandate of 5% biodiesel by 2013 and 20% by 2020; no longer has subsidies for ethanol (cost-competitive)
Japan	Mandate of 500 million liters per year by 2010
India	Mandate of 5% blend of biofuels in several states
China	$500 million investment per year in agricultural biotechnology; Mandate of 10% blend of biofuels by 2020
Canada	Mandate of 5% ethanol or biodiesel by 2010
European Union	Mandate of 10% by 2010; total EU subsidies of 3.7 billion Euros for biofuels in 2007

Pursuant to the Code of Federal Regulations (40 CRF Parts 79 and 80), biodiesel producers must complete and submit EPA registration forms 3520-12 (Fuel Manufacturer Notification for Motor Vehicle Fuel), 3520-20A (Fuels Programs Company/Entity Registration), and 3520-20B1 (Diesel Programs Facility Registration). Producers must also provide the following information in the production of biodiesel: (1) feedstocks used; (2) a description of the manufacturing process; (3) emissions and health effects testing on the manufacturer's biodiesel, or alternatively, proof of registration with the National Biodiesel Board (NBB) showing access to the Tier 1 and Tier 2 emissions and health effects testing data; and (4) test results from a representative sample of the manufacturer's biodiesel demonstrating compliance with the parameters specified in ASTM D 6751. (ASTM D 6751 is the primary industry standard for biofuel blends set by ASTM International.)

Finally, highway and nonroad biodiesel producers must comply with all of EPA's regulatory requirements outlined in 40 CFR Part 80, Subpart I. These regulations contain standards for such things as sulfur content, minimum cetane index, and maximum aromatics content. Subpart I also contains reporting and recordkeeping requirements for diesel fuel manufacturers.

WHY CELLULOSIC ENERGY CROPS ARE THE "NEXT BIG THING"?

According to the Department of Energy (DOE), starch-based ethanol feedstocks would not represent an unsubsidized competitor to traditional gasoline:

	Starch	Waste Cellulose	Cellulosic Energy Crops
Estimated cost ($/gallon gasoline equivalent)	3.78	1.62	0.91

In addition, the DOE is providing funding for numerous cellulosic biorefineries:

- Abengoa—up to $76 million
- ALICO—up to $33 million
- BlueFire Ethanol—up to $40 million
- Poet—up to $80 million
- Iogen—up to $80 million
- Range Fuels—up to $76 million

Potential Climate Change Regulations

Concern over climate change has come into increasingly sharp focus in recent years, leading to considerable debate in the United States over whether and how to institute climate change regulations. Although the Clean Air Act and other environmental protection laws regulate certain types of emissions and pollutants, no federal laws have been enacted with the specific, express intention of mitigating public or private companies' impact on global climate change.

However, a number of recent developments on both the federal and state levels suggest that the United States may be on the verge of significant change in this area of

The EPA and Pew Center on Global Climate Change both maintain a useful list of state and local climate change initiatives

http://epa.gov/climatechange/

http:// www.pewclimate.org/

environmental law and policy. In 2009 much discussed bills addressing climate change were introduced into U.S. Congress. The bills included a number of strategies for reducing greenhouse gas emissions in the United States, including implementing a carbon "cap and trade" system, a carbon tax, and various incentives for the use of alternative energy sources. The EPA has also come under pressure to regulate carbon dioxide following a 2007 U.S. Supreme Court determination that carbon dioxide should be considered an "air pollutant" for purposes of the Clean Air Act. Finally, some states and municipalities are imposing their own systems of climate change regulation, and a number of voluntary programs exist to encourage the reduction of greenhouse gases.

As of August 2010, there are no robust federal regulations on greenhouse gas emissions in the United States. However, the below initiatives and bills represent significant steps forward in federal climate change regulation.

CARBON TAX VS. CARBON CAP AND TRADE SYSTEM

A *Carbon Cap and Trade System* involves establishing a limit, or cap, on the total amount of greenhouse gases that can be emitted over a specific period of time. It then allocates portions of that limit to various carbon emitters through carbon allowances—essentially permits to emit a set amount of carbon pollution—so that the total of the individual emission allowances is equal to the overall cap. Carbon emitters then have two options: (1) Emitters can adjust their practices to emit less carbon pollution, thereby meeting their own carbon allowance; or (2) Emitters can buy additional allowances from other carbon emitters that do not exceed their allowances. This leaves all companies with an incentive to reduce their own emissions and it creates an incentive for green companies to increase profits through the sale of unused allowances.

A *Carbon Tax* program attaches a tax to CO_2 (or other greenhouse gas) emissions that result from the burning of fossil fuels. A carbon tax both creates an incentive for companies to lower their own emissions and increases competitiveness of low-carbon technologies and renewable resources that do not produce harmful emissions. A number of European countries, including France, Finland, the Netherlands, and Italy, have implemented carbon taxes of some kind. Some states and municipalities within the United States have also introduced carbon taxes in recent years. For example, the Bay Air Quality Management District in the San Francisco Bay Area now charges businesses 4.4 cents per ton of carbon dioxide they emit.

Climate Change Response Council

In September 2009, U.S. Secretary of the Interior Ken Salazar launched the Department of the Interior (DOI) Climate Change Response Council to coordinate the DOI's response to the impacts of climate change. As part of this effort, eight DOI regional Climate Change Response Centers and a network of Landscape Conservation Cooperatives will be established in order to address climate change throughout the country. The Council will also oversee the DOI Carbon Storage Project, through which the DOI is developing methodologies for carbon storage, and the DOI Carbon Footprint Project, meant to reduce the DOI's own carbon footprint. More information on the DOI Climate Change Response Council is available here: www.doi.gov/climatechange/.

EPA Final Mandatory Reporting of Greenhouse Gases Rule

In September 2009 the EPA issued the Final Mandatory Reporting of Greenhouse Gases Rule, which requires the reporting of greenhouse gas emissions from large sources and suppliers in the United States. The rule is intended to collect accurate and timely data regarding greenhouse gas emissions to inform future policy decisions. Although this rule does not regulate the amount of greenhouse gas emissions produced by large sources, it may be a step toward future EPA regulation of such emissions. More information on this rule can be found here: www.epa.gov/climatechange/emissions/ghgrulemaking.html.

Recent Proposals

In 2009, both the U.S. House of Representatives and the U.S. Senate began considering proposals to regulate greenhouse gases. The bills under discussion were the Clean Energy Jobs and America Power Act of 2009 (Kerry-Boxer Bill, Senate Climate Bill) and the American Clean Energy and Security Act of 2009 (ACES, Waxman-Markey Bill, House Climate Bill).

In the summer of 2009, the House of Representatives passed the House Climate Bill. The bill proposes national regulation of greenhouse gases through a cap and trade system. Regulated entities may purchase and sell permits which would allow for the emission of greenhouse gases. Over time, the permissible emissions by those entities shall be decreased periodically until 2050. The House Climate Bill also requires electric utilities to meet 20% of their electricity demand by 2020 through renewable energy generation and energy efficiency. The bill creates also subsidies for clean technologies.

The Senate Climate Bill was not passed through the Senate but Senate leaders have suggested that climate regulations will be a part of the 2010 Senate calendar (although this has not occurred as of August 2010). The initial bill offered in 2009 proposes a mandate to curb the nation's greenhouse gas emissions by 20% from 2005 levels by 2020. However, many of the details behind the bill were not initially provided and are expected to be part of further negotiations.

PRICING CARBON

What is the estimated impact of a cap-and-trade system or carbon tax on average Americans?

ELECTRICITY

- 13.8% increase in electricity prices—Jefferies & Company
- 36% to 65% increase in electricity prices by 2015 and 80% to 125% by 2050—Charles River Associates

GASOLINE

- 12 cents per gallon increase in price of gasoline during the first year of cap-and-trade; 15 cent increase by 2020—Point Carbon
- 26 cents per gallon increase by 2030 and 68 cents increase by 2050—Environmental Protection Agency

> **OVERALL COSTS PER HOUSEHOLD**
> - Additional annual costs of $960 or 2.8% of gross income for middle class families—Congressional Budget Office (assuming a 15% reduction in carbon emissions)
> - $3100 in additional costs for each household (according to House Republican leader John Boehner) while John Reilly (one of the authors of the MIT report on which these figures are based) believes that number is much too high, offering a figure of $340 in additional costs for each household. *Both* of these estimates are based on same data from the same MIT Joint Program on the Science and Policy of Global Change's 2007 report "Assessment of U.S. Cap-and-Trade Proposals."
> - $800 to $1300 in additional costs annually per household by 2015, rising to $1500 to $2500 by 2050—Charles River Associates.

What Would SEC-Mandated Change Disclosure Mean?

In 2009, the National Association of Insurance Commissioners (NAIC) mandated that insurance companies disclose to their respective state regulators the financial risks such insurance company foresees from climate change and the steps required to address those risks. Some experts have noted that the NAIC policy represents the precursor to Securities and Exchange Commission (SEC) climate change disclosure requirements. The SEC has been considering various proposals and approaches to require disclosure to investors of climate change risks, and in 2009 SEC Commissioner Elisse Walter said it was time for the SEC to consider specific steps to govern what climate change-associated risks must be disclosed.

As of August 2010, there were no specific SEC regulations for climate change risk disclosure. However, under regulations S-K, publicly traded companies are required to disclose material information, which certain companies have determined would apply to climate change and other related environmental risk. It is unclear as to the approach that the SEC will most likely take and whether its disclosure requirements may tie into systems providing voluntary disclosure opportunities such as the Carbon Disclosure Project.

What are the types of risks related to climate change a company may be likely to disclose?

- Litigation risk
- Regulatory risk
- Market risk
- Physical or property risks

Will the disclosure requirements be detailed enough to provide investors with real information (related to actual costs, specific corporate actions, or emissions amounts) or could the requirements result in substantial business costs for reporting of the disclosure requirements?

Other Federal Regulations

Energy and environmental regulation has developed significantly in the United States over recent years. It is likely that a great deal of further action will be taken on both the state and federal levels in the years ahead. A few possibilities for future federal regulation are summarized here.

Federal Renewable Energy Portfolio Standard

A renewable portfolio standard (RPS) system requires a certain percentage of electricity sold by utilities to retail customers to be generated by renewable energy. Most U.S. states either have, or are considering implementing, RPS at the state level. However, as of August 2010, the United States has not yet implemented a national standard.

During the 2008 Presidential Campaign, now-President Barack Obama called for 25% of electricity used in the United States to come from renewable sources by 2025. Essentially, President Obama was calling for the creation of a federal RPS. Since becoming President, bills have been introduced into both the House and Senate that would create an RPS. However, attempts to create a federal RPS have been met with criticism over whether such a standard could account for the regional difference in renewable development potential and existing renewable generation capacity among different states.

Renewable Energy Payment System

Under a renewable energy payment system, or "feed-in tariff," renewable electricity producers are paid a fixed, above market-rate tariff for electricity that is fed into the grid. The cost of the tariff is then spread across consumers of electricity. The system is meant to give homeowners and businesses an incentive for producing green energy.

Feed-in tariff systems are becoming popular in Europe, where they have been introduced in a number of countries with success. Germany's feed-in tariff system, enacted in 2003, is often looked to as a model of this kind of system. As of June 2008 Germany had provided 14% of its electricity supply from renewable sources—much of it from solar cells. France, Spain, and Australia have also implemented successful feed-in tariffs in recent years.

In the United States, several municipalities and states have taken action toward starting a local feed-in tariff system. In the spring of 2009, Gainesville, Florida became the first city in the United States to introduce higher payments for solar power-generated electricity. At the state level, Hawaii, Washington, Oregon, California, Maine, Vermont, and Wisconsin are examples of states that have taken steps toward introducing feed-in tariffs. Although the Renewable Energy Jobs and Security Act, a bill implementing a national feed-in tariff system in the United States, was introduced into Congress in 2008, the bill never became law. A national debate continues over whether such a system could and should be implemented at the federal level.

State Programs

Despite a lack of strict federal enforcement of carbon emissions, many states have introduced regulations, climate action plans, and voluntary programs that regulate greenhouse gas emissions at the state level.

Throughout the United States state governments have enacted their own laws meant to regulate energy use, encourage energy efficiency, and promote environmental protection. Many new and expanded regulations have been passed in recent years, and state laws affecting the cleantech industry are likely to continue to expand quickly in the near future. States—as well as counties and local municipalities—have also

For more information about state-specific regulations, see the Database of State Incentives for Renewables & Efficiency

http://www.dsireusa.org/

created incentive programs, loans, grants, and tax credits aimed at encouraging efficient, responsible energy use.

California, Oregon, and Massachusetts are three states that have led the charge in enacting state regulations, creating state programs, and participating in joint efforts to support energy efficiency and renewables. Examples of such efforts taken by each state are described below.

California

California has been a leader in encouraging and requiring energy efficiency and environmental protection efforts in both the public and private sectors. As early as 1991 California mandated efficiency standards to guide the purchase of fuels and vehicles for state fleets. The 2006 legislation passed in the state now sets milestones for increasing alternative fuel use in state fleets by 2012, 2017, and 2022. More recently, California has set a high bar for fuel efficiency standards in cars produced in California.

The state has established a number of requirements and standards affecting California energy use and utilities. In 1998, the California Energy Commission created its Emerging Renewables Program, which began offering cash incentives to promote the installation of grid-connected small wind and fuel cell renewable energy electric-generating systems. In 2002, California adopted a renewable portfolio standard program. This program requires utilities to obtain 20% of their electricity from renewables by the year 2010, and 33% by 2020. California also requires all utilities to offer net metering for solar and wind systems and allows commercial, residential and industrial property owners to be eligible for a property tax exemption for solar energy systems.

For more information about California's regulations, visit http://www.dsireusa.org/

California has also passed several laws aimed at reducing California's greenhouse gas emissions. Those laws included the Global Warming Solutions Act, Assembly Bill 32, California Senate Bill 1368, and Senate Bill 107. Senate Bill 1368 directs the California Energy Commission to set a greenhouse gas performance standard for electricity procured by local public utility companies. California has also joined with six neighboring states and four Canadian provinces to establish the Western Climate Initiative, an effort to introduce carbon trading in the region by the year 2012.

The California Assembly Bill 32 (a.k.a. the Global Warming Solutions Act) set a greenhouse gas reduction goal into California law. The bill directs the California Air Resources Board to develop an action plan to reach a target goal of reducing greenhouse gas emissions so that the amount of California's 2020 emissions does not exceed the amount of the state's 1990 emissions. The Act also creates a mandatory emissions-reporting system and helps create incentives to businesses to reduce greenhouse gas emissions.

ALTERNATIVE FUELED CITY FLEETS IN U.S. CITIES

1. Las Vegas, NV (63%)
2. Honolulu, HI (51%)
3. Kansas City, MO (45%)
4. Albuquerque, NM (42%)
5. Dallas, TX (39%)
6. Denver, CO (31%)
7. Phoenix, AZ (28%)

8. Los Angeles, CA (25%)
9. Seattle, WA (25%)
10. Portland, OR (25%)

Source: SustainLane City Rankings 2006–2007.

Oregon

Oregon is another leader in taking state action encouraging energy efficiency and renewables. The 2001 Oregon Sustainability Act created a broad framework for encouraging sustainability efforts throughout Oregon. The Act followed a goal set by Oregon's former Governor John Kitzhaber to reach statewide sustainability by the year 2025. The Act authorized the creation of a Sustainability Board, which is composed of representatives with knowledge of the business and small-business sectors, natural resource conservation, sustainable development, and health or economics.

For more information about Oregon's regulations, visit

http://www.sustainableoregon.net

Under the broad structure created by the Sustainability Act, Oregon has several state programs regulating CO_2 emissions, offering broad-based net metering, encouraging alternative fuel use and development, and offering tax incentives for renewable energy systems. For example, Oregon offers Small-Scale Local Energy Project loans to support alternative fuel production. The state's Department of Energy also offers tax credits to businesses and individuals that invest in alternative fuel or hybrid vehicles, solar electric and thermal systems.

Further, in 2007, the Oregon Legislature passed the Oregon Renewable Energy Act, which establishes a renewable energy standard in the state. Under the standard, Oregon's largest utilities must acquire 25% of their electricity from renewable energy sources by 2025. Smaller utilities must meet targets of 5–10% renewable energy by 2025.

Massachusetts

Massachusetts is another state that has been a leader in enacting regulations surrounding energy efficiency and renewables. For example, Massachusetts was the first state to provide a state gas tax exemption for biofuels via the Clean Energy Biofuels Act. Massachusetts is also one of the first states to strictly regulate greenhouse gases through its Global Warming Solutions Act (GWSA). The GWSA mandates an 80% reduction in greenhouse gas emissions by 2050 from 1990 levels.

For more information about Massachusetts' regulations, visit

http://www.mass.gov/dep/energy.htm

The state's Green Communities Act sets a renewable energy portfolio standard requiring state utilities to derive 15% of their energy resources from renewable sources by 2020, with an additional 1% per year thereafter. Further, the state's Green Jobs Act establishes preliminary funding for a Clean Energy Seed Grant Program. The Program is meant to stimulate clean energy research and new venture creation. It also authorizes preliminary funding for a Green Jobs Initiative to launch job-training programs at schools and training organization and authorizes exploration for funding of a Clean Energy Fellowship program.

Still more legislation promoting energy efficiency and renewables in the state includes the Massachusetts's Oceans Act, which provides a regulatory framework for developing wind, wave, and tidal power generation resources in Massachusetts's state waters; the Commonwealth Solar initiative establishing a rebate program to support adoption of solar power; and a Department of Public Utilities Decoupling Order providing utilities with incentives to use energy efficiency measures.

HOW DO I FIND MORE INFORMATION ABOUT MY OWN STATE?

A resource for finding contact information for the state entities regulating energy use in your state is found at http://www.naseo.org/members/states/default.aspx.

More detailed information on specific laws and regulations in each state is available on the Database of State Incentives for Renewables & Efficiency is found at: http://www.dsireusa.org/.

15

Grants, Loans, and Other Green Government Funds

Economist Lord Nicholas Stern estimates that almost $400 billion of the approximately $2.6 trillion in economic stimulus allocations announced as of the end of 2009 by G20 nations are earmarked for clean technologies such as renewable energy, improved electrical grids, and cleaner cars, with investments of over $52 billion being made by China and over $80 billion by the United States. At the start of 2010, cleantech grant experts Skipso has identified at least $26 billion in specific grants and other programs already available for green businesses. With the wealth of funding available and to be made available in the coming years, understanding government funding is a must for any green business.

Identifying Cleantech Grants and Loan Programs

With numbers like $400 billion being discussed, it seems like finding money for a clean technology business or project should be simple, right? Not so fast. Governments have long histories of providing substantial obstacles to access funding opportunities. In addition, some of these earmarked funds have not yet been "spent" or even allocated, and some of the funds may only be tangentially related to clean technology. However, there are still billions of dollars already available. One of the first steps is identifying the funding source and putting the company in the best position to be awarded the grant or loan.

Cleantech.com provides the U.S. Government Stimulus Portal (http://cleantech.com/research/government_stimulus_portal/) tool that provides a searchable source for federal and state government funding opportunities including grants, loan guarantees, tax credits, and other financing programs. Another of the largest resources for locating grants available

U.S. Government Stimulus Portals

http://cleantech.com/research/
government_stimulus_portal/

http://www.cooley.com/
cleantechstimulusportal

to green businesses across federal, state, nonprofit, academic, and international boundaries is Skipso (http://www.skipso.com). As of the start of 2010, Skipso listed nearly 600 grant, showcase, prize, and other programs aimed at various sectors of clean technology and green businesses. In addition, the United States funnels most of its grants through its Grants.gov Web site which lists thousands of grant programs (many of which are tied to clean technology programs).

Government Grants for Green Initiatives

Skipso (www.skipso.com)

Locate grants, showcases, prizes, and other programs in one of five major categories with subcategories for each grant category:

- Renewable energy (270 grants)
- Energy efficiency and distribution (220 grants)
- Clean products and production (50 grants)
- Waste and pollution (100 grants)
- Water management (70 grants)

Grants.gov is a central storehouse for information on over 1000 federal government grant programs and provides access to approximately $500 billion in annual awards. Grants.gov was established as a governmental resource in 2002 in order to improve public access to government services. Not all grant applications can be submitted through the Grants.gov Web site, but those that require a separate submission process or have more information will provide you with Web site links.

According to the Grant.gov Web site, "A federal grant is an award of financial assistance from a federal agency to a recipient to carry out a public purpose of support or stimulation authorized by a law of the United States. Federal grants are not federal assistance or loans to individuals." There are 26 federal grant-making agencies that use Grants.gov including the Department of Agriculture, the Department of Energy, NASA, the Environmental Protection Agency, the Department of Transportation and the Small Business Administration, all of which make some grants to green businesses. In addition, Grants.gov allows you to segment grants into one of the 21 categories of government grants including Business and Commerce, Energy, Environmental Quality, Natural Resources, the Recovery Act, and Science and Technology, each of which may have specific applicability to green entrepreneurs.

How to Apply for a Government Grant via Grants.gov

1. *Identify a grant opportunity.* Initially, you may review the grant programs currently soliciting applications. However, organizations should also consider setting up email notifications for new grant opportunities as they are posted to the Grants.gov site.

2. *Register your organization in order to access and submit grants.*

To find information on Recovery Act funds allocated by the states visit

http://www.recovery.gov/Transparency/info/Pages/State_Territory_Recovery_Sites.aspx

3. *Download an application package.* Once you have located a grant opportunity, check to see if it is available to apply online through Grants.gov. You will need to enter the Funding Opportunity and/or CFDA number to access the application package and instructions. In order to view application packages and instructions, you will also need to download and install the PureEdge Viewer. This free program will allow you to access, complete, and submit applications electronically and securely.

4. *Complete the application process.* This process can be a time-consuming one, and reviewers suggest that ensuring an application is complete, correct, and without errors is crucial to receiving grant funds. There are consultants and advisors who can assist a company applying for grant money.

5. *Track your progress.* You can log-in to the Grants.gov site to track the status of the application.

While the lure of grant money is tempting, remember that many other companies, organizations, and individuals are also looking to the government for funding. In the fiscal year 2009, over 300,000 submissions to grant programs were made—a growth of 50% over the prior year. For many of the stimulus grants, the grants were oversubscribed. For example, for the Department of Energy (DOE) smart grid stimulus grants, 100 projects were selected, but the grants were oversubscribed 4:1 in terms of the number of applications and 5:1 in terms of the dollar amount of applications.

GRANTS.GOV ORGANIZATIONAL REGISTRATION CHECKLIST

The process to register your organization with Grants.gov to submit applications for federal grant programs can take as few as three days or as long as three weeks, depending on the type of organization and if all the steps are met in a timely manner.

1. *Obtain Data Universal Number System (DUNS) number.* Visit Dun & Bradstreet at http://fedgov.dnb.com/webform/displayHomePage.do. You will receive same-day confirmation of your DUNS number online.

2. *Register with Central Contractor Registration (CCR).* You can apply online by going to http://www.ccr.gov. As long as you have an EIN number, the registration process takes one to three business days to complete; otherwise you need to obtain an EIN number prior to starting the process. If your company is already registered with CCR, you must renew your registration annually which may take up to five business days. Additional information by downloading the CCR handbook at http://www.bpn.gov/ccr/doc/UserAccount.pdf.

   Note: When your organization registers with CCR, you must designate an E-Business Point of Contact (E-Biz POC). This person will identify a special password called an "M-PIN." This M-PIN gives the E-Biz POC authority to designate which staff member(s) from your organization are allowed to submit applications electronically through Grants.gov. Staff members from your organization designated to submit applications are called Authorized Organization Representatives (AORs).

3. *Authorized Organization Representatives (AORs) must complete profiles and obtain usernames and passwords.* AORs should login to their profiles at Grants.gov and can obtain their username and password immediately.

4. *E-Business Point of Contact (E-Biz POC) approves AORs.* Only AORs approved by the E-Biz POC can submit grants on behalf of the organization.

5. *AORs are approved to submit grants.*

   Note: If you are an individual applying for a grant on your own behalf and not on behalf of a company, academic or research institution, state, local or tribal government, not-for-profit, or other type of organization, refer to the Individual Registration. If you apply as an individual to a grant application package designated for organizations, your application will be rejected.

DOE Advanced Research Projects Agency-Energy

One of the new DOE programs for funding innovative clean technology research and development is the Advanced Research Projects Agency-Energy (ARPA-E). ARPA-E was modeled after a similar program in the Department of Defense, the Defense Advanced Research Projects Agency (DARPA). Through technology funded by DARPA technological, innovations such as the Internet and the stealth technology found in the F117A and other modern fighter aircraft were created, and the aim of ARPA-E is the same type of technology leaps that may not have been otherwise funded through traditional commercial mechanisms.

For more information about the ARPA-E Program

http://arpa-e.energy.gov/

The mission of ARPA-E is to fund "projects that will develop transformational technologies that reduce America's dependence on foreign energy imports; reduce U.S. energy related emissions (including greenhouse gasses); improve energy efficiency across all sectors of the U.S. economy and ensure that the United States maintains its leadership in developing and deploying advanced energy technologies." ARPA-E was originally authorized under the America Competes Act of 2007, and was later allocated $400 million in federal funding through the Recovery Act.

The ARPA-E awards can be utilized for a portion of the allowable project costs in the range $500,000 to $10 million. If an applicant is exclusively a university, college or other educational institution, at least 10% of the total project costs must be supplied by a non-federal government source, with at least 20% for all other applicants. For example, a $1 million research project without university partners would require the project to have at least $200,000 in funding from other sources including loans or equity investments.

In October 2009, ARPA-E distributed $151 million in funding to small businesses, education institutions, and large corporations. Examples of funded projects included liquid metal grid-scale batteries; research of bacteria used in direct solar hydrocarbon biofuel production; CO_2 capture using artificial enzymes; and low cost crystals for LED lighting. A second round of funding was authorized for projects in the areas electrofuels, batteries for electrical energy storage in transportation, and innovative materials and processes for advanced carbon capture technology. It is anticipated that this program will continue to provide funding for advanced research projects in subsequent funding rounds.

USDA Business and Industry Loan Guarantee Program

The U.S. Department of Agriculture's Rural Development program offers business and industry guaranteed loans to improve, develop, or finance business, industry, and employment—and improve the economic and environmental climate—in rural communities. Many different entities are eligible to apply for these loans, including for-profit and nonprofit businesses, individuals, corporations, and partnerships.

For more information about the USDA Loan Guarantee Program

http://www.rurdev.usda.gov/rbs/busp/b&l_gar.htm

DOE Loan Guarantee Program

In order to help push clean technologies ahead, the federal government recognized the need to fill the gap between early stage commercialization and mass-market adoption. The loan guarantee program has been received with open arms by many green businesses because it potentially provides a funding source to help build commercial facilities for clean technology businesses that may be past the appropriateness of venture finance, without the ability to obtain acceptable commercial loan terms for the cutting-edge technologies, and unable to access the public markets for funding. According to the DOE Web site:

> The U.S. Department of Energy's Loan Guarantee Program paves the way for federal support of clean energy projects that use innovative technologies, and spurs further investment in these advanced technologies.
>
> Established under Title XVII of the Energy Policy Act of 2005, the Secretary of Energy is authorized to make loan guarantees to qualified projects in the belief that accelerated commercial use of these new or improved technologies will help to sustain economic growth, yield environmental benefits, and produce a more stable and secure energy supply.

In the first round of loan guarantees, the DOE evaluated loan guarantee preapplications for projects that employed technologies in the following areas: biomass, hydrogen, solar, wind and hydropower, advanced fossil energy coal, carbon sequestration practices and technologies, electricity delivery and energy reliability, alternative fuel vehicles, industry energy efficiency projects, and pollution control equipment. The first round of loan guarantees drew nearly 150 applications, of which the DOE had identified 16 projects to fund from that group by early 2009. Since then, the DOE released new solicitations for the program making a total of more than $30 billion in loan guarantee authority available. Depending on the ongoing success of the program, additional authority may be granted in the future.

For more information about the DOE Loan Guarantee Program

http://www.lgprogram.energy.gov

Loan guarantees are not an appropriate tool for research and development or pilot-scale projects. Instead, the DOE is looking to support products that are commercializing new clean technologies. Specifically, the loan guarantee funds are applicable for projects that avoid, reduce or sequester air pollutants or anthropogenic emissions of greenhouse gases and employ new or significantly improved technologies as compared to commercial technologies in service in the United States at the time the guarantee is issued. To qualify, technology must have been successfully demonstrated both at the pilot and demonstration scale and be ready for commercialization. Support is limited to technologies that are installed in no more than three commercial projects in the United States which have been in operation for no more than five years. Technologies that have been commercialized outside the United States are eligible, but the proposed project must be located in the United States.

For access to key documents and webinars on the DOE Loan Guarantee Program visit

http://www.lgprogram.energy.gov/keydocs.html

The 2009 DOE Loan Guarantee Program set out seven submissions periods beginning on September 14, 2009 and ending on August 24, 2010—however, in the event the $8.5 billion in loan guarantee authority was fully obligated prior to the termination date, then such late submissions will not be approved. In the event a company passes the eligibility test, it then will be subject to a full review of technical and financial merit. Loan guarantees are only applicable to the following project types: manufacturing projects or stand-alone projects. And these projects must fall under one of the nine categories including the following:

DOE Loan Guarantee Program Application Recommendations

http://www.lgprogram.energy.gov/SS-LG-Apps.pdf

Category 1: Alternative fuel vehicles

Category 2: Biomass

Category 3: Efficient electricity transmission, distribution and storage

Category 4: Energy efficient building technologies and applications

Category 5: Geothermal

Category 6: Hydrogen and fuel cell technologies

Category 7: Energy efficiency projects

Category 8: Solar

Category 9: Wind and hydropower

Once the DOE has identified and selected a company for a loan guarantee, it will negotiate a term sheet with applicants. The term sheet is referred to as a "conditional commitment" upon execution. Following negotiation and execution of a term sheet, the DOE will engage with the company to negotiate a full guarantee agreement. Thereafter, the parties will finalize the project's other equity and debt agreements, execute the loan guarantee agreement and close the transaction. The process is a lengthy one by some commercial standards, but has received considerable pressure by politicians for expediency.

The DOE has compiled a list of features that have distinguished strong applications from the weaker ones. With respect to financial attributes, stronger applications tend to have third-party supply and off-take agreements in place, detailed construction budgets, accounting for all resources necessary for project completion, reviews of required permitting and environmental requirements, clear intellectual property rights, analysis of the market and competition, and sources of equity for the business. With respect to technical attributes, stronger applications had detailed engineering reports, pilot program results, clear description of technology and its advantages, and a strong staff make-up.

Additionally, many successful companies have hired consultants to assist in the application process, given the complexities and political realities of applying for the loan guarantees. Applications can be thousands of pages and can require input from a variety of internal and external stakeholders. Some experts have advised that it is important to think critically about involving a variety of partners from a variety of key geographies, industries and backgrounds in order to make the application process most compelling.

Small Business Administration Loans

The Small Business Administration (SBA) does not make loans to small businesses and startup companies. However, the SBA does offer programs whereby the SBA will act as a guarantor for loans made to small businesses through qualified lenders in order to allow businesses to obtain loans they may not have otherwise been able to obtain without the backing of the federal government. More information is available in Chapter 27 "Obtaining a Business Loan" on SBA loan programs.

In response to the economic slowdown in 2009, the SBA created a new program entitled America's Recovery Capital Loan (ARCL) Program. SBA's ARCL Program provides up to $35,000 in short-term relief for viable small businesses facing immediate financial hardship to help ride out the current uncertain economic times and return to profitability. Each small business is limited to one ARC loan. ARC loans will be offered by some SBA lenders for as long as funding is available or until September 30, 2010, whichever comes first.

SBA 504 Green Loan Program

The Energy Independence and Security Act authorized the Small Business Administration to offer energy conservation loans, grants and debentures to help small businesses develop, invest in, and purchase energy efficient buildings, fixtures, equipment, and technology. The SBA also offers pollution control loans to help eligible small businesses assist in the planning, design, or installation of a pollution control facility.

The SBA 504 loan program (sometimes called a Development Loan) is designed to assist small businesses with financing of fixed assets, including the purchase of buildings, land, and certain types of equipment. Private institutions will provide the financing for the assets through Certified Development Companies (also known as CDCs). Loans made under the 504 program are generally made at a fixed-rate, long-term basis. The way the Development Loan works is that the private lender institution will lend the

For more information about the SBA Loans and EISA

http://www.sba.gov

company 50% of the total project, a CDC will lend 40% (guaranteed by the SBA), and the company will be responsible for the remaining 10% of the total project cost. The uses are somewhat limited—inventory, debt service, short-lived equipment, and machinery aren't eligible. However, for various projects including building, construction, and facility renovation or retrofitting, these programs will be an option.

In 2009, the 504 Loan Program was expanded to provide incentives for green businesses under the EISA authorization. Specifically, three changes were made to the program that can benefit green businesses.

- Projects that result in at least a 10% reduction in energy consumption are now eligible for the program. This includes up to $4 million in loans for the purchase, construction or retrofit of facilities with energy saving technologies (and more than $9 million if certain public policy goals apply). Examples of technologies that would apply include improved insulation, lighting, HVAC (heating ventilation and air conditioning), and energy efficient windows.

- Small businesses that are involved in renewable energy generation such as solar, biomass, hydropower, ocean thermal, geothermal, and wind can utilize the 504 Program to finance real estate purchases or construction projects. The limit on these loans is $4 million. The program is not limited to small businesses generating renewable energy as their primary business; instead it also allows participation by companies that generate renewable energy for their own energy consumption from solar panels, biomass, or other sources.

- Small businesses can also qualify for 504 Loans to complete projects incorporating sustainable facility design. The eligible projects that qualify include projects that meet Leadership in Energy and Environmental Design (LEED) standards or are certified by the Green Building Certificate Institute. Businesses can obtain loans for up to $2 million to work with qualified architects and engineers.

Import Export Bank

Companies in the importation or exportation businesses may be eligible for Import-Export Bank Programs, supported by the SBA. Loans for working capital of up to $1.1 million ($1.25 million if combined with an international trade loan) can be guaranteed by the SBA. This program is typically available only for U.S. companies that have been in business for one full year, operate at a profit, and do not exclusively rely on the loan to support the business operations.

Small Business Innovation Research and Small Business Technology Transfer Programs

Many federal agencies participate in the government's Small Business Innovation Research (SBIR) and Small Business Technology Transfer (STTR) programs. The SBIR and STTR are funding programs designed to stimulate technological innovation and fulfill the research needs of the federal government. Businesses are required to meet several criteria to be eligible for grants under either program, including U.S. ownership, for-profit status, and restrictions on number of employees.

The SBIR and STTR programs differ in two major ways. First, under the SBIR program, the principal investigator listed on the SBIR application must be employed by the small business at the time of the grant and for the duration of the project. Under the STTR program, there is no such employment requirement. Second, unlike the SBIR, the STTR program requires the small business to be engaged in a collaborative relationship with a non-profit research institution located in the United States.

To apply for a SBIR award, a company must submit an application to specific Request for Proposals released by a federal agency. The SBIR award is divided into three phases. Phase I is a feasibility study to evaluate the scientific and technical merit of an idea. Awards are for periods of up to six months in amounts up to $100,000. Phase II is to expand on the results of and further pursue the development of Phase I. Awards are for periods of up to two years in amounts up to $750,000. Phase III is for the commercialization of the results of Phase II and requires the use of private sector or non-SBIR federal funding.

The following federal agencies are eligible to participate in SBIR award programs: Department of Agriculture, Department of Commerce, Department of Defense, Department of Education, Department of Energy, Department of Health and Human Services, Department of Transportation, Environmental Protection Agency, National Aeronautics and Space, Administration National Science Foundation and Department of Homeland Security.

For more information about STTR announcements and solicitations visit

http://www.sba.gov/aboutsba/sbaprograms/sbir/announce/sbir_links.html

STTR awards are designed to provide funding to small businesses in partnership with nonprofit research institutions to move ideas from the laboratory to the marketplace, to foster high-tech economic development and to address the technological needs of the Federal Government.

Like the SBIR program, STTR awards are structured into three phases. Phase I is the startup phase for the exploration of the scientific, technical, and commercial feasibility of an idea or technology. Awards are for periods of up to one year in amounts up to $100,000. Phase II is to expand Phase I results. During this period the R&D work is performed and the developer begins to consider commercialization potential. Awards are for periods of up to two years in amounts up to $750,000. Phase III is the period during which Phase II innovation

For more information about SBIR announcements and solicitations visit

http://www.sba.gov/aboutsba/sbaprograms/sbir/announce/sbir_links.html

moves from the laboratory into the marketplace. Like SBIR awards, there is not STTR funding in this phase. The Department of Defense, Department of Energy, National Aeronautics and Space Administration, Department of Health and Human Services and National Science Foundation make awards under STTR programs.

EXPANDED GOVERNMENT FUNDING FOR NEW ENVIRONMENTAL TECHNOLOGIES

The Small Business Innovation Research program has long been designed to provide government grant funding to small businesses engaged in the commercialization of cutting-edge technologies. In 2009, the Environmental Protection Agency (EPA) and the National Science Foundation (NSF) granted nearly $200 million through SBIR programs at these agencies to businesses engaged in the commercialization of environmental technologies.

What types of businesses are eligible? Companies with fewer than 500 employees are eligible to apply to both EPA and NSF, although they can only accept funding from one source.

When are deadlines? Each year, these agencies will solicit grants and dates will vary. In 2009, the EPA's SBIR Phase I solicitation closed on May 20, 2009, and the NSF's SBIR Phase I solicitation closed on June 9, 2009. Visit the Web sites for more details.

How much are the grants worth? Each winner of an EPA Phase I aware is eligible for funding of up to $70,000. Each winner of a NSF Phase I award is eligible for funding of up to $150,000.

For information on EPA environmental technology needs and application requirements, visit www.epa.gov/ncer/sbir. For information on the NSF's SBIR Program,
visit the NSF Web site at: www.nsf.gov/eng/i8ip/sbir.

Raising Funds through Green Bonds

Clean Renewable Energy Bonds

The Clean Renewable Energy Bond (CREB) program was established by the Energy Policy
Act and is administered by the IRS. CREBs allow cities, counties, school districts, and
tribes to obtain interest-free financing for renewable energy projects. The recipient of a
CREB is only required to pay back the principal of the bond, and the bondholder receives
federal tax credits rather than traditional bond interest. CREB tax credits may be taken
each year the bondholder has a tax liability, as long as the credit does not exceed limits
established by the Energy Policy Act.

Qualified Energy Conservation Bonds

Qualified Energy Conservation Bonds (QECBs) are similar to CREBs in that they are
funds allocated to state, local, and tribal governments to finance certain types of energy
projects. The bonds may be used to fund energy efficiency expenditures in public buildings; renewable energy production; various research and development applications; mass
commuting facilities that reduce energy consumption; energy related demonstration
projects; and public energy efficiency education campaigns. The difference between
QECBs and CREBs is that QECBs are not subject to U.S. Treasury application and approval
process. Rather, bond volume is issued to each state based on state's percentage of the U.S.
population. Each state then allocates that portion to local governments based on population within the state.

16

Taxes and Incentive Programs

Tax incentives have served as the U.S. government's main tool to encourage investment in the production and distribution of energy from renewable sources. Clean energy technology is becoming increasingly important in U.S. economic, environmental and tax policy. Over $60 billion of the $787 billion Recovery Act was earmarked for energy-related investments as part of President Obama's pledge to take "bold action[s] to create a new American energy economy that creates millions of jobs for our people."

Understanding Tax Incentives

U.S. tax policy is set by Congress with the primary objectives of raising revenue and influencing taxpayer behavior. Consistent with the second objective, the tax code is often used to encourage investment by providing accelerated tax deductions and tax credits for specific business expenditures. In the case of clean technology, there are numerous tax incentives that are available to businesses that should be a part of the strategic planning process of a green business.

For more information about federal tax incentive programs, visit

http://www.energy.gov/taxbreaks.htm

There are two primary mechanisms that policymakers have put in place for green businesses: tax credits and tax deductions. So just what is the difference between a tax credit and a tax deduction? A tax credit reduces an individual's tax liability by a direct dollar-for-dollar amount. A tax deduction removes a *percentage* of the tax that is owed. This means a tax credit is generally more valuable than an equivalent tax deduction.

One of the issues for some green businesses is being able to utilize the potentially valuable tax assets, particularly if the business is operating at a loss for a number of years. As discussed in Chapter 12 on project finance, there are oftentimes ways to structure investments into a business or project to allow for the tax benefits to be allocated to parties who are better able to use these assets. In particular, there are a number of commonly used structures for wind projects to utilize that allow for the investment of capital to be made into a wind farm that will assign the tax benefits to a tax investor initially while that investor can utilize those benefits before "flipping" the structure to allow the wind project sponsor to utilize these benefits (if they continue to exist).

For more information about the Database of State Incentives for Renewables & Efficiency, visit

http://www.dsireusa.org/

Although several incentives summarized in the next section may be available to entrepreneurs already, future incentive programs are likely to develop at the federal, state, and local levels. To take full advantage of incentives available to greentrepreneurs, it is important to stay knowledgeable of ongoing developments in federal and state law and policy, to check for relevant updates on the EPA, DOE and IRS Web sites, and to seek additional information from state, municipal, and local energy regulating entities. This chapter will focus on federal tax incentive programs, but there are numerous programs in each state. The Database of State Incentives for Renewables & Efficiency provides a useful resource for finding up-to-date information on both federal and state incentives.

Production Tax Credit

The production tax credit (PTC) encourages renewable energy production by providing a tax credit for each kilowatt-hour (kWh) of wind, biomass, geothermal or hydro-based electricity produced by and sold to an unrelated party during the year. The credit was initially established by the Energy Policy Act of 1992 and has since been renewed and expanded several times, most recently by the Recovery Act of 2009. Wind, geothermal, and closed-loop biomass-produced energy is eligible for a 2.1 cent/kWh credit, while qualified hydropower, municipal solid waste, and open-loop biomass-produced energy is eligible for a 1.0 cent/kWh credit. Solar facilities placed into service after 2009 are not eligible for the PTC. The rules governing the PTC vary depending on the nature of the energy resource and of the facility used to produce energy.

For more information about the Production Tax Credit, visit

http://energytaxincentives.org/business/renewables.php

The PTC is "paid out" over 10 years beginning in the year a qualified energy production facility is "placed in service" (i.e., ready and available to begin energy production). A taxpayer must produce the energy and either own or lease the production facility to take advantage of the PTC. If a taxpayer does not have enough taxable income to fully utilize the tax credit in a given year, the PTC may be carried back one year or forward up to 20 years. The Recovery Act extends the PTC for wind-based energy property placed in service through December 31, 2012 and for all other resource-based energy property placed in service through December 31, 2013.

Investment Tax Credit

The investment tax credit (ITC) encourages the construction and/or acquisition of new property used to produce and distribute clean tech energy. The ITC is a one-time credit equal to 30% of the cost basis of the energy property placed in service during the year. The Recovery Act permits firms that qualify for the PTC to take the ITC instead. The Recovery Act extends the ITC for solar and fuel cell property placed in service through December 31, 2016, for wind property placed in service through December 31, 2012, and for biomass, geothermal, municipal solid-waste, and qualified hydropower property placed in service through December 31, 2013. A 10% credit is available for microturbine, geothermal heat pump, and combined heat and power equipment placed in service through December 31, 2016. The Recovery Act also provides investment tax credits for certain coal gasification and advanced combustion facilities.

Similar to the PTC, if a taxpayer does not have enough taxable income to fully utilize the tax credit in a given year, the ITC may be carried back one year or forward up to 20 years.

The ITC was first enacted in 1962 (but allowed to expire in 1985) and originally provided a credit of 20% per year over five years for investments in renewable energy production.

A taxpayer may not take both an investment tax credit ITC and a production tax credit PTC for a facility that could qualify for both.

WHY SOLAR IS SO "HOT" RIGHT NOW?

In 2003, the amount of worldwide energy generation from solar was negligible (0.0% of the total)—meaning that solar didn't provide any meaningful amount to the global energy usage. And, while the second half of the decade saw massive investment and excitement surrounding solar generation, it still remains a very tiny piece of worldwide energy generation. Estimates for 2010 forecast that only 0.2% of all energy generation would come from solar with a total of 32.9 GW of solar installed.

But, according to the Energy Information Administration and estimates by Jefferies, by 2030, 10.5% of our worldwide energy generation will come from solar energy.

In order to grow to 10.5% of worldwide energy generation requires substantial investment and growth, which is why business leaders from around the globe have attempted to gain a stake of that growing sector.

Renewable Energy Grants

Since many early stage businesses are unable to utilize tax credits, recent changes in the rules allow companies to elect to receive grants in lieu of the PTC or ITC. This is particularly beneficial for firms without sufficient taxable income to fully utilize a tax credit. The Recovery Act recently authorized the U.S. Energy and Treasury departments to administer a Renewable Energy Grant Program. Grants are available to eligible property placed in service in 2009 or 2010, or placed in service by the specified credit termination date. Definitions of eligible property types and renewable technologies generally include the following:

- Solar technology
- Fuel cells
- Small wind turbines
- Wind energy facilities
- Closed-loop and open-loop biomass facilities
- Geothermal energy facilities and geothermal heat pumps
- Landfill gas facilities
- Trash facilities
- Qualified hydropower facilities
- Marine and hydrokinetic renewable energy facilities
- Microturbines
- Combined heat and power systems

Vehicle Tax Credits

A number of tax credits are available for use of hybrid, electric, and alternatively fueled vehicles. Many of these credits were created or expanded by the Recovery Act in 2009. Several of these credits are summarized as follows:

For more information about Vehicle Tax Credits, visit

http://www.fueleconomy.gov

- Hybrid vehicles purchased or placed into service after December 31, 2005 may be eligible for a federal income tax credit of up to $3400.

- Qualifying electric vehicles purchased new are eligible for a one-time federal tax credit of up to $4000.

- Qualifying alternative fuel vehicles (AFVs) purchased or placed into service between January 1, 2005 and December 31, 2010 may be eligible for a federal income tax credit of up to $4000.

- Some diesel vehicles purchased or placed into service after December 31, 2005 may be eligible for a federal income tax credit of up to $3400.

Accelerated Depreciation

When businesses buy equipment, they get to write the value of the equipment off (i.e., deduct the purchase price of the equipment) a little bit at a time over the course of several years. A common and (relatively) simple way to encourage business investment is to allow businesses to purchase business-related property and write it off at a faster than usual rate, accelerating future tax benefits to the current year. Just what benefit does this provide your business? By accelerating depreciation, you are able to shorten the useful life of a piece of capital as it's recorded for tax purposes. For example, let's assume you install a small-scale wind turbine to generate electricity for your business. You expect to receive cash flows for the next 20 years from this system. However, by accelerating depreciation over a five-year period, you will have lower taxable earning for the first five years.

Accelerated depreciation is allowed for certain renewable energy technologies. For solar, wind, and geothermal equipment installed after 1986, the allowable property class is five years. Additionally, the Recovery Act included a one-time, 50% bonus depreciation for systems purchased and installed in 2008 and 2009.

In 2009, businesses may purchase and place in service up to $800,000 of qualifying property during the year and immediately write off up to $250,000 of their purchases. Because the tax incentive (referred to as a "Section 179 deduction") is aimed at helping small and mid-sized businesses, the $250,000 maximum benefit is reduced dollar for dollar for investments of more than $800,000 in qualifying property. This means that if a business purchases and places in service more than $1,050,000 of qualifying property, the business' Section 179 deduction is zero. However, "50% bonus depreciation" (i.e., an immediate write-off of 50% of qualifying property) is available for property where a Section 179 deduction is not taken.

Let's say you plan to purchase a small-scale wind unit to place on your rooftop at a cost of $400,000. Here is an example of how Section 179 and "50% bonus depreciation" can reduce your business' after-tax cost of investment in 2009:

Total Equipment Purchased	$400,000
Section 179 Deduction	(250,000)
Reduced Basis of Property	150,000
50% Bonus Depreciation	(75,000) = 50% * 150,000
Reduced Basis of Property	75,000
"Regular" Depreciation for 5-year property	(15,000) = 20% * 75,000
Basis of Property at the end of year 1	60,000

Total 2009 Depreciation Deducted	($340,000) = (250,000) + (75,000) + (15,000)
2009 Tax Savings from Deduction	($119,000) = (340,000) * 35% statutory tax rate
After-tax Cost of Equipment	$281,000 = 400,000 − 119,000

Source: http://www.section179.org/section_179_deduction.html

TAX TIP

While Congress has the ability to change the scope and amount of the Section 179 deduction at any future date, the maximum amount that can be written off in 2010 is only $134,000 (and only $25,000 in 2011). While Section 179 depreciation is limited to the amount of a taxpayer's active trade or business taxable income, businesses with bonus depreciation in excess of taxable income can use the depreciation to create a net operating loss (NOL). NOLs can generally be carried back 2 years and forward 5 years. Because businesses are allowed to carry back their 2008 and 2009 NOLs up to five (as opposed to the general two) years, thanks to a temporary provision in the 2009 Worker, Home Ownership, and Business Assistance Act, firms can use their 2009 bonus depreciation in excess of taxable income to generate an NOL, use the NOL to offset taxable income in the previous five years, and receive a tax refund from the government.

Commercial Building Tax Deduction

The Commercial Building Tax Deduction was enacted as a part of the Energy Policy Act of 2005 to improve the energy efficiency of commercial buildings. In 2008, the deduction was extended through 2013. Qualifying expenses incurred to make a building more energy efficient may be deducted by the building's owner. This deduction can apply to commercial buildings of any size, apartments of four or more stories that are being leased, and commercial energy renovations.

To qualify, the energy efficiency improvements must reduce the total energy and power costs by at least 50% compared to a reference building. Up to $1.80 per square foot of the

For more information about the Commercial Building Tax Deduction, visit

http://www.efficientbuildings.org/

property may be deducted, with partial deductions for certain improvements in interior lighting, HVAC and hot water systems, and building envelope systems. Businesses are required to show energy simulations and obtain an inspection and certification of the efficiencies by a registered engineer.

Manufacturing Tax Credit

While the Advanced Energy Manufacturing Tax Credit (MTC) authorized in the Recovery Act only accepted applications until the fall of 2009, its popularity and success has led the White House to publicly declare its plans to work with the U.S. Congress to expand the program from the original $2.3 billion authorization up to $5 billion in 2010.

The MTC was initially established as a part of the Recovery Act to encourage the development of a U.S.-based renewable energy manufacturing sector, intended to promote the building of manufacturing facilities for things like wind turbines, solar panels, alternative fuel vehicles, and other clean technologies. The investment tax credit is equal to 30% of the qualified investment required for an advanced energy project that establishes, re-equips, or expands a manufacturing facility that produces any of the following:

- Equipment and/or technologies used to produce energy from the sun, wind, geothermal, or "other" renewable resources
- Fuel cells, microturbines or energy-storage systems for use with electric or hybrid-electric motor vehicles
- Equipment used to refine or blend renewable fuels
- Equipment and/or technologies to produce energy-conservation technologies (including energy-conserving lighting technologies and smart grid technologies)

The fundamental goal of the MTC is to grow the domestic manufacturing industry for clean energy and therefore, this tax credit does not support energy generation projects themselves. Instead, the MTC is intended to promote the building manufacturing facilities that support generation and conservation.

Other Tax Incentive Programs

Biofuel Excise and Income Credits

Biofuel excise and income tax credits are available for taxpayers who use biofuel in their trade or business or who sell biofuels at retail. The credits are available in varying amounts for different types of biofuels and biofuel mixtures: ethanol ($0.51/gallon), biodiesel ($0.50/gallon), and agri-biodiesel and renewable diesel ($1.00/gallon). These credits expired on December 31, 2008 excluding the ethanol-related credit, which expires on December 31, 2010.

Value-Added Producer Grants

The U.S. Department of Agriculture's Rural Development program offers Value-Added Producer Grants for value-added agricultural products and for farm-based renewable energy. Eligible applicants include independent

For more information about USDA Value-Added Producer Grants, visit

http://www.rurdev.usda.gov/ga/tvadg.htm

producers, farmer and rancher cooperatives, agricultural producer groups, and majority-controlled producer-based business ventures. Applications and additional information can be obtained through Rural Development State and Local Offices.

Carbon Dioxide Sequestration Credit

The Emergency Economic Stabilization Act of 2008 created a new carbon dioxide sequestration credit which provides a tax credit for each metric ton of qualified CO_2 that is captured and disposed of in secure geological storage or used as a "tertiary injectant" in a qualified enhanced oil or natural gas recovery project. Taxpayers can receive a \$10 credit per ton for the first 75 million metric tons of CO_2 captured and transported from an industrial source and stored or used in the United States for use in enhanced oil recovery and a \$20 credit per ton for CO_2 captured and transported from an industrial source for permanent storage in a geologic formation.

Part IV

Green Progress (So Far)

17

Greening Your Business

Producing green energy, green products, or green services isn't enough to truly be a "green" business if you aren't also operating your business in a green manner. This chapter is all about helping you to develop a strategy to operate and act green.

The federal government resource, Business.gov, has created a list of steps that any business should follow in order to become greener. We'll use that as our guide and offer a series of tips, techniques, and activities that your business can use in order to operate "greenly" on the inside and outside.

WAVES OF CORPORATE ENGAGEMENT INTO ENVIRONMENTAL INITIATIVES

First wave: *Do No Harm*.
Second wave: *Do Well by Doing Good* (improving the bottom line through improved efficiencies).
Third wave: *Growing the Top Line through Innovation*.

Source: Joel Makower, CleanEdge.

Build Green

Green building represents one of the fastest growing segments of the construction industry. And now, with the LEED certification system, you can be certain that prospective builders meet industry standards before pursuing green certifications. Also, renovating an existing building can incorporate green building concepts and take that old building from nongreen to green (and get certified too). Rents and property values in LEED certified buildings are higher than comparable buildings and retain that value over time. More information is available on the Department of Energy Web site at: http://www1.eere.energy.gov/buildings/commercial/.

Install a "Cool" Roof

A poorly insulated roof can allow heat to escape, leading to increased energy use and heating and cooling costs. A number of steps can be taken to help "cool" your roof during warmer months. First, proper insulation can help block unwanted heat from getting into your building in the same way it can block generated heat from escaping during winter months. Further, use of reflective paints or specially designed synthetic materials on the outside of your roof can help reduce the amount of heat absorbed and transferred into your building. An

For more information on "cool" roofs, visit

http://www.coolroofs.org

219

alternative, more creative way to cool your roof could be to install a rooftop garden. In addition to being visually appealing, rooftop gardens can absorb sunlight and heat into plants, grasses and mosses rather than into the building itself.

Install Green Flooring

Carpet and vinyl account for about 70% of all floor covering in the United States. Many types of carpet are made from petroleum-based nylon. Vinyl is made from petroleum-based polyvinyl chloride. The production of these products involves use of a tremendous amount of nonrenewable resources, and neither carpet nor vinyl is biodegradable.

Several natural flooring materials exist as preferable options for those businesses that wish to reduce their environmental footprint. Green flooring can also help to reduce allergens and other levels of toxicity in your workspace and increase energy efficiency.

Here are a few examples of alternative flooring options:

- Sustainable wood flooring, such as bamboo, can work as an attractive and sustainable alternative to traditional hardwood flooring. Bamboo, a giant grass, is one of the fastest growing plants in the world, meaning it can be harvested every few years. Other forms of wood certified by the Forest Stewardship Council (www.fscus.org) meet strict criteria related to forest management. If you do choose hardwood floors, look for an FSC-approved product that allows you to trace the wood to its source.

For more information on Green Flooring, visit http://www.greenfloors.com

- Carpet or rugs made from natural materials, such as wool, are also sustainable alternatives to nylon carpet. Seagrass, sisal (a fiber from the leaves of the agave plant), and coir (fiber harvested from coconut husks) are other natural products used to make rugs and carpets.
- Natural linoleum, made from linseed oil, cork dust, wood flour, tree resins, ground limestone and pigments, is also a sustainable option far more environmentally friendly than petroleum-based vinyl. Although natural linoleum is more expensive than vinyl, it lasts 10–20 years longer than the alternative.

Buy Green Products

Experts suggest that businesses consider buying green and environmentally friendly products that are made from postconsumer, recycled materials; bio-based; nontoxic; energy-efficient rated products, such as ENERGY STAR®; renewable and recyclable; and/or locally produced, such as food that is locally grown and organic. Below are some ways to build green buying into your business.

Purchase Green Cleaning Supplies

Americans generate up to 1.6 million tons of waste per year by using common household cleaning products. Some of this waste is potentially hazardous due to harsh chemical ingredients. Using green cleaning supplies made from natural and environmentally friendly ingredients can reduce the danger posed by more hazardous cleaning products. A growing number of green cleaning products are now readily available across the country, and many of them are priced to be competitive with, more hazardous supplies. Web sites such as The Green Seal (www.greenseal.org) and Environmental Choice (www.ecologo.org) provide product certification programs that help ensure green products truly are "green."

Utilize Green Packaging Policies and Techniques

The U.S. EPA estimates about 30% of all municipal solid waste is currently generated from packaging-related material. Although several European countries have taken deliberate legislative action to reduce packaging waste and mandate recycling programs, a great deal of waste continues to be generated from excessive packaging. As a green business, there are a number of ways in which you can reduce the waste produced through packaging materials.

- *Minimize packaging.* Design packaging in such a way as to use as little material as possible. Try to avoid excessive "interior" packaging enclosed within an outer package.
- *Make smart packaging purchases.* Use packaging materials made from biodegradable materials, such as biodegradable plastics, recyclable cardboard, and materials made from other recycled items.
- *Consider the carbon footprint.* If you purchase packaging materials, learn how and where they are produced and transported. Buy packaging materials locally, where possible, and/or from those suppliers that generate the least amount of waste in producing and transporting the materials.
- *Use green labeling techniques.* The process of printing, cutting, and gluing labels can, itself, create an enormous amount of waste. Where possible, print labels directly onto your product rather than produce and adhere separate labels.
- *Use green packing materials.* Rather than using Styrofoam packing peanuts or sheets of plastic bubble wrap, consider using shredded paper documents, recycled newspaper, biodegradable packing materials (see, e.g., www.puffystuffn.com), or shredded cardboard as packing alternatives.

Use Renewable Energy

Generate Your Own Solar Energy

Solar energy is a readily available resource that can be utilized by businesses to reduce greenhouse gas emissions, conserve fossil fuel supplies, and provide a clean and renewable source of electrical energy. Although a robust solar power system may cost as much as $20,000 to install, that cost will be recovered over time in reduced energy costs. The federal government—as well as many state governments—also offer tax credits and other incentives to those businesses that install solar energy systems.

For more information on installing solar power, visit

http://www.solarelectricpower.org

Here are some additional ideas for incorporating solar energy into your business's energy plan:

- *Natural gas.* Convert all your major appliances to clean-burning natural gas or propane. By doing so your demand for electricity is limited to relatively low-consuming lighting systems and small electronics. This equipment can be powered, either completely or partially, by solar energy.

- *Storage and additional space.* Where possible, take small outer buildings and structures off the grid and install a solar power system that generates all the power required for that structure.

- *Business signage.* Convert your businesses signage to be illuminated by solar power. A small and relatively inexpensive solar panel may be all you need to keep a sign lit during nighttime hours.

- *Water pumps.* Install solar-powered water pumps. Solar energy can be used to power water pumps, landscape water features, and irrigation and drinking wells.

- *Special event lighting and power.* Use portable solar power systems, rather than generators, for any special events or activities conducted outside of the office.

Install Wind Turbines

A growing number of businesses are installing small-scale wind power generating devices in order to provide power to the business. Although wind turbines may not be effective everywhere, they can be a great way for businesses to save costs while producing zero CO_2 emissions. They can also generate electricity 24 hours a day, 365 days a year, so long as the wind is blowing. Wind turbines can be utilized most effectively in areas that have an average annual wind speed of 11 miles per hour or more. They cost a business between $6000 and $20,000 to install, but this cost will typically be recovered in 5–10 years.

For more information on installing wind power, visit

http:// www.windenergy.com

Purchase Renewable Energy Certificates to Offset Carbon Emissions

If the nature of your business involves CO_2 emissions in one form or another, one way to preserve your "greenness" is to purchase renewable energy certificates to offset those emissions. Renewable energy certificates (a.k.a. green tags, renewable energy credits, or tradable renewable certificates) are tradable energy commodities that guarantee that a set amount of renewable energy (e.g., wind or solar power) will go into the grid on the buyer's behalf. The Bonneville Environmental Foundation (BEF), is a reputable organization that sells renewable energy certificates. BEF allows its customers to purchase certificates based on a set dollar amount, or based on the average amount of CO_2 emitted during a flight, by a car, or by a home.

For more information on purchasing BEF credits, visit

http:// www.b-e-f.org

USE OF RENEWABLE ENERGY IN U.S. CITIES

1. Oakland, CA (17%)
2. Sacramento, CA (12%)
3. San Francisco, CA (12%)
4. San Jose, CA (12%)
5. Portland, OR (10%)
6. Boston, MA (8.6%)

7. San Diego, CA (8%)
8. Austin, TX (6%)
9. Los Angeles, CA (5%)
10. Minneapolis, MN (4.5%)
11. Seattle, WA (3.5%)
12. Chicago, IL (2.5%)

Source: SustainLane City Rankings 2006–2007.

Adopt Energy-Efficient Practices

Utilize Energy-Efficient Lighting Techniques

Lighting systems account for between 30% and 50% of the total annual electrical energy consumption in a typical U.S. office building. Fortunately, lighting technology has greatly improved in recent years, and a business can significantly reduce the amount of energy it consumes by taking a few relatively simple steps towards improving its lighting systems.

- *Take advantage of available daylight.* This can be done by designing buildings in such a way as to allow for maximum sunlight penetration through the building's windows or skylights, or by installing automatic sensors and dimmers that will dim or turn off lights during peak sunlight hours.

- *Use florescent, HID, and LED lighting systems.* Florescent lighting systems consume as much as 75% less energy than do older incandescent systems. LED systems can be even more efficient than florescent systems. A business can greatly reduce its overall energy consumption by replacing incandescent bulbs with energy-efficient alternatives.

  For more information on energy efficient lighting, visit

  http://www.americanlightingassoc.com
  http://www.energystar.gov

- *Use energy-efficient reflectors or louvers rather than diffusers.* Many office lighting systems utilize diffusers (plastic sheets or lenses that cover lighting to reduce the glare or intensity of naked bulbs). Diffusers are often inefficient, as they absorb much of the light intended to illuminate a workspace. Use of reflectors, clear patterned lenses, or parabolic louver shields can greatly improve energy efficiency without detracting from the appearance of a lighting system.

Make Your Computers Run Greener

Based on current trends, did you know that energy used by Information and Communication Technology (ICT) equipment is estimated to require 45% of the U.S. domestic energy use by 2020? That's right, 45%! Today, ICT equipment accounts for approximately 2% of global CO_2 emissions, and its energy use doubled between 2000 and 2005 alone. Companies offering Green IT solutions have seen rapid expansion including Verdiem, offering its energy monitoring application "Edison" that allows computer users to actively influence their CO_2 output by controlling the energy consumption of their PCs. Many of the largest users of computing power including Google, Microsoft, Sun and Intel have all embracing green as a way to lower their operating costs.

The money saving opportunities from more efficient use of ICT resources holds true for smaller companies too. What can you do to make your IT resources more green?

- *Track ICT power consumption.* You can't improve if you don't know how you are doing. Therefore, the use of energy monitoring tools in information technology departments has been a key to better management. More IT managers are now being rewarded as they find unique ways to save money through lowering energy bills.

- *Upgrade your old system.* Newer computer models and operating system have begun to operate more efficiently with power consumption. Look into upgrading sooner and factor into the buying decision the cost savings you can save by switching. Models that are ENERGY STAR qualified products meet the standards for energy efficiency.

- *Recycle your old computer.* And what should you do with that old machine? Don't just throw it in the trash as durable electronics can actually be recycled very successfully these days. While not all cities recycle durables like computers, monitors and printers, many private organizations are filling in that gap. In addition, charities such as Little Geeks Foundation (www.littlegeeks.org) will take your old computers, refurbish them and then provide them to underprivileged children.

For more information on green HVAC systems and proper insulation techniques, visit

http:// www.ceedirectory.org

http:// www.insulate.org

- *Use power management settings.* It may seem simple enough, but many users of computers don't use the tools that come pre-loaded onto every personal computer. By enabling these options on each computer, it is estimated that you can save as much as $60 per machine annually in energy-related costs.

- *Don't use a screensaver.* Sure, it seems nice to have photos of your vacation on the screen or lines and fishtanks pop up, but the reality is that these force the monitor to remain on. Instead, set your monitor to dim when idle and turn off automatically when not in use.

- *Switch to off.* Switch off your computer at the end of the day, as well as turn off monitors, printers, speakers, and other electricity hogs that are often left on overnight. These devices not only use electricity themselves but produce heat which can increase spending on cooling costs.

Make Smart Choices about Heating, Ventilating, and Cooling Your Workspace

Heating, ventilating and air conditioning (HVAC) systems consume approximately 40% of the energy used in commercial buildings. Much of this energy is wasted due to inefficient equipment and practices. By making smart choices about how you purchase, use and maintain your HVAC system, you can reduce the cost of utility bills and reduce greenhouse gas emissions.

Here are a few suggestions:

- *Weather-proof.* Use weather-stripping and caulking to minimize heat loss from windows and doors.

- *Seal and insulate air ducts.* Use mastic sealant or metal (foil) tape and environmentally friendly insulation to reduce energy loss from ducts. Also check to see that connections at vents and registers are well-sealed where they meet the floors, walls, and ceiling. If you do not have access to perform this work yourself, make a suggestion to your building's manager or maintenance personnel.

- *Use a programmable thermostat.* If you have access to your office thermostat, program the thermostat so that it is set at energy-saving temperatures during times when the office is unoccupied.

- *Use a green air filter.* A green air filter improves filtration efficiency and air quality while minimizing energy consumption and greenhouse gas emissions.

- *Buy ENERGY STAR.* Look for the ENERGY STAR label when purchasing new HVAC equipment.

For more information on ENERGY STAR® equipment, visit

http://www.energystar.gov

Green Your Office Equipment

One easy way to green your business is to make good decisions regarding the use and purchase of office equipment. Technology plays a big role in most offices, and several steps can be taken in order to ensure that technology is used efficiently.

- *Look for an ENERGY STAR label.* When buying appliances and equipment for your office, check to see if the equipment has been certified. An ENERGY STAR label indicates that a particular product is among the most energy-efficient options available.

- *Turn off and unplug machines and equipment when they are not in use.* This practice should include computers and computer monitors. Contrary to a common misconception, turning off a computer every night will not harm it.

For more information on recycling and recycling centers, visit

http:// www.earth911.com

http://www.therecyclingcenter.info

- *Laptops vs. desktops.* Consider buying laptops instead of desktop computers, as laptops are more energy efficient.

- *Buy rechargeable.* Use rechargeable batteries wherever possible, and check with your trash removal company about safe disposal of single-use batteries.

Reduce, Reuse, and Recycle

Start an Office Waste Reduction and Recycling Program

An important step towards greening your business is starting an office recycling program. Of course, before taking the time to sort recyclables from trash, it helps to first reduce the amount of trash produced.

Here are some simple tips for reducing trash and developing a recycling program:

- *Reusable dishware.* Use reusable plates, mugs, flatware, and glasses.

- *Ditch the water bottles.* Rather than stocking your fridge with bottled water, use a water purifier or water cooler.

- *Go electronic.* Avoid printing documents or other materials that can otherwise be transferred and reviewed electronically.
- *Double-sided recycled paper.* When printing is necessary, use recycled paper and print on both sides of the paper.
- *Recycling stations.* Set up recycling stations in office common areas. Educate your staff on different types of plastics, glass, and other recyclable materials as well as how they should be sorted.
- *Durables can be recycled too.* Learn where and how to recycle used office equipment, printer cartridges and toners, batteries, CDs, and other forms of technology whenever possible.

Be Conscious of Your Coffee

Although a morning cup of coffee may seem like an insignificant part of your day, a surprisingly large amount of waste is produced on a daily basis by coffee drinking Americans. According to calculations made by www.papercalculator.org and the Environmental Defense Fund, Americans will consume an estimated 23 billion paper coffee cups in 2010. The production of 23 billion paper cups requires the consumption of 9.4 million trees and 7 trillion BTU's of energy. Further, 363 million pounds of solid waste results from the production and disposal of these cups. Your business can help reduce waste produced by disposable coffee cups by encouraging the use of reusable mugs and thermoses and not utilizing single-use paper or Styrofoam cups in the workplace.

Reduce Waste from Toner and Ink Cartridges

According to a recent study, approximately eight printer cartridges are discarded every second in the United States. Each one of those cartridges can take up to 450 years to decompose in a landfill.

Businesses can avoid contributing to ink and toner cartridge waste by following a few easy steps:

- *Print less and more efficiently.* Of course, the less ink you use, the less often you will need to replace the ink and toner cartridges in your printer or copy machines. Avoid printing when electronic transmission or use of documents is a reasonable alternative option.
- *Purchase ink and toner refill kits rather than new cartridges.* Refill kits can be used for cartridges found in copiers, faxes, and laser printers. Find refill kits at www.abcink.com and www.tonerrefillyourself.com, as well as retail office supply stores.
- *Recycle used cartridges.* Search for local recycling centers that accept cartridges in your area or ask for recycling information when you purchase new cartridges. Several businesses will even pay you for your used cartridges in order to then recycle and reuse them. Examples of such businesses include www.freerecycling.com and www.tonerbuyer.com. Staples, Inc. also offers a rewards program for customers that return used cartridges to Staples' retail locations.

Conserve Water

According to the U.S. Geological Survey's latest data, the United States uses approximately 410 billion gallons of water per day. Decreasing water supply is a growing concern throughout the country, and excessive water use is harmful to the ecosystems surrounding many of the country's lakes, rivers, and streams.

For more information on water usage and conservation, visit

http://pubs.usgs.gov/circ/1344/

http:// www.epa.gov/watersense

Your business can help conserve water with a number of simple steps:

- *Regular plumbing maintenance.* Check pipes, faucets, and toilets for leaks regularly.
- *"Green" landscaping.* Consider landscaping techniques that require less watering than traditional grass lawns, and avoid overwatering areas that require irrigation.
- *Water recovery.* Avoid dumping clean water down drains if it can be used to water plants, mixed with cleaning supplies to clean your office, or used in other ways.
- *Conservation systems.* Install low-flow toilets and faucet aerators.
- *Track your progress.* Monitor your water meter and water bills and set goals to reduce your overall water consumption.
- *Purchase water-efficient products.* Buy products that have a WaterSense label indicating water efficiency.

Prevent Pollution

Find Natural Ways to Control Pests

Like chemical cleaning supplies, chemical pest control agents can contain hazardous materials that cause environmental damage. Rather than using hazardous chemicals to control mice, cockroaches, ants, or other pests, first establish an integrated pest management system.

This system can be established by following these steps:

- Create an effective barrier that prevents pests from entering your workplace. Check your office building regularly for possible points of entry for pests and be proactive in ensuring that those points are sealed.
- If pests do get into the building, eliminate all food and drink that attracts and sustains them. This may include checking for leaks or dripping water, more carefully disposing of garbage, and so on.
- Finally, if a pest problem still persists, use the least toxic product available to eliminate the pests. If possible, use mechanical traps, or ask a pest control professional about the environmental impact of his/her products.

Make a Paperless Office a Reality

Despite the considerable advancement in computer technology, email communication, Internet research, and digital storage capacity many offices continue to consume enormous amounts of paper, much of which ends up being wasted unnecessarily. According to the EPA, the average office worker in the United States uses 10,000 sheets of copy paper each year and generates approximately two pounds of paper and paperboard products every day. If use of office paper in the United States was reduced by just 10% it would prevent the emission of 1.6 million tons of greenhouse gases, an amount equivalent to taking 280,000 cars off the road. Setting a goal of going paperless—and actually sticking to that goal—is one of the ways for a company to go green and save substantial money doing so.

Follow these simple steps to reduce the amount of paper consumed in your workplace:

- *Resist the urge to print.* Some people still prefer to read a document in paper form rather than on a computer screen. Although it may be impractical to avoid all printing, transmit and read documents in electronic rather than paper form wherever possible. Reducing the brightness of computer screens, purchasing large monitors, using text applications' zoom functions, or increasing font size may be simple ways of making electronic reading easier on the eyes.

For more information on paper waste and paper recycling, visit

http://www.epa.gov/waste/conserve/materials/paper/

- *Print efficiently.* When you do print, print on both sides of the paper, use smaller margins, and take the time to format documents so they fit on as few pages as possible.
- *Don't print extra handouts and materials.* Accept the risk of having too few handouts. Many paper handouts, agendas, and meeting materials passed around during the typical workday are printed in excess, and those materials are often quickly discarded. Exchange materials electronically wherever possible allow attendees to share materials if feasible, and reuse or recycle documents when they are no longer needed.

Create a Green Marketing Strategy

A great way for businesses to prove to potential customers that they "walk the talk" is to promote themselves as green businesses by using green marketing methods. In other words, don't advertise your business commitment to reducing your carbon footprint by posting paper fliers all over town.

Here are a few tips for green marketing techniques:

- *Social media and online marketing.* Take advantage of social media platforms and online marketing techniques. Facebook, Twitter, and LinkedIn are examples of increasingly popular Internet platforms where small businesses can conduct free advertising. Set up an account for your business on any or all of those applications, and update it regularly in order to build an online network.
- *Create eco-friendly paper products.* If you do use paper mailings or other forms of paper-based advertising, use recycled paper products, soy-based ink, and other natural materials that help show your commitment to being green.
- *Network in-person with other greentrepreneurs.* Attend local social and networking events catering to other green entrepreneurs. Meeting other

like-minded business owners may be a great way to build partnerships, increase word-of-mouth advertising, and learn from others' work and ideas.

Comply with Environmental Regulations

As a green business, remember that it isn't enough to simply create green products. You too must comply with environmental regulations. The following are common categories of environmental regulations that businesses must comply with:

- Environmental permits
- Air quality regulations
- Environmental cleanup
- Ecosystem protection
- Emergency and disaster planning
- Fish and wildlife regulations
- Pesticides
- Pollutants, chemicals and toxic substances
- Storage tanks
- Toxic release inventory (TRI)
- Waste regulations
- Water quality regulations

Encourage Use of Public and Alternative Forms of Transportation

Nearly 90% of American workers living outside of major cities drive to work. Further, commuters stuck in traffic waste an estimated 30 billion gallons of gas each year. In many areas of the country, driving may be the only reasonable means available for workers to commute to work. However, a green business can be creative in encouraging and supporting workers who commute in ways that reduce the overall waste produced by automobiles.
Here are some ideas:

For more information on encouraging use of public and alternative forms of transportation, visit

http://www.erideshare.com

http://www.cleanair.org/Transportation/greenCommute.html

- *Carpool.* Set up a carpool system and offer incentives to those workers that use it, such as free parking passes or gift certificates to local lunch destinations.

- *Buy bike storage.* Install bike racks and, where possible, showers and lockers for employees who choose to bike to work.

- *Employee benefits.* Include public transportation tickets or passes in your employees' benefits packages.

- *Understand public commuter incentive programs.* Educate yourself and your employees on available commuter tax credits or other federal and state incentives meant to encourage alternative forms of transportation. The IRS Web site provides general information on commuter tax credits here: www.irs.gov/formspubs/article/0,,id=181059,00.html.

- *Lead by example.* If you are the business owner or manager, make an effort to show that you are using alternative forms of transportation to get to work.

Developing an Environmental Management Plan

Decrease Your Carbon Footprint

A business's carbon footprint refers to the amount of CO_2 emitted into the atmosphere through routine business activity. Many businesses strive to reduce their carbon footprint with an ultimate goal of becoming "carbon neutral," and incorporate these goals into an overall Environmental Management Plan. Businesses accomplish this by first taking the time to calculate their carbon footprint, accounting for emissions from factories, equipment, and vehicles used in creating and shipping the company's products. Once this footprint is measured, a business can then strive to be "carbon neutral" by taking deliberate steps to eliminate extraneous carbon emissions within its own day-to-day operations, or by paying offsetting fees or trading carbon credits with other companies. Building a plan can help a business meet environmental and carbon emissions goals.

A few Web sites offer "carbon calculators" that will assist you in making these measurements

http://www.epa.gov

http://www.nature.org

http://www.carbonfund.org

Making "Greening" a Priority during Contract Negotiations

Whether it be contracting with manufactures, shippers, property owners, maintenance, or landscaping workers, partner organizations or any other entity upon which your business relies, including green provisions in contract negotiations is a great way to ensure that the services or products that you pass on to consumers are thoroughly green. Although not all businesses or individuals will make it a key priority to be environmentally conscious, a simple request to consider reasonable green alternatives in your contracts will help the effect. If a green solution or alternative is easy, those other entities may choose to offer the same alternative or solutions to other customers or partners, thereby expanding to the growing green market. On the other hand, if the prospective business partner is unwilling to consider green alternatives, perhaps you would be better off shopping around. This approach may help you develop and expand a green network of other businesses—and eventually potential customers—who value your commitment to being green.

18

Green Certifications

William Brent of Weber Shandwick has suggested that 2010 and beyond will see a growing emphasis on standards and certifications for green businesses and technologies. In 2010 Brent expects to start to see "standards gain a higher profile—whether building codes, water or carbon labeling, unified standards for the smart grid, and so on, creating a clear marked playing field grows in importance, including communicating the rules to consumers as needed."

That said, even today, there certainly isn't a lack of green standards and certifications. For example, there is the Green Seal, EPA Designed for the Environment, EcoLogo, GreenBlue, Green Label, Environmental Choice, Eco-Label, GreenGuard, and a new addition to the list in 2009, the *Good Housekeeping* official green seal, just to name a few. The bigger issue is making sense of the myriad of seals, certifications, logos, labels, and guides—certifying the certifiers and developing standards for key products that consumers begin to trust.

The Growth of Green Standards

It isn't surprising that standards for environmentally friendly products have grown steadily since the 1960s. With greater research into the harms of chemicals, greenhouse gases, nonrecyclable, noncompostable, and numerous other anti-green actions, consumers have looked for a guide to help with their selection of the "most" green product (or perhaps the least "non-green" products).

For a list of ecolabels and certifications, visit

http://www.ecolabeling.org

From automobile standards for miles per gallon of gasoline to the recycling symbols on certain consumer packaging products to better labels and descriptions on product packaging, green standards and reporting of those standards were a logical outgrowth of the environmental regulations of the 1970s. And with the resurgence of environmental concerns in the first decade of the twenty-first century, we witnessed an increase in green labeling, certifications, and standards.

EcoLabeling.org identified more than 325 ecolabeling schemes across the globe. The ecolabels range from labels and certifications for food (90 different ecolabels) and retail goods (74) to buildings (64) and electronics (40) to textiles (40), forest products (36), energy (31), tourism (28) and carbon (15). The United States leads the way with 46 ecolabels, while Europe as a whole has 115. A few of these labels and certifications are given in Table 18.1.

For more information about consumer reports greenerchoices, visit

http://www.greenerchoices.org/eco-labels/eco-home.cfm

Consumers have expressed concern about the increase in ecolabeling and green certification programs—particularly given the inability to distinguish between the relative value of each label and program. Consumer Reports Greenerchoices has developed a tool to assist consumers to distinguish between reliable and unreliable certification programs. The program is relatively new, and not all ecolabels and certifications are currently available.

TABLE 18.1

Sample Certification and Labeling Programs

Certification or Label	Oversight Organization	Emphasis	Products Included	Certification Process
Cradle to Cradle Certification	McDonough Braungart Design Chemistry (MBDC)	General certification	Various	Independent research and analysis; levels: basic, silver, gold or platinum
Ecologo Certified	TerraChoice Environmental Marketing	General certification	Various	Third party auditors test for compliance with standards
ENERGY STAR	U.S. Environmental Protection Agency and Department of Energy	Energy efficiency	Appliances, light commercial HVAC, office equipment, lighting, external power adapters, roof products, room air cleaners, transformers, water coolers, windows and doors	Manufacturers submit data detailing how a product meets standards
EPEAT	Green Electronics Council	Green electronics	Computers and monitors	Manufacturer to provide production reports, lab analysis or other data; may be audited/verified
FLO-CERT	Fairtrade Labeling Organization International (FLO)	Certified fairtrade products	Various	Independent analysis of compliance with FLO standards
Green-e	Center for Resource Solutions	Renewable energy & greenhouse gas emissions	Green-e climate (credits); Green-e energy (renewable energy); Green-e marketplace (company purchases certification)	Independent research and analysis
Greenguard	Greenguard Environmental Institute	Indoor air quality	Adhesives, appliances, ceilings, cleaning systems, flooring, insulation, office equipment, office furniture, paint, textiles, and wallcoverings	Independent lab emissions test
Green Seal	Green Seal	General certification	Various	Products must meet Environmental Leadership Standards; application reviewed by Green Seal staff
Leadership in Energy and Environmental Design (LEED)	US Green Building Council (USGBC)	Green building	Environmentally sustainable construction and remodeling	Submission of application documenting compliance with the requirements of the LEED rating system

TABLE 18.1 (continued)

Sample Certification and Labeling Programs

Certification or Label	Oversight Organization	Emphasis	Products Included	Certification Process
SCS Certification	Scientific certification systems	Independent certification of environmental and sustainability claims	Agricultural production, food processing and handling, forestry, fisheries, flowers and plants, energy, green building	Engineers test products based on internationally recognized standards and certification programs; also tests external standards
Sustainable Materials Rating System (SMaRT)	Institute for the Market Transformation to Sustainability	General certification for sustainability	Building products, fabric, apparel, textile & flooring	Levels: sustainable, silver, gold or platinum
USDA Organic	U.S. Department of Agriculture	Organic	Food, some cleaning products and cosmetics	National Organic Program certifies agents who inspect organic production

ISO 14001 Certifications

In order for a green business to assure itself and its customers that it indeed is a "green" business, business owners can seek an ISO 14001 Certification. In short, this certification allows a business to prove that it has successfully implemented a system meant to ensure that its operations, products, and services are environmentally friendly. What follows is a brief summary of ISO standards and the ISO 14001 Certification process.

For more information about ISO 14001

http://www.iso.org/iso/management_standards.htm

http://www.epa.gov/OWM/iso14001/isofaq.htm

ISO is the International Organization for Standardization, which is a nongovernmental organization headquartered in Geneva, Switzerland. ISO coordinates a network of national standardization institutes from 162 countries throughout the world. That network promotes the development of voluntary international standards relating to a wide variety of subjects. As of November 2009, ISO has developed over 17,500 International Standards and publishes some 1100 new standards every year.

The goal behind ISO's international standardization process is to provide a reference framework and common technological language to consumers, businesses, and government entities. That reference framework can then help to facilitate trade and the transfer of technology throughout various industries. Of interest to green entrepreneurs, ISO standards allow business owners to base the development of their products and services on reference documents that are recognized internationally and that have broad market relevance. Business owners can also seek certifications that show they comply with ISO standards.

ISO 14000 refers to a series of voluntary standards that relate to the environmental field, including standards used in environmental auditing, environmental performance evaluation, and environmental management issues. Included in the series is ISO 14001, which outlines procedures and requirements for creating an Environmental Management System (EMS). By developing an EMS, a business can identify and control its environmental impact, improve its environmental performance, and implement a systematic approach to setting environmental objectives and targets.

Once a business has implemented an EMS, it can then assure both itself and its customers that it complies with that EMS by seeking an ISO 14001 certification through an environmental audit. ISO 14001 certifications can be obtained through an auditing body certified by RABQSA International or the Board of Environmental, Health & Safety. Registries containing information on certified auditors are available at www.rabqsa.com/search.html and www.beac.org/Registry-homepage.html.

ENERGY STAR®

ENERGY STAR® is one of the largest voluntary climate change programs currently in existence in the United States. A joint EPA and DOE effort launched in 1992, ENERGY STAR is a voluntary labeling program used to identify energy-efficient products and practices. The ENERGY STAR label is used on major appliances, office equipment, lighting, home electronics, new homes, commercial and industrial buildings, and more. As of October 2009, ENERGY STAR works with over 15,000 public and private sector organizations.

For more information about ENERGY STAR

http://www.energystar.gov

Leadership in Energy and Environmental Design

The Leadership in Energy and Environmental Design (LEED) green building rating system was developed in the mid-1990s by the U.S. Green Building Council (USGBC). The purpose of the LEED system was to provide a clear set of guidelines and principles to govern environmentally sustainable construction. Many experts point to the LEED system as evidence of the impact that industry-wide green standards can provide—pointing to the growth of the nearly $60 billion green building construction sector in 2010. To incentivize green building, states including Nevada and Michigan, and numerous cities and local jurisdictions have passed laws providing favorable tax treatment for new construction that complies with LEED standards.

The LEED system began as a single standard for new green buildings, but ultimately became a comprehensive system of six interrelated standards that involve all aspects of the development and construction process—including new and remodeled building. The LEED green building rating system focuses on six major areas:

- Sustainable sites
- Water efficiency
- Energy and atmosphere

"LEED"ing green building

One of the fastest growing segments of green businesses are those engaged in green building. The green building market is estimated to be worth $30–$40 billion annually by the year 2010. What exactly is green building? Well, it used to mean a lot of things–perhaps you used native woods, low flow toilets, radiant heating, energy efficient lighting, or recovered carpet. Any of those could be "green" building.

In 1998, the U.S. Green Building Council (USGBC) developed the Leadership in Energy and Environmental Design (LEED) Green Building Rating System which offers standards for environmentally sustainable construction. According to the USGBC, by the end of 2010, over 100,000 buildings were expected to be LEED certified in the United States, up from just 10,000 at the end of 2007.

- Materials and resources
- Indoor environmental quality
- Innovation and design process

In the LEED rating system for 2009, up to 100 base points are possible with an additional six points for Innovation in Design and four points for Regional Priority. Based on the certifications, a building can qualify for four levels of certification:

- Certified—40–49 points
- Silver—50–59 points
- Gold—60–79 points
- Platinum—80 points and above

U.S. Department of Energy's Voluntary 1605(b) Greenhouse Gas Registry

The 1605(b) Voluntary Registry allows organizations and individuals to record the results of voluntary measures undertaken to reduce or avoid greenhouse gas emissions. The registry was authorized by the Energy Policy Act of 1992, and it is maintained by the U.S. Department of Energy. In order to participate in the registry, interested corporations, governments, and other organizations can use a web-based tool for translating common physical indicators into a preliminary estimate of greenhouse gas emissions.

For more information about the DOE GHG Registry

http://www.eia.doe.gov/oiaf/1605

Climate VISION—Voluntary Innovative Sector Initiatives: Opportunities Now

Climate VISION was launched in February 2003 as a public–private initiative designed to reduce greenhouse gas emissions. The initiative establishes partnerships between U.S. government agencies and business associations representing 14 key industrial sectors. Participating sectors work with the DOE, EPA, and DOA in order to develop work plans to introduce greenhouse management plans into business plans and decisions.

For more information about the Climate VISION

http://www.climatevision.gov/news/index.html

Climate Leaders

For more information about Climate Leaders

http://www.epa.gov/stateply/partners/index.html

Climate Leaders, a voluntary industry–government partnership managed by the EPA, was launched in February 2002. The objective of the program is to challenge individual companies to develop long-term climate change strategies. Participating companies develop a greenhouse gas reduction target goal, as well as an Inventory Management Plan meant to help them institute best practices in measuring and reporting greenhouse gas emissions. Best practices employed for greenhouse gas reductions are also shared among participating companies via the program's Web site. Finally, those companies that successfully reach their reduction goal are publicly acknowledged by the EPA. As of October 2009, well over 200 companies of all shapes and sizes are involved in the program.

The Chicago Climate Exchange

The Chicago Climate Exchange (CCX) is a privately sponsored, self-regulating cap and trade system for six greenhouse gases. Companies that join CCX make a legally binding commitment to reduce their aggregate emissions according to a certain reduction schedule. CCX members register their emissions in a CCX registry, they are then given emission allowances according to the CCX reduction schedule, and they then use a suite of tools and service offered by CCX to manage and trade their allowances. As of October 2009, over 350 organizations from North America and Brazil are members of the CCX. Those members operate on the goal of reducing greenhouse gas emissions by 6% by 2010.

For more information about CCX

http:// www.chicagoclimatex.com

Walmart Sustainability Index

Wal-Mart Stores, Inc. (branded as Walmart) was the largest public corporation in the world in 2008 by revenue, generating an estimated $409 billion in revenue in 2009. Amazingly, Walmart's revenue is higher than the Gross Domestic Product of countries that include Indonesia, Saudi Arabia, Norway, Denmark, Poland, South Africa, and Greece. As a result, when Walmart announced the development of its Sustainability Index to evaluate the sustainability of its suppliers, many other companies immediately took notice—with many experts predicting that Walmart's approach could quickly become the industry standard. The Sustainability Index measures the sustainability of all products sold in Walmart stores in four areas: energy and climate, natural resources, material efficiency, and people and community.

What Does the Sustainability Index Do and How Does It Work?

The Sustainability Index will be implemented in three phases, which will ultimately result in a new labeling or reporting system for all products sold in Walmart stores. There are three phases to the implementation of the Sustainability Index, with the first phase to be implemented by all major Walmart suppliers before the October 1, 2009 deadline.

Phase 1—Supplier assessment: Walmart has provided a survey to each of its more than 100,000 worldwide suppliers, with the largest suppliers required to complete the survey prior to October 1, 2009.

Phase 2—Lifecycle analysis database: Following submission of the supplier data, Walmart will work with a series of partners and research institutions to create a database tracking product sustainability. This analysis will use principles of the lifecycle analysis or assessment tools (LCA).

Phase 3—Rating and labeling system: The data gathered from the Lifecycle Analysis Database will be utilized to develop tools and labeling on sustainability criteria of its products. Walmart will work with its suppliers to rank products and product categories, and utilize new labeling schemes for providing consumers with sustainability ratings.

The development of a rating and labeling system will likely take a number of years to develop and implement, but Walmart expects its suppliers to play a substantial role in this process. The Sustainability Index is likely to interplay with Walmart's Packaging Scorecard, which has been utilized by Walmart to encourage its suppliers to make more efficient choices on product packaging.

The Supplier Assessment involves 15 questions for suppliers focused on the areas of energy, GHG reduction and waste reduction, sourcing and social outreach. A few of those questions covered:

- Have you measured your corporate greenhouse gas (GHG) emissions?
- Have you opted to report your greenhouse gas emissions to the Carbon Disclosure Project (CDP)?
- What are your total GHG emissions?
- Do you know the locations of 100% of the facilities that produce your products?
- Do you work with your supply base to resolve issues found during social compliance evaluations and also document specific corrections and improvements?

Carbon Disclosure Project

Some experts have suggested that businesses may be unwilling or unlikely to publicly disclose their emissions. However, the Carbon Disclosure Project (CDP) has been effective at making that voluntary disclosure common among the largest companies in the world. CDP is currently involved with more than 3000 of the biggest corporations in the world in order to promote carbon emissions standards and reduction strategies. The CDP published its 2008 report detailing the emissions information for 1550 large corporations, which represented more than 26% of world's emissions.

One of the keys to the success of CDP in obtaining voluntary reporting comes from the support of nearly 500 of the largest institutional investors. Key stakeholders in the businesses

For more information about the Carbon Disclosure Project

https://www.cdproject.net

including boards, investors and management, regulators, and corporations themselves are looking to utilize this data to aid in decision-making with respect to public disclosure, risk mitigation, industry comparisons, litigation risk, and consumer preferences. Today, the CDP has developed what is understood to be the common reporting standard for reporting GHG emissions.

BIODEGRADABLE VS. COMPOSTABLE

One of the "hot" topics in the green labeling debate has been between describing a product as biodegradable and compostable. What do these different labels mean and how can someone understand whether their product is both, is either, or is neither?

Generally, for a product to biodegrade it will break down into its component organic parts—typically when exposed to some combination of air, light, heat, water, and microorganisms. This means that a product which contains nonorganic materials, by definition, cannot be biodegradable. Composting is one type of process that causes biodegradation—typically the process that is used to create soil, compost, or fertilizer.

Let's first start with what consumers think

Biodegradability: The American Chemistry Council asked consumers to define what they thought a biodegradable product should be. Your standard consumer said a product would be biodegradable if it was able to break down naturally (on its own) in 1 year or less and leave nothing behind, or completely disappear.

Compostability: The American Chemistry Council also asked consumers to define what they thought a compostable product should be. Your standard consumer said a product would be compostable if the product could be put back into the ground to make soil, mulch, or fertilizer that is able to be used in a garden, home, or farm.

What is the reality

The Federal Trade Commission (FTC) in its "Green Guide" has stated that if a product claims to be *degradable*, *biodegradable*, or *photodegradable*, that the materials will break down and return to nature within a reasonably short time after customary disposal.

The FTC goes on to define composting as the process of turning degradable materials into useable compost-humus-like material that enriches the soil and returns nutrients to the earth. For a product to make a claim that it is *"compostable"* the products or packages will break down, or become part of usable compost (e.g., soil-conditioning material, or mulch), in a safe and timely manner in home compost piles. For composting, a "timely manner" is approximately the same time that it takes organic compounds like leaves, grass, and food stuff, to compost. The ASTM (American Standards for Testing Materials) has set out a definition for whether or not a product is compostable. Its definition is a product that is "capable of undergoing biological decomposition in a compost site as part of an available program, such that the material is not visually distinguishable and breaks down into carbon dioxide, water, inorganic compounds, and biomass at a rate consistent with known compostable materials."

The take-away: Consumers do not necessarily grasp a full understanding of these differences and there is not a clear-cut difference in the marketplace—and many consumers simply interchange both of these concepts.

The American Chemistry Council survey did showcase one of the key differences between a biodegradable product and a compostable product (in the eyes of these consumers): A belief that the decomposition would be beneficial to the earth whereby consumers felt that biodegradable byproducts would simply disappear completely (which isn't in fact a correct assessment …).

These confusions will increase the focus on these green claims (and already have, based on recent suits by the FTC). Companies should plan on greater labeling requirements and a need to emphasize scientific research for any claims of biodegradability or compostability. The Biodegradable Products Institute (BPI) has developed a certification for compostable products based on the ASTM requirements, including a labeling scheme to aid consumers.

19

Venture Capital and Clean Technology

Venture capital has played a key role in advancing the innovation economy of the United States. According to the National Venture Capital Association, U.S. companies that received venture capital between 1970 and 2005 accounted for approximately 10 million jobs and $2.1 trillion in revenue in 2005. And, although venture capital represents a mere 0.2% of Gross Domestic Product (GDP) of the United States, venture-funded companies account for nearly 17% of GDP.

And the venture capital industry has become heavily invested in clean technology—so much so that as of the end of 2009, the sector was the largest single investment category of venture capitalists. The growth is expected to continue with clean technology investors anticipating an average of nearly 20% growth in the number of new, first round clean technology investments in 2010, according to a survey of 40 clean technology venture capital firms by Reuters.

What does this mean for green businesses? Well, it certainly isn't the silver bullet that suddenly will fund anyone with a good, green idea. Venture capital remains a source of funding for a narrow slice of businesses, green or otherwise. And while advancing technologies from solar, biofuels, green chemistry, advanced batteries, and more has become a substantial segment of the venture capital sector, perhaps as few as 1% of green businesses seeking venture capital will ever raise a single dime from institutional investors (Table 19.1). But if your business fits the model, venture capital can provide a key source of funding to grow a clean technology and green business.

Green Venture Capital Industry

Venture capital investment into clean technology companies has steadily grown in recent years. As recently as 2005, less than $500 million in venture investment was focused on clean technology annually. By 2008, that number figured had skyrocketed to more than $7.6 billion according to Greentech Media—growth of more than 1000% over that period (Figure 19.1). Estimates are that over 100 venture capital and private equity funds are now investing in clean technology companies.

For more information about venture capital and raising money from venture capitalists, go to

- Raising money from venture capital on page 336.
- Developing your fundraising strategy and plan on page 343.

And with such growth has come suggestions that the clean technology investment sector is overhyped. The struggles and failures of some heavily venture capital funded companies (biofuel company Imperium Renewables and solar developer Ausra, just to name a few) led some pundits to describe the growing clean technology investments as a bubble. And while investments in clean technology slowed in connection with the global economic downturn in 2008 and 2009, funding into the space has continued to outpace other investment sectors (Table 19.2).

TABLE 19.1

Venture Capital Investments by Sector (2009)

Sector	Investments[a]	Deals	Average Deal Size[a]
Solar	$1415	84	$16.9
Biofuels	975	44	22.2
Automotive and transport	543	29	18.7
Batteries, fuel cells and storage	455	36	12.6
Energy Efficiency and smart grid	401	34	11.8
Green buildings	143	10	14.3
Wind	142	17	8.4
Green materials	131	9	14.6
Water	130	33	3.9
Lighting	115	16	7.2
Green information technology	106	10	10.6
Environmental technology	77	7	11.0
Green finance and project development	59	2	29.5
Geothermal	35	2	17.5
Carbon markets	26	8	3.3
Tidal	22	3	7.3
Nuclear	9	1	9.0
Green consumer products	3	2	1.5
Miscellaneous	57	9	6.3
Total	$4857	356	$13.6

Source: Adapted from Greentech Media.

[a] In million dollars.

Should we expect this growth to continue or is it truly a bubble? Only time will tell. But many of the key fundamentals that venture investors look to remain. In a 2009 editorial in Greentech Media, Michael Kanellos suggested a number of reasons why he expected investment to continue:

1. Certain clean technology investments don't require substantial amounts of capital, including smart grid companies, software companies managing building controls, home retrofits and energy consumption, and solar companies such as Skyline

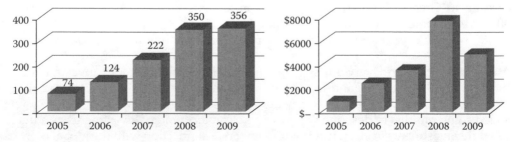

FIGURE 19.1
Clean technology venture capital investment deals and investments (in billions of dollars). (Data from Greentech Media).

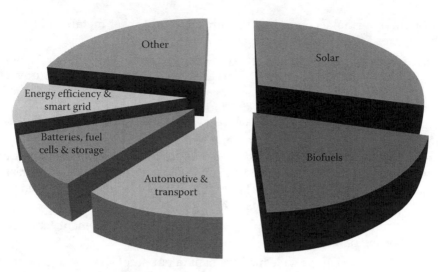

FIGURE 19.2
Venture capital investments by sector (2009).

TABLE 19.2

Sample of Recent Funds Launched or Announced Investing in Clean Technology

Firm	Approximate Fund Size[a]
Advanced Technology Ventures	$303
Altira Group	176
CMEA Ventures	400
Crosslink Capital	400
Curzon Park Capital	28
Dawson and Fielding	120
Demeter Partners	250
Element Ventures	486
Environmental Technologies Fund	218
Exel Venture Management	125
First Reserve Corp	9000
Index Ventures	440
Khosla Ventures	1000[b]
KSK Emerging India Energy Fund	199
Masdar	250[b]
Nature Elements Capital (China)	350[b]
Olympus Capital	250
Pine Brook Road Partners	1430
RockPort Capital Partners	450
Sindicaturn Carbon	280
Virgin Green Fun	199
Wellington Partners	389
Wheb Ventures	102
Yellowstone Capital Partners	50
Zouk Solar Opportunities	83

[a] In million dollars.
[b] Announced.

Solar (who now make concentrators with auto body sheet metal) and Cool Earth (who utilizes Mylar in its technologies).

2. Clean technology is like biotechnology in that it relies on the commercial development of basic sciences. As proven in other sectors, government labs, universities, and even large corporate entities such as utilities or conglomerates do not have proven track records at commercialization of technology. As Kanellos noted, "Private equity funds and government-sponsored funds, meanwhile, do best with meat 'n' potatoes concepts like retail and real estate. Look at Optisolar. The company deliberately avoided VC funds and was strictly funded by private equity. It burned through over $300 million. When it went under, a lot of the manufacturing equipment was still in the crates. The technology wasn't great. It did, though, have one really valuable asset: the rights to develop desert land."

3. Venture capital understands how to work with the management teams of new technology companies.

4. Venture capitalists can provide management talent necessary for scaling the businesses. In companies such as Aurora Biofuels and Amyris, former executives from oil companies were taken on board to run the businesses; and First Solar appointed a former Intel executive.

5. There have already been some proven successes for investments in Clean Tech. For example, First Solar had seen its stock grow 10 times over after its IPO and had its revenue and earnings exceed expectations.

6. Exits exist for clean tech companies. Leading companies will continue to integrate horizontally (adding key segments of the supply chain). Large conglomerates such as GE, IBM, Siemens, Allied Materials, Intel, and BP have each indicated they intend to make acquisitions to expand into key segments within the clean technology space. While the IPO market has slowed during 2008 and 2009 for all companies (including clean tech businesses), experts continue to suggest that clean technology businesses will have opportunities to raise funds on both domestic and international financial markets (Table 19.3).

7. Failure is tolerated in the venture investment model. The clean technology sector continues to offer substantial opportunity. While those opportunities may come with sizeable risk, selecting the right team and technology, coupled with ongoing

TABLE 19.3

Clean Technology Venture Capital Deals by Region (2008)

	Regions	No. of Deals
1.	Silicon Valley	77
2.	New England	38
3.	Northwest	26
4.	Los Angeles/Orange County	22
5.	Midwest	21
6.	Colorado	19

Source: Adapted from PricewaterhouseCoopers/National Venture Capital Association MoneyTree™ Report.

support from the investors can provide 10× returns. And, in the event failure occurs along the way, it is simply the nature of doing business.

Importance of Venture Capital for Clean Technologies

While venture capital is not the "silver bullet" for green businesses, the support of these institutional investors is very important to advance technologies that may otherwise find it difficult to obtain other backing needed to take relatively high-risk technology from the concept stage to commercialization. Like biotechnology, medical devices, software, and computer hardware before it, the investors are looking at the huge opportunity as the potential pay-off for their investments.

Venture capital is a key driver of innovation and entrepreneurship. Many of the most successful companies today simply wouldn't exist without this funding. Therefore, venture capital is instrumental for accelerating high technology businesses—a model that other countries across the globe have begun to leverage to further technology development in their local economies (Table 19.4). Below are a few of the key benefits that venture capital firms provide to their portfolio companies:

- *Money.* Cash is king and providing cash to startup businesses that, if successful, offer handsome rewards to the investors helps to align the incentives for the parties to these transactions. The reason for the crucial role of venture capital in high technology businesses is that few other sources of capital have risk tolerances that allow investment into risky, early stage companies. The venture model works because the high rewards realized from a relatively few investment "wins" offset the investments that are losers. High technology businesses require cash to develop their technologies, scale their businesses, and reach new customer streams. While there may be less expensive sources of financing, venture capital money remains an integral source of that cash necessary for many of today's clean technology startups.

- *Advisory.* The partners of venture capital firms tend to be individuals that have a mix of entrepreneurial, business, and investing experiences. This skill set often uniquely positions a venture capitalist as an excellent resource to provide operational, strategic, or practical guidance for a new startup company. In addition, as a venture capitalist will often focus on an industry or technology, the VC may be uniquely situated to share knowledge, insight, and information across multiple portfolio companies facing shared challenges and obstacles.

- *Contact networks.* Venture capitalists provide a powerful resource in their personal and firm networks. From these contacts, VCs are able to tap into potential employees and management team members, companies to be partners or customers, or can leverage market intelligence among their portfolio companies. And while a venture capitalist may be an early source of funds for the company, as the company looks to raise additional funding, many of the sources of fresh capital will result from introductions from the venture capitalist's network.

TABLE 19.4

Most Active Venture Capitalists in Clean Technology (as of December 31, 2006)

Firm	Location	Dedicated or General	Type	No. of Reported Clean Tech Investments
Nth Power	West	Dedicated	VC	20
RockPort Capital Partners	East	Dedicated	VC	19
SAM Sustainable Asset Management	Worldwide	Dedicated	VC	15
Chrysalix Energy	Canada	Dedicated	VC	14
Draper Fisher Jurvetson	West	General	VC	12
Perseus LLC	East	General	VC	12
Harris & Harris Group	National	General	VC	11
NGEN Partners, LLC	West	General	VC	11
Altira Group LLC	West	General	VC	10
E2 Venture Fund	Worldwide	General	VC	9
EnerTech Capital	East + Canada	Dedicated	VC	9
OPG Ventures Inc.	Canada	Dedicated	Corp.	9
Technology Partners	West	General	VC	9
CDP Capital—Technology Ventures	Canada + France	Dedicated	VC	8
DFJ Element	National	Dedicated	VC	8
Kleiner Perkins Caufield & Byers	West	General	VC	8
NGP Energy Technology Partners	East	Dedicated	PE	8
OnPoint Technologies	East	Dedicated	VC	8
Vantage Point Venture Partners	National + Canada	General	VC	7
BDC Technology Seed Fund	Canada	General	VC	7
Conduit Ventures	UK	Dedicated	VC	7
SJF Ventures	East	Dedicated	VC	7
Sustainable Development Technology Canada	Canada	Dedicated	VC	7
3i	Worldwide	General	VC	6
Braemar Energy Ventures	East	General	VC	6
Fonds de Solidarite FTO	Canada	General	VC	6
Hydro Quebec Capitech	Canada	General	Corp.	6
Intel Capital	Worldwide	General	Corp.	6
NEA	Worldwide	General	VC	6
Polaris Venture Partners	National	General	VC	6
Siemens Venture Capital	Worldwide	General	Corp.	6
Ventures West	Canada	General	VC	6
Angeleno Group	West	Dedicated	VC	5
Apax Partners	Worldwide	General	PE	5
Asia West LLC	Worldwide	Dedicated	VC	5
Commons Capital	East	Dedicated	VC	5
Cordova Ventures	South	General	VC	5
Cornell Capital Partners	Worldwide	General	VC	5
EcoElectron Ventures	West	Dedicated	VC	5
Pangaea Ventures Ltd.	Canada + East	Dedicated	VC	5
Sevin Rosen Funds	National	General	VC	5
Solstice Capital	National	General	VC	5
Venrock Associates	Worldwide	General	VC	5

Source: Adapted from Silicon Valley Bank, Cleantech Primer 2007.

GREEN BUSINESS VENTURE CAPITAL ODDS

How many green businesses take venture capital?

- Approximately 115 green businesses received a first investment from a venture capital firm in 2007 (not follow-on investments), according to the Pricewaterhouse Coopers/National Venture Capital Association MoneyTree™ Report.
- Approximately 131 green businesses received a first investment from a venture capital firm in 2008.

What are the odds of receiving venture capital?

- It is difficult to know the exact number of companies competing for first sequence venture funding. But, according to venture industry statistics used within the industry *only 1% to 3%* of companies seeking venture capital are successful in receiving funding.

How many other green businesses are looking for venture capital investment?

- We can estimate that between *4000 and 13,000* green businesses are competing for those approximately 130 new venture capital investments made annually into green businesses.

Should Your Business Pursue Venture Capital?

Venture Capital is not a good fit for every company (in fact, probably not for most companies). Venture capital is expensive, has a fairly limited time horizon, and tends to operate in a selected group of industries, technologies or fields. Before investing substantial time, resources, and energy into attracting venture capital, it is extremely important that an entrepreneur understand the market and the types of companies that make sense for venture capital. Countless excellent companies have been built without venture capital—either because other funding alternatives were available or because the company was not a good fit for venture funding.

To find out if venture capital might be a good fit for your business, take a test at

- Venture capital "fit" test on page 361.

However, certain high technology business ideas will require both substantial upfront time and money investments in order to perform research, development, and testing efforts. And, in many of these cases, venture capital offers the right mixture of capital and expertise to help an idea grow rapidly to a significant size. Moving a technology from a laboratory to a prototype may involve well over a year of time and millions of dollars. Finding a traditional bank loan program or living on savings alone is simply unrealistic for many startup companies. Therefore, obtaining venture capital financing is a necessity for certain businesses in the high technology arena.

Certain companies are considered to be "natural" fits for venture capital—oftentimes the business involves a former CEO or technical lead from another successful startup company. And if you aren't the founder of a company that sold for a billion dollars, how can you tell if you are a "natural fit" for venture funding? VCs are looking to mitigate their risk. Venture capital firms are looking for a management team they can trust to execute on a business plan they believe can provide high returns. Does the business have a team that can execute on a business idea with the potential for a big reward? What does the team have that mitigates the risk to the investor?

The box below describes a few of the key factors that venture capitalists point to as instrumental in their decision-making process. Management has been and continues to be a key factor for investors. Proven startup experience (and hopefully startup success) is valued by the venture community. But also note the weight the investors polled below placed on factors like market size and growth potential.

WHAT DO VCs LOOK FOR IN A COMPANY?

Profit Dynamics Inc. surveyed venture capital firms and asked them to rank five important factors that influence their decision to invest, with 5 being the highest rank you could award a factor, 1 being the lowest.

These factors were:

- Quality of the management team
- Size of the company's market
- Proprietary, uniqueness, or brand strength of the company's product
- Return on investment
- Company's potential for growth

The percentage of respondents that awarded "5" to a factor were as follows:

- Management—52%
- Return on investment (ROI)—42%
- Market size—27%
- Growth—25%
- Uniqueness—19%

In overall average score, the factors were ranked as follows:

- Management—4.1
- Return on investment ROI—3.5
- Market Size—3.3
- Uniqueness—3.3
- Growth—3.0

Source: Profit Dynamics Inc.

20

International Landscape

To say that green business is a global opportunity may understate the importance of the international landscape on the green movement. Many scientists predict an increase in global natural disasters attributable to climate change. Countries such as China and India expect to see their energy needs explode in the next 50 years. To meet those increasing demands in developed and developing nations, substantial public and private investment will be necessary. In fact, experts estimate that over $400 billion in government stimulus funds have been earmarked for clean technology projects—and only 10–12% of those funds reside in the United States. This means that huge opportunities exist around the globe for green businesses.

It is simply impossible to paint a full picture of the international landscape and opportunities in green business. So instead of trying to tackle everything international, we'll highlight a few key stories, trends, opportunities, and case studies in the international side of green.

Green Goes Global

According to a survey of U.S.-based venture capital funds investing in clean technology, over three quarters of survey respondents believe the United States will be the best market for green businesses over the next five years. Eighty-eight percent said America was the best place to base a green business in the next five years. Additionally, most of the top green entrepreneurial companies continue to come from the United States.

According to the Guardian's Top 100 Cleantech Firms (the full list is available at the end of this chapter), 55 of the top 100 cleantech firms were headquartered in the United States in 2009; 13 came from the United Kingdom; 10 from Germany; 5 from Israel; and rounding out the top 5, four came from Sweden. China did not have any companies on that list, but India did have two companies make the top 100.

But that information certainly doesn't tell the full story. Shawn Lesser of Sustainable World Capital has ranked the top 10 countries for green business in 2009, on the basis government initiatives and programs, large investment mandates, entrepreneurial innovation, and cultural and social drivers (Table 20.1). While the strong venture capital backing within the United States provides a solid base for the clean technology businesses, the mixture of government initiatives and various cultural and social drivers actually led five European nations to rank ahead of the United States, including Denmark, Germany, Sweden, the United Kingdom, and Switzerland.

For more information on selling products and services abroad, see Chapter 10, Making the Sale

- Selling your products abroad on page 127.

TABLE 20.1

Top Countries for Green Business

	Country
1	Denmark
2	Germany
3	Sweden
4	United Kingdom
5	Israel
6	Switzerland
7	United States
8	United Arab Emirates
9	China
10	Canada

Source: Adapted from Sustainable World Capital (2009).

International Treaties

Research by scientists in the 1970s and 1980s began to detail a growing problem for the earth: Global Warming. As carbon dioxide and various other gases had built up in the atmosphere since the beginning of the Industrial Revolution, global temperatures too had begun to rise. Scientists expressed concerns over the damages that could result including rising oceans, droughts and famines, hurricanes, melting glaciers, loss of wildlife, and ultimately devastation to human populations. Since greenhouse gases were not the problem of just one nation, it was recognized that the issue required a global consensus, most likely under the auspices of the United Nations.

The names of three locations have (or soon will) become synonymous with the issue of an international climate change treaty: Kyoto, Copenhagen, and Mexico. The history of the international efforts to combat climate change on a global scale began in earnest at the Earth Summit in 1992 held in Rio, which resulted in 154 countries agreeing to voluntary reductions in greenhouse gas emissions to 1990 levels by the year 2000. But, the process of enacting binding reductions and setting forth a path toward global carbon dioxide emissions cuts has been full of bumps.

Five years after the initial agreement at the Earth Summit, the Kyoto meeting was held to set out specific targets for emissions reductions. At that meeting, the Kyoto Protocol set out legally binding targets for reductions of greenhouse gas emissions for the European Union and 37 industrialized nations, including the United States. The Protocol called for an average reduction in GHGs of 5% over 1990 levels prior to 2012—with the EU agreeing to an eight 8% cut, the United States to seven 7% and Japan to six 6%.

Four years later, a treaty was agreed to on the implementation of the Kyoto Protocol, but the United States refused to ratify the treaty leading many to question the value of a treaty without the world's largest emitter of carbon dioxide. In spite of those limits, the treaty was ratified by sufficient countries to receive the consent of countries emitting at least 55% of greenhouse gases. In 2005, the Kyoto Protocol took full effect with 185 countries as signatories.

While the Kyoto Protocol represented a major step forward, the cuts that were agreed upon were relatively minor and failed to include the two largest emitters of greenhouse gases, the United States and China. As a result, it was agreed to revisit the issue of climate change in Copenhagen in 2009—a discussion which was to include both China and India, each of whom represented major emitters. While many had anticipated that the parties would agree to a climate treaty including the United States, the EU, China, Brazil, and India, such agreement was not reached. And while some progress was made including the drafting of the Copenhagen Accord, there was no legally binding agreement which came out of the Copenhagen meetings. The 2010 United Nations Climate Change Conference will be held in Mexico. The aim of this meeting is to continue to progress on the issue of climate change at this meeting, but many experts note that the failures in Copenhagen do not bode for a quick resolution of the future binding emissions cuts through the United Nations governance.

European Union

While it is easy to combine all of Europe into a single group for purposes of understanding the green business environment, in fact much of the impact still depends on the specific country in question (Table 20.2). In Germany, nearly three-fourths of all companies have some green business policies in place while in the United Kingdom, less than half of all companies have any policies in place. In France, fewer than half of companies measure environmental credentials, while over 85% of companies in Spain and Germany require these credentials.

But Europe does provide us with an example of the impact of climate change regulations. With the ratification of the Kyoto Protocol, Europe has committed to reduce the emission of carbon by 20% of 1990 levels by 2020 and some experts predict that this level could be increased in connection with a potential global climate change treaty. Europe's implementation of the world's first mandatory carbon cap-and-trade system experienced its share of problems (including the over allocation of carbon emissions credits), but also

TABLE 20.2

2008 Cleantech 100: Best in European Class

	Company	Sector	What They Do	Country
1	Odersun	Solar power	Design and manufacture of thin-film flexible solar cells	Germany
2	Deep Stream Technologies	Distribution and management	"Embedded intelligence" circuits for power management	UK
3	CamSemi	Electricals	Low cost, low power standby mode technology	UK
4	SiC Processing	Industry	Hydrocyclone technology to improve solar cell production	Germany
5	Marine Current Turbines	Marine power	Tidal turbines	UK
6	Sulfurcell Solartechnik	Solar power	Thin film solar technology	Germany
7	Pelamis Wave Power	Marine power	Wave energy technology	UK
8	Solarcentury	Solar power	Mass market solar technology	UK
9	Nujira	Electricals	Low-power mobile-phone and radio transmission	UK
10	Atraverda	Electricals	Conductive ceramics for power storage	UK

Source: Adapted from the Guardian/Library House.

has grown into a substantial market. The UK estimated the global market for carbon to be more than a $6 trillion market in its *Low Carbon Economy Report*. As companies continue to provide low carbon products and services, more businesses are making money from carbon regulations—leading to new business opportunities across the continent.

These opportunities have led to an increase in private investment. In 2009, the clean technology investment sector grew to the second largest venture capital sector for European venture capital funds trailing only biotechnology. 2008 saw more than 150 institutional investors in Europe make their first investments into a European clean technology business. And the trend looks to continue as substantial government stimulus money has been earmarked for green business opportunities throughout Europe. European nations now represent some of the leading innovation centers for technologies including wind, tidal, and solar energy generation.

China

China represents a number of things in the green business revolution. On one hand, it represents a huge opportunity for companies looking to supply clean products to meet China's growing appetite for energy, transportation, fuels, and other products and services. On the other hand, it represents a huge threat to established green businesses in the United States, Europe, and Asia. And throughout all of this runs a thread of uncertainty—will China fully embrace the green revolution or will economic development win out at the expense of environmental damage? Tom Friedman points out in his writings that China is poised to leapfrog the industrialized world based solely on the clean technology opportunity—a wakeup call that government and corporate leaders need to address.

China is the leading emitting nation for carbon dioxide—exceeding the United States (second) and Russia (third). Yet, on a per capita basis, Chinese citizens still lag behind Americans, which creates a huge potential risk as the Chinese population gains wealth and requires more resources. China holds vast amounts of coal reserves and expects to build hundreds of coal-fired power plants in the near future to meet its growing need for electricity. Estimates also show that the number of automobiles on China's roads will triple by 2020—to more than 150 million vehicles which would account for a fifth of global carbon emissions.

While these statistics could be a cause for grave concern, many instead view these as opportunities. Large opportunities exist within China for the generation of electricity from renewable sources, green transport options for their growing populace and for an increase in energy efficiency for current and future power usage. The Chinese government has already proven it is able to enact effective environmental reforms, including its efforts to clean up emissions prior to the 2008 Olympic Games held in Beijing. And efforts such as an electric vehicle initiative and others do provide opportunities for global companies to partner with Chinese businesses for green reforms.

Renewable generation will be a huge opportunity in China. In 2008, China doubled its capacity for wind energy to become the fourth largest producer of wind power. China also produces nearly one-third of all of the world's photovoltaic solar panels annually. But, it is without question that renewable generation is not enough in itself to keep up with the growing need for power in the country. Numerous old, run down coal-fired plants continue to operate in China and more coal-fired power plants are planned throughout the country. Many foresee opportunities to implement more "clean coal" technologies on both new and existing coal facilities. China does recognize these challenges, establishing the National Energy Commission in 2010 to oversee energy security and development and coordinate with international efforts.

Mass transportation remains a priority for the government, and as fast as new train lines are be built more continue to be needed. To address the growing concern of an increasing demand for automobiles, a $3 billion dollar initiative was enacted by the government and led to the purchase of nearly 15,000 electric vehicles by a dozen cities throughout China.

Energy efficiency also represents an immense opportunity. China's energy intensity (the amount of energy needed to create one unit of gross domestic product (GDP)) has fallen by 75% since 1990, now placing China in the mid-range for energy efficiency among the world's largest economies. But as more electronics, consumer durable goods, home heating and cooling, and other luxuries of a growing middle class become standard "must-haves," energy efficiency technologies will be crucial to slow the growing rate of energy consumption. The government has set specific national and local level policies to further improve energy intensity, and appears willing to provide subsidies and direct investment into these technologies.

Thinking Globally about Intellectual Property

In today's global economy, the importance of thinking internationally about your intellectual property cannot be understated—particularly for green businesses that may be looking to Europe and Asia for future product growth. International efforts oftentimes entail individual efforts in many foreign jurisdictions. In some cases, these efforts are substantial and ongoing compliance is significant. For instance, some countries will require you to have a local attorney to file your application; others may require regular filing updates and reports. Before undertaking a strategy to acquire international protections, be certain you have a process in place to manage your filings and international compliance—many times your attorney may be able to assist you with these ongoing efforts.

There are some entities that can assist with efforts by providing a streamlined process for submitting multiple applications. These include the Patent Cooperation Treaty which provides a simplified process for filing for protections in over 100 countries. In Europe, the European Patent Office (under the European Patent Organisation) which was established by the European Patent Convention of 1977 offers a single patent filing process for its member countries, while the Office for Harmonization in the Internal Market provides trademark registration in European Union member states.

International Considerations for an IP Strategy

Listed are some key steps a startup company should consider in developing their intellectual property strategy:

- *Utilizing the Paris Convention.* Most of the key nations that an investor would be likely to consider for patent filings are members of the Paris Convention. The

Convention provides entrepreneurs with one year from the date of a patent application filing in any of its member countries to effectively file on that first date in other member countries. For a U.S. company filing an application with the USPTO (without public disclosures prior to your filing), you'll have a year to file in other countries even if you subsequently make a public disclosure. Nevertheless, the costs of a broad international patenting strategy may be a hurdle to a filing in many countries. Be aware of the costs of the strategy and consider the costs to protect your key technologies in important markets.

- *Selecting a Global Mark.* A startup can save itself a great deal of effort in international mark registration by researching its mark internationally early. While you may not choose to register your mark in other jurisdictions at formation, by selecting a mark that allows for international expansion, you may prevent future issues from arising.

- *Disclosure Risks.* One of the key differences between the United States and international patent regimes is that most other countries give priority to the "first-to-file." *In the United States, the "first-to-invent" is given priority.* Therefore, be aware of the risks of disclosing your invention prior to filing in international markets. While you would not be barred following disclosure within the United States, you may be barred internationally. Putting your product on sale or marketing your product for sale could immediately bar your product from patent protections abroad—so *be certain to file in the United States prior to sales efforts if you intend to file for patents abroad.* One area that startups may not realize is that an issued patent will qualify as "publication" in international jurisdictions. Therefore, if you intend to file for patents abroad, be certain to make those filings prior to the publication of your patent in the United States.

For businesses that intend to develop an international patent strategy, be certain of the markets before spending the money to file in various international countries. Some experts suggest that the total cost to file between four and six patent applications in foreign jurisdictions may cost a company upward of $100,000 when you include translation costs, filing, fees, agent costs in appropriate countries, and so on. In the event you do not file a patent in a foreign jurisdiction, you still do have the rights to prevent the importation of foreign patented products in the United States.

Raising Money from Foreign Markets

Raising money in the public markets has continued to evolve in recent years, with a growing number of U.S. companies looking to foreign markets as a source of funding. In particular, clean technology companies have looked to the public investment markets in the United Kingdom and Canada as a source of funds.

There is no consensus about listing on a foreign market, with experts weighing in on both sides of the debate regarding the benefits and obstacles companies face in listing on a foreign public market. Even so, more U.S. green businesses have considered the Alternative Investment Market (AIM) or the Toronto Stock Exchange (TSX) as fundraising tools. The two TSX markets boast some 114 clean technology companies currently listed on its exchanges (of the approximately 3500 total companies listed). Some of these are small companies that would be unlikely to list on the large U.S. exchanges and others have turned to the TSX because of the inability to raise mezzanine funding after raising venture capital. However, between 2005 and 2009, only 14 clean technology companies became listed on a TSX exchange according to the Cleantech Group.

For many companies that are currently unable to consider listing on Nasdaq or NYSE, a foreign market such as the AIM may provide another source or method to raise money. Even if your company does not meet the traditional criteria for listing on a public stock exchange, you may still have options in this area. Increasingly, established stock markets are accommodating smaller, fast-growing companies on specially designed sub-markets. Two good examples of this trend are the TSX Venture Exchange, associated with the TSX and the AIM, a sub-market of the London Stock Exchange. Increased internationalization in markets and investing has caused stock exchanges like these to compete globally for the business of fast-growing, entrepreneurial companies. With options in different markets around the world, going public is no longer the exclusive privilege of established, traditional companies.

The AIM market and green business

Between 2005 and 2009, 49 green companies became listed on AIM, including six U.S.-based companies according to the Cleantech Group

This section discusses some of the benefits and considerations of listing on one such market, the London AIM, and briefly describes the process for listing. AIM provides an alternative for fast-growing companies seeking to list on a public market. Its streamlined regulatory scheme was designed with smaller, entrepreneurial companies in mind, and it provides an option for companies that wish to raise capital through a public market but do not meet the requirements for listing on a traditional exchange. At the same time, as a sub-market of the London Stock Exchange, AIM offers its companies an opportunity to gain visibility in Europe and around the world. Today, more companies are giving serious consideration to a foreign listing on markets such as the AIM and other alternative markets.

During 2006 and 2007, many experts suggested that the AIM market may become *the* public market for clean technology companies. Between 2005 and 2009, 49 green companies became listed on AIM, including six U.S.-based companies according to the Cleantech Group. However, in 2008 and 2009, just four clean technology companies were listed on the AIM market due to a variety of reasons including the macroeconomic slowdown and issues experienced by the initial companies listing on the exchange with raising additional funds outside of the AIM.

Introduction to the London Alternative Investment Market

The AIM was created as a sub-market of the London Stock Exchange in 1995, and has rapidly grown, attracting over 2000 companies and raising over £20 billion. The primary attraction for companies has been AIM's simplified regulatory environment, which has made going public possible for many smaller, fast-growing businesses that would not have met traditional market-listing criteria. AIM has attracted companies from around the world, particularly in the wake of the passing of Sarbanes-Oxley in the United States in 2002, which added restrictions on listing publicly in the United States.

Benefits of Listing on the AIM

Listing (or "floating," as it is also called on AIM) can provide a number of benefits for a company. The most obvious benefit is rapid access to capital for further growth. Access to capital can be had first when a company lists initially on AIM, and later through further capital raisings.

Additionally, there are a number of secondary benefits to listing on AIM. It is a global market whose membership is increasingly international. Listing on AIM provides visibility on the world stage, and signifies that a company has a particularly increased presence in Europe. AIM might be an appropriate market for a U.S. company that is seeking to go public and is contemplating expanding in Europe in the near future.

Going public on AIM also provides opportunities to repay investors and employees for their support. Listing allows early stage investors a chance to realize their investment, and also rewards employees with stock option plans or other stock-based incentives.

Considerations, Costs, and Downsides of Listing on the AIM

At the same time, listing on the AIM is a major commitment that entails costs and potential downsides. Any prospective company should carefully consider what an AIM listing will require and evaluate the risk before proceeding.

First, be prepared for the regulation and closer scrutiny that comes with having shares traded on a public market. Management must be comfortable with the AIM's requirements for communication, which seek to ensure that the market is appraised of the company's financial status and business prospects and thereby allow investors to make informed decisions on the value of the company's shares. The AIM requires a listed company to promptly notify the market of any development that could impact the company's share price. Additionally, the AIM requires a listed company's financial statements to conform to internationally recognized accounting standards, such as the U.S. GAAP or the UK GAAP.

Second, any prospective company should also be prepared for regular trips to London for meetings with analysts and investors. Frequent trips can lessen the time to focus on business and customers in the United States.

Third, it is important to remember that while part of the AIM's attraction for U.S. companies lies in the avoidance of strict U.S. regulations such as Sarbanes-Oxley, U.S. companies that conduct an IPO on the AIM are still subject to some U.S. securities laws. The SEC considers U.S. securities rules and regulations to have "extraterritorial reach" in many situations. These extraterritorial regulations include restrictions on certain transfers of shares (depending on whether the shares were issued in the AIM IPO or before) and the possibility that a U.S. company will inadvertently become a "reporting company" under U.S. regulations (requiring registration with the SEC and periodic reporting). These restrictions can increase the cost of the AIM listing and may negatively affect the liquidity of a company's shares on the AIM.

Finally, be aware that the AIM has been criticized in the past for being too lax with regard to the requirements it makes of its listed companies. Admittedly, some of this criticism has come from executives of rival exchanges. The initial success of the AIM silenced some of these critics, and the AIM itself has announced that it intends to continue strengthening its monitoring of listing companies. Nevertheless, consider whether any residual doubts about the regulatory regime of the AIM would impact business or ability to raise money through an AIM listing.

**2009 Global Cleantech 100: Recognizing the Most Promising Private Clean Technology
Companies around the World**

Agriculture

Exosect	Natural pesticides	Winchester, UK
Jain Irrigation Systems	Land Management	Jalgaon, India

Air and environment

Luca Technologies	Emissions control	Colorado, USA
Neosens	Monitoring/compliance	Toulouse, France

Energy efficiency

Albeo Technologies	Lighting	Colorado, USA
Alertme	Buildings	Cambridge, UK
BridgeLux	Lighting	California, USA
CamSemi	Other	Cambridge, UK
ChromoGenics Sweden	Glass	Uppsala, Sweden
ClimateWell	Buildings	Hägersten, Sweden
Cpower	Other	New York, USA
d.light design	Lighting	New Delhi, India
EnOcean	Buildings	Munich, Germany
EPS Corporation	Other	California, USA
Ice Energy	Buildings	Colorado, USA
Novaled	Lighting	Dresden, Germany
Nujira	Other	Cambridge, UK
Powerit	Buildings	Seattle, USA
Serious Materials	Buildings	California, USA
SynapSense	Other	California, USA
Ubidyne	Other	Germany
Verdiem	Other	Washington, USA

Energy generation

AltaRock Energy	Geothermal	California, USA
Amyris Biotechnologies	Biofuels	California, USA
BioGasol	Biofuels	Ballerup, Denmark
BrightSource Energy	Solar	California, USA
ChapDrive	Wind	Trondheim, Norway
Chemrec	Biofuels	Stockholm, Sweden
Cobalt Biofuels	Biofuels	California, USA
Concentrix Solar	Solar	Freiburg, Germany
Coskata	Biofuels	Illinois, USA
G24i	Solar	Cardiff, UK
Gevo	Biofuels	Colorado, USA
GreenVolts	Solar	California, USA
Hellatek	Solar	Dresden, Germany
Infinia	Solar	Washington, USA
IQWind	Wind	Bazra, Israel
LS9	Biofuels	California, USA
Marine Current Turbines	Hydro/Marine	Bristol, UK
Mascoma	Biofuels	New Hampshire, USA

Nordic Windpower	Wind	California, USA
Odersun	Solar	Frankfurt, Germany
Pelamis Wave Power	Hydro/Marine	Edinburgh, Scotland
QuantaSol	Solar	Kingston, UK
Sapphire Energy	Biofuels	California, USA
Solaire Direct	Solar	Paris, France
Solarcentury	Solar	London, UK
SolarCity	Solar	California, USA
SolarEdge Technologies	Solar	California, USA
Solazyme	Biofuels	California, USA
Solel Solar Systems	Solar	Beit Shemesh, Israel
Solexant	Solar	California, USA
Solyndra	Solar	California, USA
Sulfurcell Solartechnik	Solar	Berlin, Germany
SunEdison	Solar	Beltsville Maryland, USA
Tigo Energy	Solar	California, USA
Xunlight	Solar	Ohio, USA
Zeachem	Biofuels	Colorado, USA
Ze-Gen	Other	Massachusetts, USA

Energy infrastructure

eMeter	Transmission	California, USA
Enphase Energy	Management	California, USA
GridPoint	Transmission	Virginia, USA
Power Plus Communications	Management	Manheim, Germany
ResponsiveLoad (RLtech)	Transmission	London, UK
Silver Spring Networks	Smart Grid	California, USA
SmartSynch	Transmission	Mississippi, USA
Tendril Networks	Transmission	Colorado, USA
Trilliant	Transmission	California, USA

Energy Storage

A123 Systems	Advanced batteries	Massachusetts, USA
ACAL Energy	Fuel cells	Cheshire, UK
Bloom Energy	Fuel cells	California, USA
Boston Power	Advanced batteries	Massachusetts, USA
Deeya Energy	Advanced batteries	California, USA
Electro Power Systems	Fuel cells	Turin, Italy
EnStorage	Fuel cells	Zichron Yaacov, Israel
Imara	Advanced batteries	California, USA
Pentadyne Power	Hybrid systems	California, USA
ReVolt Technology	Advanced batteries	Zurich, Switzerland

Manufacturing/industrial

Metalysis	Smart production	Rotherham, UK

Materials

Xylophane	Biofuels	Gothenburg, Sweden

Recycling and waste

MBA Polymers	Recycling	California, USA
Recupyl	Recycling	Grenoble, France
SiC Processing	Recycling	Hirschau, Germany

Transportation

Achates Power	Vehicles	California, USA
Adura Systems	Vehicles	California, USA
Better Place	Fuels	California, USA
Fallbrook Technologies	Vehicles	California, USA
Fisker Automotive	Vehicles	California, USA
Tesla Motors	Vehicles	California, USA

Water and wastewater

Aqwise	Wastewater treatment	Herzliya, Israel
Arvia Technology	Water treatment	Liverpool, UK
Danfoss AquaZ	Water treatment	Nordborg, Denmark
Epuramat	Water treatment	Luxembourg
HydroPoint Data Systems	Water conservation	California, USA
Inge	Water treatment	Griefenberg, Germany
Microvi Biotech	Wastewater treatment	California, USA
MIOX	Water treatment	New Mexico, USA
NanoH$_2$O	Water treatment	California, USA
Oasys	Water treatment	Massachusetts, USA
Ostara Nutrient Recovery Technologies	Wastewater treatment	Vancouver, Canada

Source: Adapted from the Guardian and Cleantech Group.

21

Growth Opportunities

As green businesses continue to grow from the niche categories into multimillion dollar revenue companies, certain greentrepreneurs will need to consider the possibility of future exits, fundraising, and growth. Of course, if your business is not a good candidate for public offerings and mergers, then these growth opportunities may not be on your mind. However, if you business is looking for venture capital, needs to utilize project finance or perhaps is looking for opportunities abroad, the needs of your business may obligate you to look at these opportunities.

Let's be clear here that doing an IPO is a bit like finding that proverbial needle in a haystack these days. Hundreds of green companies have taken venture capital in the past decade and only a handful of those companies have been listed on the U.S. public markets. A few more have been listed on the AIM or TSX (discussed in Chapter 20). And still others have been acquired or sold to another business. So the idea that any green business is a viable public company or merger candidate is by no means a given.

For more information about certain growth opportunities, see Chapter 31 on M&A and IPOs, and go to

- Deciding to Pursue an IPO or Merger on page 389.
- Background on Mergers and Acquisitions (M&A) on page 392.
- Background on Initial Public Offerings (IPOs) on page 397.

This chapter examines the opportunities for acquisitions and mergers in the green business space and offers details about just what it takes to go public or be acquired these days.

Growth, Liquidity, and Exit Events

For certain companies taking funds from institutional investors, the concept of a liquidity or exit event will be important. For most new companies, predicting a future event such as an initial public offering on a U.S. or foreign market or a future merger or acquisition event is simply impossible. However, these events represent important considerations for a company, particularly for a private equity-backed or venture-backed company. Investors understand that investing in a private company offers them the opportunity to supply necessary capital for a company they hope will produce exceptional gains in the future. Yet, these investors also understand that it is very difficult for an investor to recoup these gains if the company stays as a private entity without available liquidity in the company's stock.

Owning stock of a privately held company is very different from holding stock in IBM, Disney, or Ford. In order to avoid registration with the SEC and other state regulators, private companies must place restrictions on the transfer or sale of their stock. While this helps private companies avoid the expensive process of registering with the SEC (as well as providing ancillary benefits of a more manageable capitalization structure of the company), these restrictions will prevent investors, executives, and other employees from selling their ownership of the company. Therefore, after a startup business reaches a certain level of sustained growth and consistent results, many key stakeholders may begin to be looking for an opportunity to gain liquidity in their shares of the company. For this reason, even an early stage investor will want to be certain that the company has considered potential options for a future liquidity event.

Not all merger and acquisition activity results in the startup company being acquired by a larger company—there are oftentimes cases where a medium-sized startup company will acquire or merge with another medium- to small-sized competitor with the aim to better compete in the marketplace. And, oftentimes, these events will not represent a liquidity event for any shareholders, but instead signal a new direction and could be coupled with an additional infusion of investment dollars or new technologies and products.

Both an IPO and a sale of the company offer founders, employees, and investors liquidity in their investments and new sources of capital for the company. Determining whether such a major transaction is in the best interest of your company and which type of transaction to pursue depends on a number of short and long-term strategic, financial and other considerations. Once a company has decided which transaction is in its best interest, many of these same considerations will affect the timing and, in the case of a sale of the company, the structure of the transaction. There is no one-size-fits-all decision tree, and things can change quickly as your financial projections, capital needs, and market conditions change as well. This is a major decision for the company and the entire process will take a significant amount of attention from the company management which may be a distraction from normal business operations and growth.

PUBLIC MARKETS

While your initial vision may be of brokers frantically trading your shares on the floor of the New York Stock Exchange, there are actually several markets in which you can choose to "go public."

- *New York Stock Exchange (NYSE).* The NYSE is the largest stock exchange in the world by volume of dollars.
- *National Association of Securities Dealers Automated Quotations (Nasdaq).* Nasdaq is the largest stock exchange in the United States by number of companies and is a popular exchange for IPOs since the listing fees are generally cheaper than in the NYSE.
- *American Stock Exchange (AMEX).* Generally smaller companies are listed here due in part to more liberal listing requirements.
- *Foreign Stock Exchanges.* If you are an international company, you might also consider foreign stock exchanges such as the London Stock Exchange (LSE) (and the LSE Alternative Investments Markets (AIM)), Toronto Stock Exchange (TSX), Tokyo Stock Exchange, and other more regional markets.

As the company considers the options for growth, many times the opportunity will come down to a form of a liquidity event: going public with an initial public offering (IPO) or selling the company under a merger or acquisition (M&A). In each case, the company will be looking for additional capital (provided by outside investors or through the synergies of combining with another company) and the added benefit of some level of liquidity for the stockholders.

As a company begins to reach this level of maturity, new issues and considerations will arise. These efforts may require retaining an investment bank or financial advisor to assist with sourcing new opportunities. The company will probably need to hire staff with experience in public company compliance, particularly in the accounting and finance staff. The company may desire to hire additional sales and marketing staff to showcase higher growth. These new considerations will require changes within the organization and necessitates advanced efforts.

Clean Technology M&A and IPOs

One of the reasons that venture capitalists have been willing to make clean technology the largest investment sector has been the fairly robust market for initial public offerings and merger and acquisition activity within the green business space (Table 21.1).

According to the Cleantech Group, there were an estimated 505 clean-technology M&A transactions globally in 2009. Those 500 deals amounted to approximately $31.8 billion in transaction values. While macroeconomic factors may have led to decreases in other sectors, in North America, Europe and Israel, the number of clean technology M&A deals was up from 2008 (although the total transaction value of those deals decreased over 2008). China saw M&A activity increase to all-time levels with 29 M&A cleantech transactions totaling $5.5 billion.

In 2008, clean technology companies raised an estimated $5.1 billion in 16 IPOs, according to Cleantech Group data. The public offering market in 2009 struggled, but advanced battery-maker A123 Systems completed its IPO in late 2009 and saw a positive return on the first day of the offering. Late 2009 and early 2010 saw green businesses including Solyndra, Dago New Energy, Tesla, and JinkoSolar file paperwork with the SEC to list on an exchange. Some 215 clean technology companies made a public offering on one of the various global markets in the past decade—including 190 of those just in the last five years of the decade according to the Cleantech Group. Of those 190 companies listing worldwide, 53% were in the energy generation sector.

CLEAN TECHNOLOGY COMPANY IPOs

There are several public markets that companies can utilize to raise funds, including regional markets around the globe. The most common major markets for clean technology companies to raise funds have been the New York Stock Exchange (NYSE), the Nasdaq, the American Stock Exchange (AMEX), the London Stock Exchange and its submarket the AIM as well as listing over the counter (OTC). Across all those markets, only 120 green businesses "went public" on one of those markets between 2005 and 2009. The breakdown of those listings are below:

- NYSE—15 listings
- Nasdaq—33 listings
- AMEX—3 listings
- Over-the-counter—2 listings

TABLE 21.1

Largest Global Clean Technology M&A Deals of 2009

Acquiring Company	Target Company	Amount ($)	Deal Type	Description
Bord Gais (Ireland)	SWS Natural Resources (Ireland)	720 million	Acquisition	Irish energy provider Bord Gáis completed the purchase of SWS Natural Resources, one of the largest wind generators in Ireland, in December 2009.
Statkraft (Norway)	StatoilHydro (Norway)	741 million	Divesture	Norwegian energy companies Statkraft and StatoilHydro joined forces to develop a 315 MW, offshore wind farm off the coast of Norfolk in the UK. The wind farm will consist of 88 turbines and is planned to start production.
Germany1 Acquisitions Limited (Germany)	AEG Power Solutions (France)	775 million	Acquisition	Germany1 Acquisitions Limited, a special purpose acquisition company, acquired 100% of the shares of AEG Power Solutions, a powercompany with a product line that includes solar inverters.
Great Lakes Hydro Income Fund (Canada)	Brookfield Renewable Power (Canada)	809 million	Divesture	The Great Lakes Hydro Income Fund bought the Canadian renewable power generation assets of Brookfield Renewable Power. The portfolio includes 15 hydroelectric plants and a soon-to-be-constructed wind power project.
Suez Environment (France)	Agbar (Spain)	1.3 billion	Acquisition	France's Suez Environment, a water and waste management company, acquired Barcelona, Spain-based water company Agbar. The deal makes Suez a leading player in Spain's water sector.
Bunge (USA)	Moema (Brazil)	1.4 billion	Acquisition	Bunge, an agricultural company that supplies raw materials and services to the biofuels industry, entered into an agreement to become the 100% owner of Brazillian sugar producer Moema.
Statkraft (Norway)	Sodra (Sweden)	1.5 billion	Joint Venture	Swedish forestry and timber product company Södra and Norwegian energy company Statkraft signed a renewable energy agreement in the third quarter that initiates a range of investment projects and cooperative agreements in the energy sector, including the expansion of wind power activities and the supply of district heating, biofuels and electricity.
Mainstream Renewable Power (Ireland)	FPC Services (USA)	1.7 billion	Divesture	Mainstream Renewable Power, a renewable energy company based in Ireland, acquired a portfolio of wind farm projects from Illinois wind farm developer FPC Services. The portfolio has a potential capacity of 787 MW.
GCL-Poly Energy Holdings (China)	Jiangsu Zhongneng Polysilicon Technology Development Co. (China)	3.4 billion	Acquisition	Chinese power company GCL-Poly Energy Holdings acquired Jiangsu Zhongneng Polysilicon Technology Development, China's largest polysilicon and solar wafer maker.
Panasonic (Japan)	Sanyo Electric Co. (Japan)	4.6 billion	Acquisition	Panasonic acquired a majority stake in Sanyo Electric., the world's largest rechargeable battery maker. The takeover makes Panasonic a dominant player in the fast-growing market for hybrid car batteries.

Source: Adapted from CNBC.

- Toronto Stock Exchange (both markets)—14 listings
- London Stock Exchange (LSE)—4 listings
- LSE AIM Exchange—49 listings

Source: Cleantech Group LLC.

A number of sizeable clean technology companies are expected to consider listing on an exchange in the near future including biofuel company Codexis, smart grid network companies Silver Spring Networks and Trilliant, solar thermal company BrightSource Energy, lithium-ion battery maker Ener1 Inc., Chinese solar companies Dago New Energy and Trony Solar Holdings, LED lighting company Bridgelux, solar companies Nanosolar, SunRun, and Solar City, and solar inverter innovator Enphase, among others.

TOP POTENTIAL ACQUIRERS OF CLEAN TECHNOLOGY COMPANIES

One of the reasons new investors have come into the clean technology space is the availability of exit opportunities—including by acquisition. In 2005 and 2006, 45% of the acquisitions of clean technology companies had buyers who were already fully or partially active in the clean technology space, while 42% were by acquirers who had not previously been active or exploiting the clean technology space, according to research by the Silicon Valley Bank.

Michael Kanellos of Green Tech Media has developed the following list of companies poised to make a splash acquiring green startups.

1. General Electric
2. Siemens
3. Applied Materials
4. Taiwan Semiconductor Manufacturing Co.
5. Valero
6. Toshiba
7. Philips
8. Cisco Systems
9. IBM and Intel
10. SunPower

How to Prepare Your Business

As a green startup company continues to mature, most companies will reach a stage whereby a substantial infusion of cash is necessary to continue the growth of the company. High tech companies may reach a point where they will begin to consider an "exit strategy" for the founders and investors—a point where the company will move from a privately held entity or independent entity, to a company whose stock will trade on public markets or will operate as a part of a larger entity.

At this stage, companies will begin to consider a number of options. In most cases, companies will look for an infusion of cash from an initial public offering on traditional domestic stock markets such as Nasdaq or the NYSE. Other companies may pursue an alternative fundraising course from an international stock market such as the TSX or the AIM. Still other companies will entertain acquisition and merger offers, opting to leverage the size and capital of a larger entity with greater resources.

The Next Cleantech IPOs?

- Codexis
- Silver Spring Networks
- Trilliant
- BrightSource Energy
- Ener1 Inc.
- Dago New Energy
- Trony Solar Holdings
- Bridgelux
- Nanosolar
- SunRun
- Solar City
- Enphase
- Tesla

Preparing for a major transaction such as an IPO or a sale of your company is like preparing to sell a house: you'll attract more buyers and obtain a better return if you engage in some house cleaning and maintenance prior to the transaction. In particular, companies that have been engaged in excellent corporate governance throughout their existence will find this process much less tedious (and probably much less expensive). For your company, this means resolving any potential problems with corporate or financial matters, management structure, employment issues or anything else which may deter potential buyers or investors.

Underwriters and investors in an IPO and potential acquirers in a sale of the company underwriters and investors will review: the company's corporate organization, quality and experience in management, internal financial and accounting controls, growth trends and potential, the health of the company's industry, and the company's competitive position within that industry. While some factors such as industry health are beyond the control of management, the company can be proactive in shoring up these other factors prior to the IPO process.

Either before or during the transaction, underwriters, investors, or the acquiring company in the case of the latter will perform a thorough legal and financial review of the company through the "due diligence" process. You can anticipate potential problems in due diligence by reviewing these items with the help of your legal counsel in preparation of an IPO or a sale. This will allow the IPO or sale transaction to go more smoothly once underway and may provide the company a more favorable valuation by investors or potential acquirers.

Wherever possible, the company should take the necessary steps after its review to rectify any correctable problems you have encountered. This could mean resolutions of the board of directors ratifying prior actions or hiring outside counsel, auditors or consultants to get the company in order. While this may seem burdensome, the problems will be revealed during the due diligence process and addressing as many as possible before IPO or the negotiations process is in the best interest of your company.

KEY CONSIDERATIONS BEFORE A TRANSACTION

- *Corporate matters*—Are the company's organizational documents (articles of incorporation and bylaws) up to date and reflective of the current organizational structure? Are the number and type of issued shares of the company's capital stock consistent with the authorized shares in the articles of incorporation? Have all security issuances been properly approved by the board of directors and other shareholders where appropriate? Are all stock records and minute books up to date? Is the company in good standing in all jurisdictions it conducts business? Does the company have a complex capital or organizational structure that would be difficult for an investor or potential acquirer to understand? If so, is there a compelling tax or business reason for the structure?

- *Financial matters*—Are the company's financial statements correct, up to date and properly audited? Has the company properly filed all necessary federal, state, and foreign tax returns? Are there any liens, encumbrances, mortgages, or other charges on the personal and real property of the company?
- *Management and operations*—Are the company's business plan and financial projections accurate and up to date? Does the company have internal controls and is the company in compliance with those internal controls? Does the company have rights in all the intellectual property it is using, either by patent, trademark, license, or otherwise?
- *Employee matters*—Is the company in compliance with all relevant labor and employment laws? Are all employee confidentiality, intellectual property assignment or non-competition agreements signed and current? Has the company stock option plan and each individual employee stock option grant been properly approved by the board?
- *Insurance*—Are all company insurance premiums, including workers' compensation and directors' and officers' liability insurance, current and sufficient in coverage for the company's needs?
- *Litigation*—Is there any current or pending litigation (consider breached contracts, employment disputes, etc.) that can be resolved prior to the sale of the company?

In the case of an IPO or acquisition by a public company, the company may additionally consider:

- *Board of Directors*—The NYSE, Nasdaq, and SEC place a number of requirements on the composition of the board of directors of public companies with respect to the number of independent directors and the composition of compensation and audit committees. Making some of these changes prior to being subject to the requirements may help ease the transition.
- *Changes in Management*—If the company is lacking in board members or officers with public company experience, it is advisable to hire an executive with that experience to help guide the company through the IPO process. This is particularly useful for a chief financial officer, given public company financial reporting requirements.

Part V

Green Business Fundamentals

22

Market Research and Business Planning

Market Research

One of the keys to building a successful business is displaying an understanding of the business and operating environment of the company—including the competition, the market, the industry, the technology, the key companies operating in the sector, the partners, and the history involved. According to professional investors, several of the most common mistakes they see in business plans stem from a failure to understand their competitors, the market, and the opportunity. Therefore, beginning to identify and investigate this information early will identify the key aspects that will drive the business.

YOUR MARKET RESEARCH CHECKLIST

- Understand the market and industry
- Utilize publicly available research data
- Read industry trade press
- Research trade associations
- Research public company competitors
- Research private company competitors
- Review competing intellectual property

Markets and Industry

Early stage businesses are usually unable to spend large amounts of money on market research and industry analysis reports. Plus few prospective investors will expect detailed market analysis you may see from a well-funded public company. Investors instead expect a startup company to utilize available research and information to identify market estimates to use in its decision-making.

You may decide to start with resources such as Hoover, D&B, Hill Search, and Integra (links and more information below) which provide research reports on specific markets and industries. Industry trade associations oftentimes prepare market research on the specific industry for the benefit of its members. Finally, examine research in industry trade press analyzing various aspects of an industry or market.

USING U.S. GOVERNMENT RESOURCES

- *Bureau of Labor Statistics (bls.gov)*. Identifies trends in hiring by occupation, demographic, geography, and several other criteria.

  Example: If your green product will be targeted at nurses, utilizing the BLS data, you would be able to show that there were 2,505,000 registered nurses in 2006 and that the BLS projects that these numbers will increase to 3,092,000 by 2016, an increase of 23.4%.

- *Census Bureau (census.gov/econ/www/index.html)*. Provides data on the total number of establishments, employment, payrolls, and number of establishments by nine employment-size classes by detailed industry.

  Example. If you plan to market your green product or service to physician's and dentist's offices and will market only to Chicago (your headquarters) in the first year, you can use this information to show that in 2005 there were 6843 physician's offices and 4531 dentist's office in the Chicago metro area.

RESEARCHING INDUSTRY TRADE ASSOCIATIONS

Freely available resources
- *Internet Public Library, Associations on the Net*. Offers a searchable database of various trade associations that maintain Web sites. Access at: ipl.org/div/aon/
- *American Society of Association Executives*. Provides a searchable database known as Gateway to Associations for locating trade associations. The data may be filtered by state, city, country, association type, and keyword. Access at: asaecenter.org/Directories/AssociationSearch.cfm
- *Inc.com*. Provides a listing of industry trade associations broken down into various categories. Access at: inc.com/articles/2001/02/22070.html

Fee-based resources
- *Directory of associations*. Ability to download information to create mailing labels. Prices range from $165 to $695 for access. Access at: marketingsource. com/associations/
- *Encyclopedia of associations*. Leading resource with information on over 100,000 associations world-wide. Fee for use, but may be accessed at various public and university libraries. Access at: gale.cengage.com

Competitors

Any startup company is likely to face a difficult competitive landscape—with competition from larger companies, more established ventures, or international challengers. Underestimating or failing to identify those competitive forces can spell disaster for any business. As a result, one of the key areas any new business needs to focus on is research into the competitive landscape. By understanding how your business stacks up against competitors in the markets, you can identify the opportunities and the primary challenges the business may face.

WHAT IS THE MOST COMMON MISTAKE ENTREPRENEURS MAKE WHEN COMPLETING THEIR BUSINESS PLAN?

- Stating the company had *no competition, or underestimating the strength of competitors*—32% of respondents
- Not clearly explaining the opportunity—27%
- Disorganized, unfocused, or poor presentation—12%
- *Miscalculation of market share and market size*—9%
- Failing to describe a sustainable competitive advantage—9%
- Failing to address the risks of a venture, and failing to contain a contingency plans for coping with the risks—9%

Source: Profit Dynamics Inc.

Public Companies

Research into public company competitors will generally prove easier than uncovering information on private company competitors. Public company disclosure requirements offer an insight into the products, research and development, pricing, sales, gross margins, and other key metrics of the company. Services such as Google Finance, Yahoo! Finance, the SEC Web site, and other online research tools can provide access to publicly disclosed information made available in annual reports (10Ks) and quarterly reports (10Qs). Additionally, many of the research tools identified below for private companies may also provide additional research information for public companies.

Private Companies

Private company research is more difficult to obtain than public company information. However, several sources do offer information on private companies. Many of these resources are available for a fee; however, many public or university-sponsored libraries provide free access to these resources.

ONLINE RESOURCES TO RESEARCH COMPETITORS AND MARKETS

- *D&B's International Million Dollar Database (dnbmdd.com).* Provides information on approximately 1,600,000 U.S. and Canadian leading public and private businesses. Fee for access to the database; however, many libraries have free access to the databases available for library patrons.
- *Hoovers (hoovers.com).* Provides information on public and private companies from sources such as D&B and others. Subscriptions from $75 per month.
- *Hill Search (hillsearch.org).* Provides detailed data on public and private companies in North America and access to current and archived articles from the national and regional newspapers, industry journals, trade magazines, and newswires. $59.95 per month; $650 annually.
- *Integra Information (integrainfo.com).* Provides various private company research information. Costs range from $9.95 to $200 per report, with subscription fees available for access to multiple reports.

Intellectual Property

For any new green business building its strategies around a patent portfolio, the company should research competing intellectual property to identify both companies and technology that will provide challenges to the startup's strategies.

ONLINE RESOURCES TO RESEARCH INTELLECTUAL PROPERTY

Delphion (delphio.com) provides a fee-based web service for searching, viewing, and analyzing patent documents. It provides access to the following sources:

- United States patents and patent applications (full text searching capable)
- European patents and patent applications
- PCT application data from the World Intellectual Property Office
- Patent abstracts of Japan
- Business research about patent applications

The *U.S. Patent and Trademark Office* (USPTO) Web site (uspto.gov/patft/) provides free search functionality of its "Issued Patents" (full image search from 1790 to present; full text search 1976 to present) and its "Published Applications" (2001 to present) databases.

The *U.S. Copyright Office* (USCO) Web site (loc.gov/copyright/) provides free search of its records since 1978.

The *Canadian Intellectual Property Office* (CIPO) Web site (strategis.ic.gc.ca/sc_mrksv/cipo/) provides free search functionality of its CIPO's patent database for records since 1979.

The *European Patent Office* (EPO) Web site esp@cenet® (european-patent-office.org/espacenet/info/access.htm) provides access to documents in any official language of the EPO member states.

Most countries have some web-based services for country-specific research on intellectual property in that specific country. This may involve some searching around (some countries do not have an easily identified Web site or a way to find information on the country's online databases), but with a bit of searching you can oftentimes find a way to access the information you want.

Basics of a Business Plan

According to Professor Scott Shane of Case Western Reserve University, companies that have a formal business plan in place are more likely to survive when compared with those that don't engage in formal planning. While there is no single approach that is the "right" one, most startups tend to engage in some form of business planning even if it never produces a formal business plan for the organization. Oftentimes taking those informal plans and transforming them into a formal plan is less onerous than you might think.

> ### YOUR BUSINESS PLAN CHECKLIST
> - Consider purchasing business plan software
> - Start with a business concept summary
> - Undertake sufficient market research
> - Prepare full business plan
> - Prepare executive summary at the end
> - Keep your business plan confidential

What Are the Three Main Types Of "Business Plans"?

A business plan is used for various purposes and will be prepared for various audiences. Listed below is a summary of three typical types of business planning, a startup business may utilize:

- *Business concept summary*

  The business founder's first business-planning efforts are focussed on laying out the reason for creating the business. This is sometimes called the "back of the napkin" business plan, where you lay out the high level plan and the basic assumptions behind it. Answer questions like: What is the market potential? What niche will our product fill? Who are the competitors? What is the "homerun" potential?

- *External business plan*

  Depending on the needs of the third party (investors, banks, board members, executives, etc.), the company will prepare specific planning documents and resources to convince an outside party to provide funds or other resources for the business. How does your investment fit our needs? What are the key metrics we judge our business by? Why is this approach the "right" move for the investor?

- *Internal (or strategic or operational) business plan*

  The founders and management team will undertake a comprehensive evaluation of the business and prepare a thorough plan to be used to guide the business. What are key the milestones? What are the hiring needs? What will the product development timetable be? What will the financial picture be today, this year, and three to five years from now?

The Business Plan

The business plan will lay out the product or service, the market and competitive landscape for the product or service, and the approach the company will take to tackle that market. And all of this should fit together into a story of a founding team with a unique idea to tackle a problem. The key components of a business plan are:

- Executive summary
- Product, service, or technology
- Market opportunities

- Competition
- Marketing and sales
- Financing and liquidity plan
- Management

Preparing a business plan for green business does involve certain differences from other industries. Many green businesses will be unprofitable for a substantial period during product or technology development. As a result, the focus of the business plan will be on the sizeable opportunity for long-term profit growth upon technology development. As a result, there is oftentimes less of an emphasis on short-term sales efforts and a greater emphasis on market opportunities in the long run. Additionally, expect to include a greater emphasis in the plan on topics such as development timetables and financing needs, protection of the technology, and retention of key research and development personnel.

Building Your Executive Summary

This section will focus on preparation of an executive summary business plan—the one to two page summary of the business for investors and outside parties. For more information, see the "Template Executive Summary Business Plan" that follows this section.

There are a number of resources and software packages designed to help an entrepreneur build a full business plan, both for external and internal purposes. For more information on various software packages for business planning purposes, please see the section "Business Planning Software."

Executive Summary

The executive summary of the business plan is often deemed to be the most important section of the business plan. An investor that receives dozens of business plans in any given month may only skim the full business plan (if they even read the full plan), instead focusing on the executive summary—and it is your job to hook them on the opportunity in those few minutes.

Usually, the executive summary is the final section written and should be limited to one to two pages in length. Focus on what is most important, but be certain that the executive summary adequately explains the business and the opportunity in the marketplace. The primary objective of the executive summary is to offer sufficient information to the reader, who may only read this section to make the initial decision to either reject or accept the plan. If the plan does not have a clear, concise, and compelling summary of the business concept, a plan will likely find its way into the reject pile.

WHAT TO INCLUDE IN YOUR EXECUTIVE SUMMARY

- An *introductory paragraph* to draw in the reader and identify the key opportunity the product or service will fill
- A *company overview*, including information on the organizational form, the date of formation, stage of funding/development, location, and milestones achieved to date

- A description of the *core proprietary or licensed technological* strengths
- A summary short-, medium-, and long-term *operational goals, implementation, and growth strategies*, and plans to attain and sustain market leadership
- A list of the *factors necessary for the business to succeed*
- Information on the *market*, the *market size*, and targeted or likely *customers*
- A clear description of the *key benefits, features, and strengths* of your product, service, or technology
- Profiles of *key management* and advisors
- Information on the *financing needs* of the business, including amounts raised thus far, future funds being raised, and the use for the funds
- A description of the most likely *exit strategies* for the company
- *Financial highlights* and projections
- *Summary* of the technology, team, and market opportunity

Keeping Your Business Plan Confidential and Proprietary

Since getting potential investors to sign a nondisclosure agreement (NDA) is rarely an option, you should take steps to provide some protection on your business plan itself. Use of the language "confidential and proprietary" may not provide the same level of protection as an NDA, but some courts have held that in a scenario where a person or persons were aware that they had been exposed to a trade secret, they would be barred from using it or disclosing it to others without permission. Take steps where possible to protect your proprietary information, and remember that when you cannot get an NDA signed to trust your gut—make sure to ask around before disclosing to a potential investor if you are wary of their reputation.

SAMPLE OF BUSINESS PLAN LANGUAGE

GREENSTARTUP BUSINESS PLAN

CONFIDENTIAL INFORMATION

This business plan is the property of GreenStartup, Inc. and is strictly confidential. It contains information intended only for the person to whom it is transmitted. With receipt of this plan, recipient acknowledges and agrees that:

(a) in the event recipient does not wish to pursue this matter, this document will be returned to the address listed above as soon as possible,

(b) the recipient will not copy, fax, reproduce, divulge, or distribute this confidential plan, in whole or in part, without the express written consent of GreenStartup, Inc. and

(c) all of the information therein will be treated as confidential material with no less care than that afforded to your own company confidential material.

This document does not constitute an offer to sell, or a solicitation of an offer to purchase.

Version: _____
Copy number: _____
Provided to: _____
Signature: _____
Company: _____
Date: _____

Principal Contact:

Name
Title
Address
Phone, Fax, Email

Giving a Potential Investor the Information Necessary to Make a Decision

According to Cindy Nemeth-Johannes, these are the things a venture capitalist wants to read in your executive summary (abcsmallbiz.com/bizbasics/gettingstarted/vencap_eval.html):

- Here's our estimate of the market.
- The reason why you're going to be a market leader.
- How will you benefit your customers? How much will you benefit them? Why will they choose to do business with you?
- How are you going to prevent competition from going after your market and your customers?
- What does your management team look like? What have they already done? Do you have any advisors who will carry a lot of weight?
- How much money you're looking for? What you're going to do with it? What is the current financial situation.
- What have you already accomplished? List up to eight things that you've done and don't restrict yourself to the product itself. Sure, they want to know that you can make the product work, but they also want to know what parts of your team are in place and whether you've come to agreements with any strategic partners.
- What are the milestones you intend to accomplish?
- When do you plan your exit strategy? What do you think the company will be worth then? Keep your numbers realistic or they'll figure that you're just a dreamer, not a doer.

A FEW PRACTICALITIES OF YOUR BUSINESS PLAN

- On the *cover page*, include the following information: (i) company name and logo; (ii) contact information; (iii) date of preparation; and (iv) confidentiality clause.
- Triple check for any *misspellings*.

- Limit the use of *technical vocabulary* or provide appropriate definitions.
- Use *graphs, pictures, images, and charts* to make the business plan more appealing to the eye.
- Provide detail on all the *assumptions* used in your business plan.
- Use *bold, italics, and underline* to highlight text.
- *Extract key provisions* and highlight in sidebars.
- Be sure that the business plan is *pleasing to the eye*.

Template Executive Summary Business Plan

Company name

Contact name
(Address: Street, Town, State, Zip)

Phone: Fax:

Email: Web address:

MANAGEMENT:	
CEO	**Business description:** *Briefly describe the general nature of your company. From this section the investor must be convinced of the uniqueness of the business and gain a clear idea of the market in which the company will operate.*
VP Sales/Mktg.	
VP Product Development	
CTO	**Company background:** *Provide a short summary of your company background.*
CFO	
Etc.	
Industry: Such as cleantech, software, biotech, etc.	**Management:** *List senior management and prior experience.*
Number of employees: #	
Bank: Such as Silicon Valley, Comerica etc.	**Products/services:** *Convey to the investor that the company and product truly fill an unmet need in the marketplace. The characteristics that set the product and company apart from the competition need to be identified (competitive advantage).*
Auditor: Such as PricewaterhouseCoopers, Ernst & Young, etc.	
Law firm: Such as Cooley LLP, etc.	
Amount of financing sought:	**Technologies/special know-how:** *In this section, highlight whatever aspects of your product that may be protected by current IP or patent law. Provide evidence of how your offerings are different and will be able to develop a barrier to entry for potential competitors.*
i.e. $4 million equity	
Current investors: ($Amt. invested) Any venture capitalists, private investors, or personal Funds	
Use of Funds: Such as product development, marketing/sales, distribution, etc.	**Market:** *Provide a clear description of your target market, and any market segments that may exist within that market. Include potential market size and growth rate. Also, mention your revenue model in this section.*

Distribution channels: *Indicate which channels will be used to deliver your products/service to your target markets (i.e. direct salesforce, VARs, channel partners, etc.)*

Competition: *List any current or potential direct and indirect competition. Briefly describe the competitive outlook and dynamics of the relevant market in which you will operate.*

Financial projections (unaudited):

	2011	2012	2013	2014	2015

Revenue:
EBIT
(dollars in thousands):

Business Planning Software

Table 22.1 is a review of some of the most popular business planning software tools.

TABLE 22.1

Business Planning Software Tools

	Software Publisher	Web site	Rating[a]	Relevant Software Titles	Price ($)[b]	Notes
1	Palo Alto	paloalto.com	4.5	Business Plan Pro 11.0 Business Plan Pro Premier Edition 11.0 Marketing Plan Pro 11.0	99.95 199.95 179.95	Provides largest suite of resources; best sellingsoftware
2	Business Resource Software	brs-inc.com	4+	Plan Write for Business Plan Write Expert Edition Plan Write for Hi-Tech Marketing	119.95 219.95 299.95	Offers unique business model evaluation function to identify the strengths and weaknesses of the plan; offers remote collaboration tools
3	Fundable Plans	fundableplans.com	4	Fundable Plans	39.95	Lots of links to appropriate industry-specific information
4	Jian	jian.com	4	BizPlan Builder 2008 BizPlan Financials 10	99.77 69.95	Contains numerous templates

TABLE 22.1   (continued)

Business Planning Software Tools

	Software Publisher	Web site	Rating[a]	Relevant Software Titles	Price ($)[b]	Notes
5	PlanWare	planware.org	4	PlanWrite PlanWrite Expert Edition Plan Write Hi-Tech Marketing Planner	19.95 219.95 289.95	Best software to develop a business plan in an international format
6	PlanMagic	planmagic.com	4	Business 10.0	99.95	Specific plans for hotels, bars, coffee bars, restaurants, resorts, construction, retail, bed and breakfasts
7	SmartOnline	smartonline.com	4	SmartOnline	49.95[c]	
8	Individual Software	individualsoftware. com	3.5+	Professional Business PlanMaker Professional 2008	49.95	
9	KMT Software	send. onenetworkdirect. net	3.5+	OfficeReady Business Plans 2007	9.95	
10	Nova Development	novadevelopment. com	3.5+	Business Plan Writer Deluxe 2006	99.95	
11	Socrates	jdoqocy.com	3.5+	Winning Business Plans	29.95	
12	Atlas Business Solutions, Inc.	abs-usa.com	3.5	Ultimate Business Planner 4.0	9.00	
13	Business-Plan-Success.Com	business-plan-success.com	3.5	Business Plan Success 5.0	39.99	
14	My Business Kit	mybusinesskit.com	3.5	Complete Business Kit	9.95	
15	NetEkspert	store.esellerate.net	3.5	iPlanner 2007	39.95[d]	Only does financials
16	VPS Pro	vpspro.com	3+	VPS Pro	85.95[d]	
17	Adarus	adarus.com	3	Adarus Business Plan	55.95	
18	Village Software	villagesoft.com	3	Business Plan FastPlan	99.95	

[a] Rating is based on a scale of 1 to 5, with 5 being the highest and 1 the lowest. Ratings provided by Home Office Reports (homeofficereports.com/Business%20Plan.htm).
[b] Prices as of January 2008.
[c] Price per month.
[d] Price per year.

23

Forming the Business

Selecting Your Company Name

Selecting the name of the company can be difficult, given that you want to find an available name that has a domain name, no competing trademarks, may translate into other languages, and something that speaks to your product. Based on information from Scott Trimble from Halfagain, with this section provides some insight to new businesses when considering their name. Halfagain LLC is a creator of Search and Affiliate Marketing software based in Portland, Oregon.

YOUR NAMING CHECKLIST

- Check domain availability
- Do some focused brainstorming
- Do a synonym search
- Use a word-combining tool
- Find name and word lists
- Find "puns" and "plays-on-words"
- Examine industry trends and competitor names
- Identify industry lingo
- Try metaphorical naming
- Use the USPTO site to check for any trademarks

Basic Stuff

- Let it be easy to pronounce and spell.
- Make it memorable.
- Don't pigeonhole yourself (Being too specific in the naming of your company or product (example: Dave's 256k Flash Drives Inc. or Portland Flooring Inc.) can hinder growth later).
- Go easy on the numbers.
- Don't use names that could have a negative connotation in other languages (Baka Software Inc. sounds okay in the United States, but won't fly in Japan).
- Stay away from negative connotations.
- Make sure your name doesn't alienate any group (race, religion, etc.).

- Search for existing trademarks on potential names.
- Make sure the domain is either available or purchasable in the aftermarket. Use your favorite registrar or use a bulk domain checker.

Domain Availability

Domain availability is possibly the biggest hang-up to ever happen to naming. Sure, you can come up with great potential names, but can you come up with great potential domains that are available? I won't spend much time on this because it's pretty simple. If you're creating a name for a product or business that will require a .com, be patient, keep trying and you'll start to get a feel for names that are more likely to be available than others. I've also listed some tools below that will help immensely with this.

Focused Brainstorming

Every book out there prescribes brainstorming. However, instead of just sitting back and trying to come up with *any* words that describes your business, focus your brainstorming to answering a set of questions. Answer each by making as long a list or words and phrases as you possibly can. Remember, the longer and more abstract your list, the better off you'll be. So go wild …

- What does your product do?
- What does your industry do, what's its purpose?
- What is your product's benefit to the consumer?
- What will happen for them?
- What will they get?
- What are the "ingredients" that go into your product or service?
- How are you different from the competition?
- What makes you unique?
- What's the lingo in your industry? What are the expressions that are unique to your offering and business?

… add your own as you see fit.

Synonym Search

It's pretty simple, really. Take each and every one of the words you brainstormed above and plug them into a thesaurus like Thesaurus.com. Run through each entry, keeping the words you like, trashing the ones you don't. Put these into a new list, paying attention to name possibilities.

Word Combining + a Cool Name-Combining Tool

After you've done some focused brainstorming and/or a synonym search, try word combining. Pop *all* of your words into a word combiner like My Tool (www.my-tool.com), tweak it's settings to reflect what you want it to show and combine. Depending on how many words you put into the system, you may get a massive list returned to you. To weed through them quickly, you can then hit the button at the bottom and check each domain for availability.

Name and Word Lists to Get Your Juices Flowing

Plenty of great product, company, and Web site names have their roots in other, irrelevant names. Look up "list of _____" in Google or Bing and you'll get more than you can handle. For example:

- Geologic periods
- Fruit or food names
- Types of dinosaurs
- Kinds of rocks
- Latin or Greek roots
- Place names
- Historical figure names
- Zoological names
- Botanical names
- Math or engineering terms
- Astronomical terms
- Animal, fish, or bug names

Think about this abstractly also. If your product is new and unique, what foods or plants have fresh connotations, and so on.

Punning and Plays on Words

A new beer called Tricerahops was recently released. Check out how you can create a name like that. Cruise your focused brainstorm and synonym lists for words that describe/define your product. In this beer example, we might find hops—one of the main ingredients in beer. Then, we can look through lists of animals, foods, places, and so on and see if we get any good combinations, where the words fit seamlessly. In this case, they chose a dinosaur name "Triceratops" and simply changed one letter. Here's an even easier way of doing it …

Groovy Word Tool

Use this tool More Words (www.wordlab.com) and search for any words that contain _____. You can search for anything—search for words that contain "top," or words that have a double "e." Virtually any sound or letter combo you want to find in a word, this site will do it for you.

Meaningful or Not?

For example, Dave's Rocket Repair Inc. has meaning, while Simble Inc. does not. Some say creating a name with built-in meaning is a must—new companies or products need to seem familiar and safe. Others say nonmeaningful names are the best—the name is completely yours, free of meaning (which you can then define), plus, newly coined word names connote innovation.

The jury, as they say, is out. Some things to keep in mind though:

- Newly coined words *can* convey meaning. The most championed of these may be Acura, which was formed from the morpheme "Acu" and finishing with suffix "ra." Acu as a root connotes accuracy or precision, which fits nicely for a luxury car line.
- The creator of the Acura name (Ira Bachrach of NameLabs) is purported to have a list of thousands of combinable morphemes.

A Truly Killer Naming Tool

Word Lab and specifically the Word Lab Tools pages. This Web site is considered to be one of the most powerful naming tools out there. With an absolutely massive list of company names, a morpheme name creator, name builder, and so on, this site is the juggernaut of idea generators.

Metaphorical Naming

Often called metaphorical or lateral naming, it's a branch from the focused brainstorm and often, the coolest names come from this method. It'll take a more creative, abstract frame of mind, so whatever you need to do to break out of your linear comfort zone, do it.

So, after you've changed into your tie dye and stared at your Led Zeppelin poster for a while, grab your focused brainstorm. Here we're going to center on the question—"What does your product, business, or industry *do*." You're going to sequentially take each of the words and phrases you came up with, and come up with *other* things in life that do these things too.

Let me repeat (or rewrite, as it were) that. You're going to take what your business does, and come up with other things in life that do the same thing. Make a list of everything you come up with.

Here are some examples:

- I have a *software company* and our newest product's function is to copy files (pretty high tech, I know). So I ask, "What else in life copies things?"

  A copier—too logical.

  A cell—might work, but a little "out there."

  A mime—a ha!

  Why not call the new software product—Mime.

- My *marketing company* helps its clients' voices get heard above the competition. So, what else gets voices heard or makes things louder?

  A bullhorn.

  A volume dial.

  An Amplifier—a ha!

  Why not call the company Amplify Interactive (happens to be a real company here in Portland). Volume Media wouldn't be bad either.

Misspellings

Misspellings of commonly used words can get you in familiarity's proverbial backdoor. Example—netflix.com. It's familiar, short and you instantly know what they do. Though, if looking for an available domain, you'll have to use some fancy combinations because common misspellings are already registered.

Industry Lingo

Each industry has its lingo and you may have noticed that many taglines come from this, or more distinctly, those words and expressions that are used by your consumers.

For example, I've just developed the perfect fish hook. It never, and I mean *never* lets a fish go. A common expression in fishing when you feel a fish take your bait is "fish on." This great expression combined with something else, might make a nice tagline for my fail-safe hook. How about—*Fish on … never off.*

Ask Your Friends, but …

Ask your friends' opinions, but take them with a grain of salt. First of all, your pool of test subjects is probably pretty small, leaving your results (ratio of yays to nays) with little accuracy. Second, consider whether your friend is in your target market. If they're not, they may not "get" a name which might be perfect for your market. Finally, people in general side with what's familiar. Finding your Web site, seeing an advertisement, or having a friend suggest your product can have the unique ability of making your product's name sound good. The name or names you ask your friends to grade won't have the benefit of this.

How Is the Competition Named? What Are the Trends?

Many new businesses have made the mistake of not checking the competition first and creating a name, only to find out the new name is *just* like a competitor's. Time wasted. A general rule is to find out how competitors are naming themselves and simply be different. Stepping out of the box is always a bit of a gamble, so make sure you're different in what will be seen as a positive way.

Name Rhyming

Rhymed names are memorable and can work, as long as they're not too cute or overboard. Rhyme Zone (www.rhymezone.com) is fantastic for finding words that rhyme. More Words (http://www.morewords.com) can also be good for this.

Web 2.0 Name Generators

Some people have found success through name generators, but others haven't had much luck. One example is Web 2.0 Name Generator (http://benjamin.hu/w2namegen.php). If you have a few extra minutes though, try popping some of your synonyms into the interface and seeing what it comes up with. At the very least it might give you some ideas and get your wheels turning.

Don't Put *Too* Much Stock in Your Name

They're certainly important, but naming can also be overemphasized. There are plenty of highly successful businesses and products out there with bad names. So, take your naming, like your friends' opinions, with a grain of salt. And, as with everything, the more you stress about obtaining perfection, the less likely you'll come up with that killer name that seamlessly fits your offering.

Check for Potential Trademarks

If you really like the name, don't forget to do an initial search to see if you can trademark your name—and check to see if someone else is using it. Don't assume that just because the domain name is available or that the name is available to be registered in the state that you are free to use the company or the domain name. It may seem like a lot of work to hunt through trademark databases to confirm a name is available, but the alternative is picking a great name, starting to build your business only to receive a "nasty-gram" from the lawyers of another company who claim you are violating their trademark.

Domain names and the use of meta-data on your Web site have been the source of serious headaches for today's technology companies. Companies have been found to be liable for infringement if they use a domain name or corporate/product name that is close to or similar to an existing trademark. If your site sells services or products that would confuse or mislead a consumer with another trademarked product, your company may be held to be liable.

Domain names are also protected under the Anticybersquatting Consumer Protection Act and could lead to criminal penalties for your business. In certain circumstances, the registrar company may even be able to have your domain name transferred over to the owner of the similar mark (even if your mark is stronger in the marketplace). Before using a domain name for your Web site or using a corporate/product name, be sure to conduct a trademark search (or work with an attorney to have this done). And remember to review both state and federal databases.

HOW CAN I FIND IF A TRADEMARK IS AVAILABLE?

Begin with a preliminary search of your mark within the United States.

- *United States*: U.S. Patent and Trademark Office Web site: http://www.uspto. gov. Once on the site, click on "Trademark" and then "Search."

It may be helpful to obtain a full search from a company that provides a full-service trademark search (this will prevent you from spending the time and money if your mark is unable to be protected). You may want to share this information with your attorney.

Check out international conflicts. For searching trademarks in other countries you can use the following free public databases:

- *Australia*: http://www.ipaustralia.gov.au/trademarks/search_index.shtml
- *Benelux*: http://www.internetmarken.de/bxmarks.htm

- *Canada*: http://strategis.ic.gc.ca/cipo/trademarks/search/tmSearch.do
- *European Union*: http://oami.eu.int/search/trademark/la/en_tm_search.cfm
- *Hong Kong*: http://ipsearch.ipd.gov.hk/tmlr/jsp/index.html
- *India* (status only using application number): http://www.skorydov.com/tmr/Status.asp
- *Ireland*: http://www.patentsoffice.ie/eRegister/Query/TMQuery.asp
- *Japan*: http://www3.ipdl.jpo.go.jp/cgi-bin/ET/ep_main.cgi?992599551313
- *New Zealand*: http://www.iponz.govt.nz/search/cad/dbssiten.main
- *United Kingdom*: http://www.patent.gov.uk/tm/dbase/

GREEN BUSINESS IDEAS

EMERGING TECHNOLOGIES AND OPPORTUNITIES IN BIOFUELS

Biofuels

- Biocrude
- Biobutanol
- Biodiesel from waste feedstocks
- Cellulosic ethanol—feedstocks including sugarcane, energy cane, switchgrass, algae, and even trees
- Bio-based jet fuel
- Biomass-to-liquids
- Specialized traits for the creation of "targeted energy crops"
- New feedstocks

Ancillary technologies and businesses

- Distribution
- Storage
- Refining technologies
- Engines and vehicles

Choosing an Entity

At the onset, it may seem like there are lots of choices to consider—and there may be. You can select from various entity types (a limited liability company or LLC, a corporation, a partnership, or more), you can select a state to incorporate in, and you can even get more complicated by creating subsidiaries. And while some startups will decide to create very complex structures with entities of various organizational forms located in various jurisdictions (such as setting up an entity in the Cayman Islands or splitting the company into a subsidiary to hold the intellectual property and license it to an "operating" subsidiary), you've decided to take a simpler approach. So, what next?

YOUR ENTITY CHOICE "RULES OF THUMB"

- Talk to an attorney, accountant or other advisor before forming the entity (to get some basic questions answered)
- Consider a Single Member LLC for a solo, income-replacement company
- S-Corporation for a solo, highly profitable company
- Consider a Pass-through entity such as an LLC or S-Corporation for a company that is not certain to pursue venture capital
- Consider a C-Corporation for a company looking to pursue venture capital in less than two years
- Incorporate in Delaware for a high-growth company
- Incorporate in the state where the headquarters will be if you are less certain of business growth plans
- Incorporate in the Caymon Islands if you've got sophisticated investors who can gain tax benefits from your location

The first step to form a business entity is matching the right business entity with your aims and goals of the business. To help with that effort, below is a list of five basic concepts to consider. Once you've had a chance to review these general concepts and "rules of thumb," you can begin to get more insight into some of the implications of your choices.

RULE #1

If your business is a one-man or one-woman company and you are making (or plan to make) a profit in line with salary at your prior job, consider forming a single-member LLC.

A single-member LLC is like being a sole proprietor for tax purposes (profits and losses will pass through the LLC directly to your tax return), but it will give you protection from certain liabilities. An LLC generally has the simplest tax and accounting rules.

RULE #2

If your business is a one-man or one-woman company and you are making big profits (in excess of a typical salary you'd earn working for someone else), consider forming an S-corporation.

The benefit of forming an S-corporation in this scenario is that this structure will save you on self-employment taxes. You can form the S-corporation by incorporating as a standard C-corporation (or LLC in certain cases) and then making a filing with the IRS. The accounting and taxation are more complex in this setup.

RULE #3

If you are unsure if you'll ever pursue venture capital financing and, if you were to pursue it, believe you'll wait a couple of years, consider forming either an S-corporation or an LLC.

The most important and immediate benefit to incorporating or forming an LLC is liability protections. Both entities offer that protection and allow profits and losses to pass-through to the individual for tax purposes. This approach gives you flexibility to change down the road with limited time and expense.

RULE #4

If your business growth strategy involves raising money from investors (angels and VCs) in the upcoming 6–24 months, consider forming a C-corporation.

An outside investor such as a VC will usually desire to purchase the stock of your company. They will also require that the profits and losses remain with the company and not pass-through to the owners, eliminating the ability to be an LLC or S-corporation.

RULE #5

If you are planning to create a high growth company in need of substantial outside funding, consider incorporating in Delaware.

A majority of venture-funded companies are incorporated in Delaware. There are a number of benefits including greater comfort among investors, well-established corporate laws, and fairly low incorporation fees. If you believe your company will become a venture-backed company, Delaware represents a logical choice. On the other hand, if the plans for your company are less clear and your growth horizon is longer term, plus you are planning to transact business primarily in your home state, consider incorporating (or forming your LLC) in your home state.

Summary of the Rules of Thumb

These five broad concepts should offer you some "rules of thumb" for your choices (Table 23.1). Obviously, your choice of entity is important, but unlike a diamond, it isn't forever. Ultimately, you should make your entity and jurisdiction choices based upon research and critical thinking. But you should not let the decision allow you to lose sleep. And when in doubt talk to your advisors, attorneys, and accountants who can provide further guidance.

Initial Business Filings

Once you have incorporated your business or formed an LLC, you aren't done there. Be sure to comply with filings required by the federal government (EIN), the filings required in states where you are doing business, as well as the various local and state licenses and permits required to operate a business.

TABLE 23.1

Comparison of Entity Types

Characteristics	Sole Proprietorship	C-Corporation	S-Corporation	Limited Liability Company (LLC)
Formation	No state filing required.	State filing required.	State filing required.	State filing required.
Duration of existence	Dissolved if entity ceases doing business or upon death of the sole proprietor.	Perpetual	Perpetual	Dependent on the requirements imposed by the state of formation.
Liability	Sole proprietor has unlimited liability.	Shareholders are typically not responsible for the debts of the corporation.	Shareholders are typically not personally liable for the debts of the corporation.	Members are not typically liable for the debts of the LLC.
Operational requirements	Relatively few legal requirements.	Board of directors, annual meetings, and annual reporting required.	Board of directors, annual meetings, and annual reporting required.	Some formal requirements but less formal than corporations.
Management	Sole proprietor has full control of management and operations.	Managed by the directors, who are elected by the shareholders.	Managed by the directors, who are elected by the shareholders.	Members have an operating agreement that outlines management.
Taxation	Not a taxable entity. Sole proprietor pays all taxes.	Taxed at the entity level. If dividends are distributed to shareholders, dividends are also taxed at the individual level.	No tax at the entity level. Income/loss is passed through to the shareholders.	If properly structured there is no tax at the entity level. Income/loss is passed through to members.
Pass through income/loss	Yes	No	Yes	Yes
Double taxation	No	Yes, if income is distributed to shareholders in the form of dividends.	No	No
Cost of creation	None	State filing fee required.	State filing fee required.	State filing fee required.
Raising capital	Often difficult unless individual contributes funds.	Shares of stock are sold to raise capital.	Shares of stock are sold to raise capital.	Possible to sell interests, though subject to operating agreement restrictions.
Transferability of interest	No	Shares of stock are easily transferred.	Yes, but must observe IRS regulations on who can own stock.	Possibly, depending on restrictions outlined in the operating agreement.

WHAT ARE THE COMMON FILINGS A BUSINESS WILL NEED TO MAKE?

- Obtain your Employer Identification Number (EIN)
- Qualify as a foreign entity (a company incorporated elsewhere) in a state where you are doing business
- Visit "Permit Me" at Business.gov to find state and local required licenses

Employer Identification Number (EIN)

Any employer must have federal employer identification numbers to complete its federal and state tax returns. For your federal EIN number, it can be obtained by filing a Form SS-4 with the Internal Revenue Service. This can be filled out online in well under an hour at https://sa.www4.irs.gov/modiein/individual/index.jsp. (Note: The online filing service is not available 24 hours per day.)

Some states will also have similar filing requirements. In California, employers are required to fill out Form DE-1 with the California Employment Development Department and to register with the California Employment Development Department 15 days after becoming subject to the California Unemployment Insurance Code. A California employer is subject to this provision if it has had one or more employees and paid wages for employment in excess of $100 during any calendar quarter. In New York, employers must fill out Form NYS-100.

To find out some of the state filings that may be required, specifically whether your state has an employer identification number, visit Business.gov. This site provides a service called "Permit Me" that provides links to appropriate state licensing authorities. Simply enter the state and the type of licensing, and it will provide you links to the appropriate state and local authorities.

Qualifying to Do Business in Another State (Other than the State of Incorporation)

Your business will be incorporated in one state and may do all your business in the same state. If that is the case, then you probably won't need to "qualify" or "register" in any other states. However, if you incorporated your business in Delaware, but have your headquarters in California or another state, then you'll probably need to qualify as a "foreign corporation" (a corporation or business not incorporated in that state).

If your entity is considering transacting business in any other state, it should determine, prior to doing business, whether the company must be qualified or registered to do business in such other state. Most state laws provide that a "foreign corporation" (i.e., a corporation incorporated under the laws of another state) may not "do business" within the state unless it qualifies under appropriate statutory provisions. The scope and extent of the company's activities will govern whether qualifications will be necessary.

WHEN DO YOU NEED TO QUALIFY IN A STATE?

Each state has different rules and requirements for when it is appropriate to register your business in that state as a foreign corporation or foreign LLC. However, certain activities will generally require you to register:

1. You do a substantial amount of your normal business in that state.
2. You have an active office, employees, or other facility in the state.
3. You manufacture, build, or assemble products in the state.

Remember that you could also be required to register in the event your business meets the other local requirements of the state.

Although the penalties for failure to qualify vary from state to state, the following penalties may be applicable: a denial of the right to enforce contracts in the state courts; voidability by the other parties of all contracts entered into in such state during the period when the company was required to qualify but did not; monetary fines levied against the company; monetary fines levied against agents or officers of the company; and personal liability of the officers and/or agents of the company for the company's acts in the state.

Even if qualification is unnecessary, the company may be obligated to pay corporate income and other taxes (including sales and use taxes) as a consequence of operating in a state. For this purpose, "operating" in another state may include very limited and tenuous contacts. The states are becoming increasingly aggressive in treating foreign corporations as subject to their taxing jurisdiction based on more activities within their borders. If the company employs persons located in other states, it may be subject to employer wage withholding requirements, worker's compensation requirements and other regulatory requirements. Further, if the corporation owns real or personal property in other states, it may be required to pay property taxes in such states.

State and Local Filings

Many trades, professions, businesses, and occupations are regulated by state law, which will often require that corporations meet various qualifications before granting certain certificates of registration or business licenses. Many cities also require that corporations doing business within the city limits obtain a local business license.

WHERE CAN YOU FIND THE STATE AND LOCAL LICENSES REQUIRED FOR YOUR BUSINESS?

Visit Business.gov which has a service called "Permit Me" that provides links to appropriate state licensing authorities. Simply enter the state and the type of licensing, and it will provide you links to the appropriate state and local authorities.

In addition, most states have a Web site dedicated to doing business in the state that contains this key information.

A couple of examples:

- Washington state has a Master Business License that is a single, simplified application for applying for many state licenses, registrations and permits, and some city licenses.
- California has the California License Handbook, published by the California Department of Economic and Business Development, which lists sources of California licensing requirements, the applicable regulatory agencies and details the licensing process. The California Permit Handbook addresses permits that businesses may be required to obtain, such as certain environmental permits.

TABLE 23.2

Sample State Filing Fees and Expenses (in Dollars) (as of February 2008)

	CA	CO	DE	FL	IL	MD	MA	NY	TX	WA
Incorporation (Corp.)										
Filing fee/state fee	100	125	89	70	150	120	275	125	300	175
Minimum franchise tax	800		35		75	300				59
Annual report filing fee	25	100	25	150	25	125	125	9		10
Total (estimated)	**925**	**225**	**149**	**220**	**250**	**420**	**400**	**134**	**300**	**244**
Fees for expedited filings	*350*	*150*	*119+*	*70*	*100*	*50*	*4.5% fee*	*25+*	*25*	*20*
Formation (LLC)										
Filing fee/state fee	70	125	120	125	500	100	500	200	300	175
Minimum franchise tax	800		200		75					59
Annual report filing fee	20	100		139	25		500	325+		10
Total (estimated)	**890**	**225**	**320**	**264**	**600**	**100**	**1000**	**575+**	**300**	**244**
Fees for expedited filings	*350*	*150*	*40+*	*N/A*	*100*	*50*	*4.5% fee*	*25+*	*25*	*20*
Foreign qualification (Corp.)										
Filing fee/state fee	100	125	160	88	200	100	400	225	750	175
Minimum franchise tax	800				75	300				59
Annual report filing fee	25	100			25	125	125	300		10
Total (estimated)	**925**	**225**	**160**	**88**	**300**	**400**	**525**	**525**	**750**	**244**
Fees for expedited filings	*350*	*150*	*119+*	*70*	*100*	*50*	*4.5% fee*	*25+*	*25*	*20*
Foreign registration (LLC)										
Filing fee/state fee	70	125	130	125	500	100	500	250	750	175
Minimum franchise tax	800				75					59
Annual report filing fee	20	100			25		500	325+		10
Total (estimated)	**890**	**225**	**130**	**125**	**600**	**100**	**1000**	**575+**	**750**	**244**
Fees for expedited filings	*350*	*150*	*40+*	*N/A*	*100*	*50*	*4.5% fee*	*25+*	*25*	*20*

Leaving Your Employer

The following section provides considerations before and after departing your employer to begin your new green business.

YOUR DEPARTING CHECKLIST

- Obtain copies of all employment documentation
- Check for noncompetition restrictions
- Check for employee nonsolicitation restrictions
- Check for customer nonsolicitation restrictions
- Avoid a Breach of Duty of Loyalty
- Do not Expose Trade Secrets

Are There Any Restrictive Agreements in Your Employment Documentation and Do You Have Copies in Your Files (Including Confidentiality Agreements, Noncompetition Agreements, Nonsolicitation Agreements, etc.)?

Soon-to-be founders sometimes overlook or "forget about" restrictive agreements, so it is important to be particularly diligent in performing a search on any possible restrictions. Restrictive covenants are often found in confidentiality and invention assignment agreements, offer letters, employment agreements, and stock or option agreements.

Have You Identified Every Contractual Limitation You Would Need to Observe?

For example, if you are bound by a confidentiality agreement, how does the agreement define "confidential information"? What is the scope of any noncompetition agreement? What is the scope of any nonsolicitation clause? When do these limitations expire and what are their limitations?

Are You Bound by an Employee Nonsolicitation Agreement?

Your new green startup company should never initiate contact with your former coworkers about any job prospects if you or any other members of your founder team are under a nonsolicitation agreement. Disputes regarding employee nonsolicitation clauses often focus on who initiated contact and when. If the candidate initially contacted the company about employment prospects, then generally there will be no violation of the nonsolicitation clause. As a result, it is critical to maintain documentation establishing how the initial contact was made and when (e.g., an email inquiry from the candidate, an internet application, and written notes of a telephone inquiry).

Are You Bound by a Customer Nonsolicitation Agreement?

Your new startup company should not initiate contact with the customers or clients of your former employer if you are subject to a nonsolicitation agreement for the customers of your prior employer. You may announce your new business and your participation, but be cautious about additional solicitation. If a customer does approach you (and not the other way), be certain to document the conversations and keep records as to the formation of the relationship.

Have You Breached Your Duty of Loyalty to Your Current Employer Prior to Departing?

For example, you should not solicit other employees to leave, undertake any work for the new company during work hours, or inappropriately take or gain access to any information. If any such improper activity has occurred, you should promptly consult legal counsel to evaluate whether and how the situation may be corrected, and whether it remains feasible to proceed with forming your business and hiring other employees.

Have You Been Careful to not Reveal Any Trade Secret Information?

You should identify, at a high level, the areas of trade secret information that may be considered sensitive and avoid tasks that would jeopardize the trade secret. You should also explore the time duration of the sensitivity, and plan the timing of forming your business in conjunction with these durations. You should then evaluate whether any work duties you had contemplated undertaking in your new startup could create a risk of using or disclosing such sensitive information in the course of the new job.

24

Founders

Founder Decisions and Agreements

Once you've established the founding team, too often founders forget to have the tough discussions a founding team needs to have. Don't leapfrog these discussions and jump headfirst into the challenges of starting a business. That initial meetings between the founders of a company will oftentimes be rosy and focused on the good times ahead. However, the realities of founding a business are that there will be tough choices challenges are inevitable, and change is likely to be required. Perhaps a founder will depart, the business model will change or the founders will need to give the business a loan. Recognizing the likelihood of these events, founder teams should discuss and address key issues that will inevitably face the business.

WHAT SHOULD COFOUNDERS DISCUSS BEFORE STARTING A BUSINESS?

- *Decision making*: Titles, day-to-day responsibilities, and major decisions.
- *Employment*: Roles in the company, employment terms, quitting your day job, and terminating a founder.
- *Ownership*: Founder investments, allocating stock, vesting, and transfer restrictions.
- *Documentation*: Work with your attorney to document all of these decisions (a handshake agreement won't be enough if conflicts arise).

Decision-Making Matters

I'll Be CEO, CMO, CSO, Director of Business Development, CFO, and Head Cheerleader

The founders should discuss what specific role individuals will play and what title they will hold. In the early stages, each founder will most likely be charged with accomplishing everything from technical development efforts and fundraising to recruitment and office administration. As the entity continues to move along its developmental path, roles will need to be better defined. As such, the founders should discuss both the current scope of activities for each founder as well as the plan—once the company need to expand receives funding or begins to employ additional employees. Founders should be aware that job titles assigned initially to members of the founding team are likely to change and oftentimes the company and its investors decide that adding experienced executives are necessary down the road. Disputes may arise related to the role a founder will play with the company. Spend time discussing and deciding if founders will be joining full-time or will continue to work for their current employers and work part-time on the business. Be aware that individuals may have different expectations of their role and responsibilities to the company in the early stages of founding a business.

Who Decides "What's for Lunch?"

Early on in the life of a startup, the founders may all be heavily involved in all aspects of decision-making in the business. As the business begins to grow, the founders and the company will find that it is most efficient to determine that one of the founders will serve as president and/or chief executive officer and that certain founders will have decision-making authority over decisions in their respective area of expertise.

Who Decides "If We Buy the Whole Pizza Parlor?"

For events that are outside of the day-to-day realm including incurring debt, adding additional investors, approving an acquisition, merger or sale, or entering into certain significant agreements, the founders may agree that such decisions require approval of a majority or a supermajority of the stockholders or the board members (e.g., 67%, 75%, or 100%). If the founders hold stock, they will also be responsible for electing the board of directors.

Employment Matters

"What's My Role?"

Founders should also determine the role each founder will play in the new entity and at what point founders will be full-time employees of the new company. In some circumstances, a founder may serve as simply an initial investor, a board member, an advisory board member or serve as a part-time employee. Founders should be certain to discuss the best course of action for each founder with respect to their role with the new entity and the timetable with which they will assume that role.

"How Many Days of Vacation Do I Get?"

Seems like an odd thing to consider, but founders should discuss the questions related to traditional employment issues: salaries, vacation time, severance, noncompetition agreements, and nonsolicitation agreements. Be aware that your investors may require you to modify certain terms in your employment agreements or offer letters as part of their commitment to provide funding.

Quitting Your Day Job

For many startups, it isn't possible for some or all of the founders to leave their day job in the early days of the business. While everyone may want to jump into the startup with both feet, Ramen noodles may not be for you. So the issue of leaving your current employer raises two potential questions: (1) Do you plan to leave your current employer, and if so (2) when? Founding teams may quickly realize that various founders have differing expectations for leaving their current employers. Sometimes a founder may wish to wait six months or a year before departing. In other cases, a founder may not wish to depart the company until funding is received and simply work part-time until that point. As such, it is important for the founders to determine if each founder will depart their current employer.

Pink Slips

Termination issues are difficult for founders to consider and discuss since no team ever thinks they will have to fire one of their cofounders. Unfortunately, these issues do happen and are made to be much more difficult by failing to address the issue before it becomes a problem. The founders should discuss up front how they plan to handle termination of a founder including what the process will be, what is required to initiate termination and what type of compensation will be provided upon termination.

Ownership Matters

Filling the Company Coffers

The founders should discuss how they plan to handle cash investments into the business. Even if the company does not intend to pay salaries to the founders initially, there will still be startup costs that cannot be avoided. The founders should budget for these costs and address plans for handling these amounts. Some companies have successfully financed this prefinancing startup phase on credit cards; others have used savings; whiles still other companies may look for angel funding to bridge the company's needs until it can successfully raise sufficient startup capital. If the company does need cash to be contributed by the founders, these investments may come as a result of a direct infusion of cash into the company's bank account or with the payment of certain startup costs by the founders. The company should determine how it will handle these investments into the business, when the individuals will be able to be compensated and how they plan to compensate for these payments.

Some companies will leave these amounts outstanding on their books and reimburse the founder upon a successful financing; others may issue promissory notes for this initial capital or these expenses which will be convertible into stock at a discount; and others will treat these amounts as a loan to be repaid when appropriate.

How Much of the Company Do We Own?

Each member of the founding team will contribute something to the venture—from contributions of a business plan, intellectual property, cash or commitments to provide key services for the company during the startup phase of the company. In exchange for these contributions, the company will typically issue each founder common stock in the company. Once issued, unless the founders agree differently, the stock will be owned in its entirety. For this reason, to protect the company and the other founders from a scenario where a founder departs and owns a large portion of the company, many new companies will include vesting on the founders' shares. In these early discussions, you may not need to decide the number of stock to issue to each founder; instead you can decide the percentage of the initial company to allocate to each person (e.g., you may choose to allocate 25% of the company to each of four founders). At a later point when you are prepared to sign founders stock purchase agreements and issue stock, you can determine how many shares you want to authorize in the company, how many shares you want to allocate to your initial stock option plan for future hires, and the number of shares to be issued to the original founders.

EVEN FOUNDERS' STOCK STILL MUST COMPLY WITH SECURITIES LAW

You've decided to give away a single share of stock in your new company to all the members of your collegiate alumni association in order to "spread the word." Could this cause a problem?

Perhaps. Before undertaking any sale or issuance of stock, you should ensure that sale (or gift in the case above) will not require you to register the securities or will be in violation of securities laws. For most private company startups, your stock sales can be exempted from federal and state securities laws. But this requires you to be cautious in your issuance and sale of stock. Simply because your company is small, new, and isn't selling for large dollar amounts, don't forget about the applicable securities laws and your compliance.

For each issuance of stock, or any other security, be sure to check that this sale or issuance does not violate any federal or state securities laws. Each issuance should be made under an applicable federal and state exemption—usually these are not difficult for startups but require continued diligence. And when in doubt, talk to a securities lawyer.

Some of Your Stock Now … and Some of Your Stock Later …

Many founders will choose to add repurchase provisions to the founders stock. This methodology is called "vesting." Companies may choose to apply vesting on all or part of the stock issued to its founders. One of the key mistakes that many startups make when they choose to add vesting to the shares issued to their founders is to miss the § 83(b) election (discussed below). Be aware that you only have 30 days to make the election upon issuance of the shares—and the IRS isn't forgiving to you if you miss it.

THE MISSED § 83(b) ELECTION

You issued stock to the founders subject to vesting terms, but you forgot to file § 83(b) election. If a founder is issued stock and the stock is subject to a substantial risk of forfeiture (per the IRS rules), then the stock purchase isn't complete until this risk of forfeiture is gone. Once your founders stock has vested, the risk is deemed to be gone and the IRS judges the stock purchase to be complete. At this point, according to the IRS, the difference between the original price you paid (let's say $0.01 per share) and today's fair market value after the vesting has run (let's say $10.00 per share). This difference ($9.99 per share) would be taxed as ordinary income. By filing a timely 83(b) election, you are able to avoid this problem.

When you sell this stock at a later date (after it has appreciated greatly), the appreciation would be taxed as ordinary income at almost twice the rate if you'd filed the 83(b) election and the gain was taxed as long-term capital gains! In our example, you'd only pay long-term capital gains rates on the $9.99 per share gain.

In the company's early stages, it is easy to miss 83(b) election filings with the IRS—so make sure that this responsibility is delegated to someone. A § 83(b) election must be filed no later than 30 days following the transfer of property. (Income Tax Regs. § 1.83-2(b).) When you issue stock to founders (subject to vesting) make sure to make this filing or have your attorney do the filing for you.

Note: If your company allows early exercise of options (which is oftentimes done for the advantageous tax treatment), you also will need to file timely 83(b) elections in this case.

"No, You Can't Sell Your Stock"

After you've issued stock to the founders, usually it can be freely transferred to someone else. This means your stock can be split up and possibly distributed to 20 people or even to a competitor. In particular, for startup companies relying on certain exemptions to issue their securities (so they do not need to register with the SEC), this can represent a potentially serious problem. Because of these potential problems, most attorneys will advise you to place restrictions on issued stock. Founders should discuss what type of limitations to place on the ability to transfer their stock. Generally, the founders will agree on a right-of-first-refusal provision which will allow the company to match the price any third party agrees to purchase the stock from the transferred stock. Therefore, if a founder wishes to sell the stock for $100 for all his or her shares, the company would have the right to match that price and keep the shares held by the company. Other companies will agree to have the company, the board of directors, or an independent valuation specialist calculate a repurchase price in the event a founder wishes to sell the stock and the company agrees to purchase it. Additionally, the founders may agree to allow certain transfers to be made by the other founders. In some cases, a founder may wish to transfer his or her stock into a trust fund for his family or directly to a family member for tax purposes. In these situations, if the shares are transferred to a permitted third party, the third party would be bound by the transfer restrictions to prevent further transfers.

Getting It All on Paper

Now that you've agreed to all these provisions, how do you make it all "legal?" This is usually where the lawyers come into play. Oftentimes there is not a single "founders' agreement" that will lay out the terms of your business in a single document. Rather the list of topics of discussion will tend to be handled through a variety of separate documents and agreements. For instance, you would traditionally handle stock ownership issues in your founders' stock purchase agreements, a stockholders' agreement or an LLC operating agreement, including vesting and transfer restrictions; employment issues such as salaries, termination, titles, vacation, severance, noncompetition, and nonsolicitation would all be a part of your employment agreements or offer letters; and confidentiality, invention assignment, and confidentiality are usually addressed in either the employment agreement or a specific confidentiality or invention assignment agreement.

In most cases, your attorney will be able to assist you with drafting certain documents for each of your founders including employment agreements or offer letters, confidentiality and invention assignment agreements, stock purchase agreements, stockholders' agreements, and other operational documents such as your bylaws and your formation document (articles or certificate of incorporation). These are the key documents (among others) where you'll handle each of the topics above.

Founding Team Questionnaire

The following is a sample of some of the issues founders may want to discuss before making the decision to cofound the business. Founders may decide to fill this out individually and then discuss the answers as a group, or may choose to fill out the document together. Some early stage discussions might not involve some of these questions while other companies might have other issues not contemplated here.

Founder name: _____

Personal goals with company: _____

Expectations: _____

When do you plan to "startup"? _____

Business

Name of the company: _____

Business form (Corp/LLC/Partnership; State): _____

Purpose and aims of the company: _____

_____

Initial location of the business: _____

• Would you be open to considering relocating the business?   Y/N

What milestones do you have for the business? _____

What is the fundraising strategy for the business? _____

Roles in the Company

Business lead (i.e., CEO/president): _____

Financial lead (i.e., CFO/controller): _____

Technical lead (i.e., CTO): _____

Marketing/sales lead: _____

Initial directors: _____

Will any positions be full-time initially?   Y/N

• When do you expect to have or be full-time employees? _____

What are the salaries? _____

• Do you expect to defer payment?   Y/N

• Vacation, severance, health insurance? _____

Decision-making

• What decisions required unanimity among the founders? _____

- Who is responsible for day-to-day decisions? _____
- What are typical day-to-day decisions? _____

Current Status

Current job: _____

Have you discussed the new business with your current employer?	Y/N
Have you shared all the employment agreements with your current employer?	Y/N
Is there any overlap between the activities of the new company and your employer?	Y/N

Average hours worked: _____
Current employment salary and benefits: _____
How many hours per week do you expect to contribute to the business initially? _____

Ownership Interest

Do you expect to contribute any cash initially?   Y/N

- If yes, how much. _____

Do you expect other founders to contribute any cash initially?   Y/N

- If yes, how much. _____

How will future cash contributions be handled? _____
_____

How do you expect the equity and ownership of the company to be structured? _____
_____

Do we have vesting on the ownership? _____

- If yes, what are the terms. _____

Founder Financial Status

Assets: _____
Liquid assets: _____
Liabilities _____
Credit history: Excellent/Good/Fair/Poor
Expected money to be contributed to the business in Year 1: _____
Any other personal tax planning issues to consider for company structure? _____

Exit Strategy

How do you value the shares/units upon exit of a founder? _____
Transfer:

- Can you transfer part of your ownership or only the entire stake?   Y/N
- Can you transfer to your spouse/children?   Y/N

Departure of a founder:

- "Shoot out" provision   Y/N

  Party receiving notice must elect either to purchase shares of other party or sell its shares to that party

- Staggered exit provision   Y/N

  Party leaving may be "bought out" over a period of time for cash flow and tax purposes

- "Bring-along" provision   Y/N

  If either party is transferring shares, they must require third party purchaser to offer to buy also the other party's interest at the same price per share

- "Drag-along" provision   Y/N

  If selling parties have at least a certain percentage being purchased 50%/60%/80% plus, they can obligate other party also to also transfer its shares to the same purchaser

What happens in the case of:
        Death/serious illness _____
        Divorce _____
        Sale of the business _____

What happens in the case of a deadlock or unresolved dispute (last resort: the right of either party after a minimum deadlock period for either party to call for liquidation of the company)? _____

Other Matters

Intellectual property
        Contributing IP to the company? _____
        Entering invention assignment agreements?   Y/N
Will the founders execute noncompetition agreements?   Y/N
        What terms? _____
Will the founders execute nonsolicitation agreements?   Y/N
        What terms? _____
Will the founders execute confidentiality agreements?   Y/N
        What terms?_____
Life insurance policies for the founders?   Y/N

Board of Directors

The board of directors is responsible for overseeing the management of the company. The directors represent the stockholders to ensure that the company acts in the best interest of its stockholders. But more than just providing oversight, a board provides assistance for a new company and may offer their experience to assist new entrepreneurs in their business development.

Roles of the Board of Directors

The board of directors serves to provide strategic thinking based on an "outsider's" perspective. It is important to consider each appointment to your board as an important resource and asset for the company. Board member may be able to provide you with key contacts and access to individuals to assist in the development of your business. For high technology companies in particular, research has shown that your board plays an important role in successful innovation and commercialization efforts.

For a newly formed or emerging company, the board of directors can also act to assure key constituents of the company—investors, employees, and customers. Some venture capitalists and other sophisticated investors will consider the directors as part of the knowledge base of a company and may evaluate a company based on the composition of the board of directors. In addition, some investors feel that a board which has at least some directors that have no direct relationship with the company will be better able to protect stockholders' interests with constructive, unbiased guidance.

The role of the corporate board of directors has changed in the past century. Corporations had been understood to be primarily governed by the will of stockholders and the board of directors was the mechanism that acted out that will. By the early twentieth century, that perception had changed and the role of the board began to increase as corporations provided greater direct decision-making authority to the board elected by the stockholders. This role has continued to evolve and today, more than ever before, the board of directors serves a critical role in corporate governance and decision-making.

Boards of directors have seen the importance of their position within corporations increase dramatically, particularly since the inception of Sarbanes-Oxley and its creation of a growing emphasis on effective corporate governance. And while this growing scrutiny and regulation is overtly affecting public companies, many private companies are placing important oversight roles in the hands of their boards at earlier stages than before. And while the legal role of the board to provide oversight and make certain decisions is a key part of the board's duties, a board of directors plays multiple roles for a startup company. Understanding these different roles represents an important first step in working effectively with a board and its directors.

For companies that are organized as an LLC, oftentimes the LLC will define a board of managers in their operating agreement. Generally, much of the information about a corporation's board of directors is applicable for an LLC's board of managers.

BUSINESS OPPORTUNITIES IN THE ENERGY STORAGE SECTOR

- Transportation and vehicles (e.g., automobile batteries; hybrid engines; other vehicles including military, scooters, boats, etc.)
- Consumer products (e.g., electronics; medical equipment; lighting; tools and appliances)
- Computers and communication
- Industrial (e.g., instrumentation; tools; portable power generation; elevators)
- Renewable energy generation (e.g., storage of off-grid or off-peak wind, solar, and other generation)
- Backup power
- Military electricity storage
- Utility storage

Types of Directors

The board of directors will usually be made up of some composition of the following types of directors:

1. *Inside directors.* These are individuals that are most likely part of the day-to-day operations of the company or were at one point involved in the operations. This would include having the CEO or a founder serving as a director.

2. *Outside directors.* The directors are individuals that do not have any role with the organization other than as director and do not serve as an officer of the organization. These individuals may be current or former executives of industry companies, professionals from accountancy, finance, academia or legal careers, or may have experience as a former entrepreneur of another successful company. These individuals are valued for their contacts and outsider perspectives.

3. *Investor directors.* As part of their role as a source of financial capital for your business, investors may serve as directors on your board. They too can provide similar qualities as from other noninvestor outside directors through contacts and prior experience. However, it is also important to note that these investor directors are also responsible for monitoring their investment, and may not have their interests completely aligned with the company at all times.

4. *Board observers.* Due to constraints of size or maintaining a balanced board, you may wish to select individuals to serve as board observers. These individuals will not vote in matters, but will be able to participate in discussions and provide guidance. These individuals can serve a valuable role and help the functioning of the board.

Board Meetings

Private companies have varying frequencies for board meetings. Some companies will have regular monthly meetings, while others will hold them quarterly or even less frequently. While initially startups may choose to have regular monthly meetings with their outside directors, meeting frequency tends to decrease over time. According to research by Noam Wasserman, the frequency of board meetings decreased as the number of rounds of financing increased, as revenues in the current year increased over the prior year, and as the number of years since incorporation increased.

The structure, content, and time of a board meeting will usually be determined by what is going on with the company. However, since boards of directors are often made up of busy individuals, try to set an agenda to keep people on track. This should include:

- *Operating and financial reports.* These reports are an opportunity for the directors to meet and interact with the members of the management team. Approximately 25% of the time of meetings will typically be dedicated to operational and financial matters.

- *Tactical and strategic issues.* This portion of the meeting should generally not include the management team (other than management directors) and should be discussion among the board members themselves. Approximately 50% of the time of meetings.

- *Administrative matters.* This is the formal approvals and more routine legal matters. Approximately 15% of the time of meetings. Oftentimes the board may handle administrative matters as the first order of business since these "routine" matters can oftentimes be forgotten or pushed off, but are actually very important for effective corporate governance.

- *Review of meeting.* This time will be used to discuss the meeting itself and set the agenda for the following meetings, or highlight items that the directors need in response to questions raised at the meeting. Sometimes the outside directors may ask the management directors to depart. Approximately 10% of the time of meetings.

Compensating Directors

The most typical form of compensation for directors is equity. This can come in the form of stock, warrants, restricted stock, stock options, phantom stock, or stock appreciation rights. The most common form of equity-based compensation is stock options. According to a study by the National Association of Corporate Directors of small market technology companies, nearly 80% of companies offered an initial option grant and ongoing annual option grants to directors.

Most public companies will offer cash retainers or bonuses to directors. However, directors in smaller, private companies may not receive cash payments for service on the board or for board and committee meeting attendance. Despite this, some outside directors of private companies will require these payments in exchange for their participation.

- *Board retainer.* For small, public companies, an annual board retainer will traditionally range from $10,000 to $20,000 annually. Oftentimes committee chairpersons receive an additional premium of $5000 to $10,000. In some cases, board members will receive annual retainers that exceed $50,000. These fees have increased in recent years due to the increased workload, a heightened perception of risk, competition for directors, and separate retainers that may be paid to board chairs of board committees.

- *Meeting fees.* Public company boards often compensate board members for meeting and committee attendance. In some cases, private companies will also compensate board members for meeting participation. Generally the fee for public companies range between $1000 and $2000 for a regular board meeting and may be higher in the event that a meeting requires a longer time period or involves a separate committee meeting. For nonpublic entities that do not compensate board members for meeting attendance, some will provide reimbursement for meeting attendance.

- *D&O insurance.* As the risk of liability has increased for directors, many directors today will require that the company purchase director and officer ("D&O") insurance policies.

KEY ROLES OF THE BOARD OF DIRECTORS

There are five key roles a board of directors play within the company, and particularly with respect to a green or high technology startup company.

Legal role played by the board of directors is determined by federal and state laws (primarily state laws governing the state of incorporation, but other laws may also be implicated), rules and procedures set forth in your corporate articles or certificate of incorporation and corporate bylaws, and additional rules and statutes from the Securities and Exchange Commission, Financial Industry Regulatory Authority, or other entities that regulate certain stock exchanges.

Representative role relates to the relationship between the corporate stockholders and the board of directors. As discussed previously, the board of directors historically had been elected by the stockholders as representatives and acted out the will of the stockholders. And while that relationship does remain in some senses, particularly in that the stockholders continue to elect the board, the power has shifted in many cases to the hands of the directors. Even so, it is still important to recognize that the board does represent the interests of the stockholders and is responsible for providing oversight in the interest of all stockholders of the corporation. This relationship can be at issue in cases where a stockholder (oftentimes a holder of a sizeable stake of stock) believes the board is not acting in the best interests of the stockholders, but courts have tended to give substantial deference to the board.

Strategic role largely stems from the role the board plays in corporate decision-making. Due to the fact that a board will be responsible for approving decisions from potential acquisitions to option grants to new hires, the board will hold an important role with respect to certain strategic decisions. The challenge that some corporations face is to ensure the balance between the board and management in these strategic decisions. However, there has been research that suggests there is a positive relationship between a board's involvement in corporate strategy and the performance of the company.

Advisory role for a board of directors is most often seen in the earliest stages of corporate development. Early investors in the company, oftentimes who represent venture capital funds or other serial startup investors, will provide capital for the corporation, but may also view their role to guide and assist the company in its earliest stages. The experience many of these directors will have of working with numerous companies in a similar market, at a similar development stage, or facing common market conditions gives the board the ability to provide advice and insights to the company.

Service role of the board and its members comes as a result of the fact that many small and early stage companies will not pay their board members until the company matures or will only pay a nominal fee for board service. Compensation will usually be in the form of stock options, reimbursement for travel expenses, or stipend for meeting attendance. At least in some senses, the board and its members will be providing service to the company and its contributions will be aimed at adding long-run value to the entity.

25

Employees

Recruiting Employees

Most successful organizations will employ a variety of the methods discussed in this section depending on the time in the life of the organization, the job title, the timetable for hiring, and other factors.

> **YOUR RECRUITING CHECKLIST**
>
> - Referral search
> - Retained search
> - Contingency search
> - Contract staff search
> - Traditional advertising

Referral Search

How It works

By leveraging your network, you may be able to find direct referrals for talented new hires. Utilize your current employees, your board, your advisors, your accountant, your attorney, and other service providers. Reach out to industry trade associations and other entrepreneur support organizations.

Costs

None, other than your time.

Risks

Depending on the size and scope of your network, you may be casting your net too narrowly. Be careful with particularly important positions that you do not miss opportunities to reach beyond your network to bring in new talent.

Retained Search

How It works

Typically the use of a third-party retained search is used for high-level employees such as the CEO, president, CFO, or any of your vice presidents. The search firm will likely manage the majority of the process from finding candidates to initial screenings to setting up interviews with the company. This is a full-service process and the firm will likely coordinate areas such as travel, reference checking, and evaluation processes. The process will likely take between 60 and 90 days.

Costs

The placement fee will be negotiated, but expect between 25% and 50% of the budgeted amount of the employee's first year salary and bonus (including stock compensation). Typically, you will pay a portion up front and will be responsible for the remainder upon final placement. You will also be responsible for other standard expenses such as travel, copies, and so on.

Risks

Some search firms do not work with startup or emerging companies. Be certain that any organization you work with has experience with emerging companies and in your industry. Your search firm should create a job description and a list of companies that are most likely to have the applicable talent. Be certain to review the description and the list of companies to prevent the search firm from reaching out to companies you may want them to avoid (such as a potential competitor, a business partner, or a customer).

Contingency Search

How It works

For contingency searches, you will only pay the search firm upon successful placement of a candidate (unlike the retained search process). The agency will be less likely to prepare job descriptions, to determine target companies, or to provide reference checking. In addition, they oftentimes do not provide a first screening and may provide you with a much higher volume of candidate resumes.

Costs

The placement fee will be negotiated, but expect between 10% and 25% of the first year salary and bonus of the employee. However, you will not pay until the search is successful.

Risks

Contingency searches may not provide you with the highest caliber of potential hires as a greater number of unemployed individuals or others are looking to leave their employer will likely come through this process. If you would like the search firm to contact potential hires at specific companies or in specific jobs via phone or email, this may be an additional charge.

Contract Search Staff

How It works

At the early stages of your organization, you may find it is necessary to hire an individual or team to provide contract search staffing. This may involve employing a search coordinator on a limited, part-time basis to handle your internal search efforts.

Costs

Contract employees to provide search servicing may likely cost between $25 and $100 per hour depending on the experience level and support required.

Traditional Advertising

How It works

Utilizing traditional advertising in local newspapers, placement Web sites (including places such as CraigsList), and industry periodicals.

Costs

Individual advertising rates will apply. These can run from the hundreds of dollars for an individual ad to thousands for a multiple ad listing.

EMBEDDED WATER

Embedded water or virtual water is defined as the amount of water that is used in the production and trade of a product (it is similar to the concept of embedded carbon). While an 8 ounce cup of coffee may only have 8 ounces of water in your cup, the concept of embedded water tracks the amount of water used to produce the coffee beans that produce the coffee.

- Cup of coffee: 37 gallons of water
- Single hamburger: 634 gallons of water

Hiring an Employee

Hiring employees implicates a number of practical considerations from payroll and taxes to confidentiality and invention assignment. Therefore, startup companies should be aware of the necessary restrictions and requirements that are undertaken with a new hire.

YOUR HIRING CHECKLIST

- Should this person be an employee or an independent contractor?
- Do I use an offer letter or employment agreement to document the employment terms?
- What are the restrictions I need to put on the employee, such as noncompetition and nonsolicitation provisions?
- How do we ensure our intellectual property is protected?

Employees or Independent Contractors

For early stage companies, hiring full-time employees may represent a substantial undertaking from a financial and operational perspective. An independent contractor is responsible for payment of the taxes that an employee would typically pay: payroll taxes, FICA, Social Security taxes, and unemployment tax. Therefore, many companies will initially decide to utilize their team of founders and retain independent contractors on a project-by-project basis, for specific tasks, or on a less than full-time basis.

If you decide to utilize independent contractors, be aware that simply calling someone an independent contractor may not be enough in the eyes of the Internal Revenue Service to avoid paying certain taxes. The IRS recognizes that companies that rely on independent contractors have lower effective tax rates, and therefore the IRS carefully scrutinizes these relationships to determine whether the relationship is, in fact, an employment relationship and subject to taxes. There are multiple factors the IRS examines in these scenarios, but factors include the level of control the company has over the contractor's work, the number of other clients the independent contractor has, and the employment-like activities of the contractor (e.g., Does he or she have a desk? Does he or she have an identical company name badge? Does he or she have a company email address?). If you are uncertain about these issues you should discuss them with an attorney familiar with employment matters.

Independent contractors may be important resources for your company. Be aware that those that are engaged with your company should sign agreements governing nondisclosure of confidential information and assignment of inventions.

In the event you do hire any employees, the company is responsible for payment of taxes and proper employee withholdings. Be aware that many companies get themselves into trouble due to improper management of employment taxes and withholdings. You should carefully structure your accounting system to track payroll, taxes due, and withholdings. Many companies will outsource their payroll and employee taxes to a third-party provider.

Offer Letters and Employment Agreements

Offer letters and employment agreements represent a written understanding of the key terms of the employee–employer relationship. These agreements represent an important tool for clarifying the understanding of the parties and ensuring that the parties accept any restrictions or responsibilities.

Today, the trend is for companies to utilize offer letters, although some companies and some employees will insist on an employment agreement. The difference is a technical one: an offer letter is an offer for an employment contract that is accepted upon the employee beginning work, while an employment contract is accepted upon signing of the parties. Generally, an employment agreement is perceived to be a bit more formal, but both agreements will offer sufficient evidence of the employment relationship.

Restrictive Agreements

Companies will often include certain restrictions on their employees in their ability to join competing ventures, to solicit your current employees after they've departed, or to solicit your company's customers after they depart. The enforceability of these provisions depends on the state where the employee resides and you should only include the provisions if you have researched their enforceability in the state.

- *Noncompetition.* A noncompetition agreement is an agreement between an employer and employee in which the employee agrees not to pursue a similar profession or trade in competition with the employer. Generally, these provisions will have a limited time period on the employee effective after departing the company. Many states enforce agreements preventing employees from working for competitors, so long as the restriction is based on legitimate business need, and is reasonable in scope, duration, and geographic limitation. In California, noncompetition provisions are generally not enforceable. When evaluating the enforceability of a noncompetition agreement outside of California, it is important to review legal authorities of the relevant state carefully.

- *Employee nonsolicitation.* These nonsolicitation clauses are aimed to prevent a mass exodus from one company to follow a former employee by limiting the ability of a former employee to recruit former colleagues. Yet, despite the strong public policy favoring employee mobility, most courts (including California courts) have upheld reasonable restrictions on a former employee's ability to solicit his former employer's employees or customers. Many companies will include nonsolicitation language in employee agreements to restrict the ability of a former employee from solicitation efforts.

- *Customer nonsolicitation.* Reasonable agreements not to solicit customers are often enforced by courts, but the use of such agreements has been narrowed by many states. These restrictions should be limited to a particular time period, and the company should recognize that these provisions will only forbid solicitation by the former employee but will not prevent customers from working with the departed employee.

Protecting Intellectual Assets

For most green companies, disclosure of confidential or proprietary information, invention produced by employees, and employee use of trade secrets represent potential risks to the company. Therefore, it is important that the company take actions at the time of hiring, throughout the employee's tenure, and following termination in order to protect the company.

Many companies will develop an agreement that includes matters of confidential information, invention assignments, works made for hire, and may include references to noncompetition or nonsolicitation restrictions (which are discussed in the prior section). These terms and restrictions can be included in a single agreement (sometimes called a Confidential Information and Invention Assignment Agreement) or may consist of a number of separate agreements. In these cases, an employer may utilize confidentiality or nondisclosure agreements to protect confidential information and an intention assignment agreement to assign rights to employee-developed inventions.

Employee Compensation

Hiring an employee involves a substantial investment—not only in a salary for the employee, but costs ranging from bonuses and commissions for sale people to various benefit plans, employment taxes, and costs to provide equipment and technology to the employee.

> ### TYPICAL COMPONENTS OF EMPLOYEE COMPENSATION/COSTS
> - Salary
> - Bonus
> - Commissions
> - Taxes
> - Insurance
> - Retirement plans
> - Equity compensation

Employee Salaries

Salaries and bonus programs for startup companies vary based on the region, the industry, and the position. Startup salaries tend to be somewhat lower than established companies, but the salary is balanced by equity grants and the opportunity for the company to produce a sizeable appreciation in stock value.

Visit CompStudy.com to find out detailed information used to set salaries, bonuses, equity compensation and more for specific positions. The Web site offers guidance based on the size of your organization, the fundraising status, and the title or position of the employee.

Industry trade associations and local entrepreneurial support organizations may provide anecdotal evidence of salary ranges. You may be able to research comparable salaries on various job-posting Web sites such as Monster.com or CareerBuilder.com. In addition, PayScale.com and Salary.com provide detailed salary and benefit reports you can purchase based on the position, the region, the size of the organization, and various other criteria.

The chart in Figures 25.1 and 25.2 provide some representative salary and bonus amounts segmented by position and industry. While there is no segment for clean technology itself, usually a company can identify with either an information technology sector or a life sciences sector.

Employee Taxes and Benefit Programs

The cost to add each new employee represents more than just the salary you've agreed to in the offer letter. You'll be responsible for costs from taxes and benefits to rent and equipment.

The Department of Labor provides information on the costs an average employee costs to the employer. According to the DOL, an average employee costs $25.93 per hour when you factor in costs of salaries, benefits, and taxes. While these figures represent useful information, you should note that these numbers represent a broad range of employees across all industries in the U.S. economy (Table 25.1).

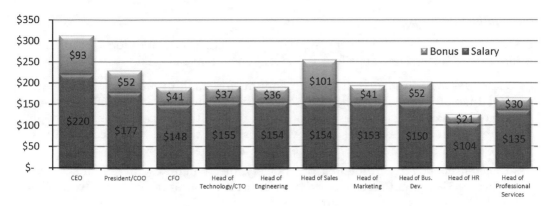

FIGURE 25.1
Cash compensation (in thousands) in the information technology sector. (Data from 2006 Compensation & Enterpreneurship Report, www.compstudy.com)

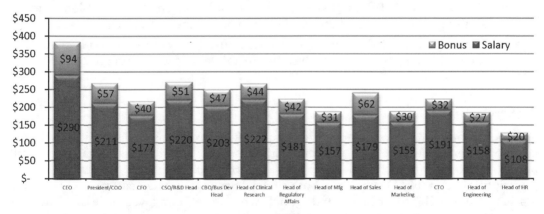

FIGURE 25.2
Cash compensation (in thousands) in the life sciences sector. (Data from 2006 Compensation & Entrepreneurship Report, www.compstudy.com)

TABLE 25.1

Private Industry Employer Compensation Costs

Employer Cost	Cost per Hour ($)	Total Costs (%)
Wages and salaries	18.32	70
Paid leave benefits	1.77	7
Supplemental pay	0.78	3
Insurance benefits	1.97	8
Retirement and savings	0.88	3
Legally required benefits	2.21	9
Total	25.93	100

Source: Adapted from U.S. Department of Labor's Bureau of Labor Statistics (June 2007).

Common Benefit Plans

In general, providing certain benefits will cost your company an additional 20% to 40% of the employee's salary. Here are a few of the key benefit plans offered:

- *Health insurance premiums.* In 2006, premiums for small group plans (plans with less than 50 people), averaged $3732 for a single coverage and $9768 for family coverage, according to the research of America's Health Insurance Plans. Although many small businesses will only cover a portion of the costs, these could still be in excess of $3000 for single coverage and $8000 for family coverage.

- *Dental insurance.* According to a 1999 Society for Human Resources Management survey, 83% of small businesses that offer health insurances cover between 25% and 50% of dental insurance premiums.

- *Vision insurance.* According to the same Society for Human Resource Management study, 58% of small businesses offering health insurance also offer vision insurance. The costs for vision insurance on the employer are generally very low—typically ranging from $50 to $100 per employee annually.

- *Life insurance and long-term disability insurance.* Expect costs for providing life insurance to cost between $150 and $250 annually per employee and long-term disability insurance to cost between $200 and $350 per employee each year.

- *Retirement plans.* A 401(k) plan will generally be funded by employee withholdings although the company will bear the administrative costs of the program.

Employment Taxes

The specific amounts for certain required employment taxes for social security, Medicare, unemployment insurance, and workers' compensation, described as "legally required benefits" above, can be found on various government Web sites. Federal Insurance Contributions Act (FICA) consists of both a Social Security (retirement) payroll tax and a Medicare (hospital insurance) tax. Individuals who are self-employed have their own tax that is similar to FICA referred to as the Self-Employment Contributions Act. Federal Unemployment Tax (FUTA) finance the administrative costs of unemployment insurance. Finally, each state will have an individual state unemployment tax.

You can find updated information and rates for the following:

- *FICA*, www.ssa.gov
- *Medicare*, www.ssa.gov
- *FUTA*, www.workforcesecurity.doleta.gov/unemploy/
- *Individual state unemployment taxes*, www.toolkit.com/small_business_guide/ (summary of individual state rates)

Equity Compensation

Many companies will attempt to structure general rules to allocate initial hire stock grants that depend on the seniority of the individual. Companies will identify the general tiers of the organization which would range from the CEO and the senior vice presidents through the entry-level employees. In general, the number of options granted to a new hire will

TABLE 25.2

Example of Option/Share Distribution Plan

Tier/Position	Option/Share Grant	Percentage[a]
Chief executive officer	600,000	3.000
COO/CFO/senior vice president	200,000	1.000
Vice president	100,000	0.500
Director	50,000	0.250
Manager	25,000	0.125
Staff	10,000	0.050
Entry level	5000	0.025

[a] Assumes 20,000,000 shares of stock issued and outstanding (including option pool).

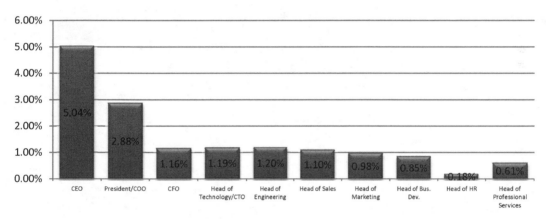

FIGURE 25.3
Equity/option grants at hire (nonfounders) in the information technology sector. (Data from 2006 Compensation & Entrepreneurship Report, www.compstudy.com)

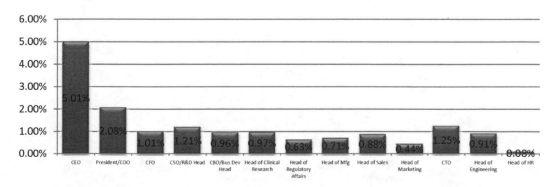

FIGURE 25.4
Equity/option grants at hire (nonfounders) in the life sciences sector. (Data from 2006 Compensation & Entrepreneurship Report, www.compstudy.com)

decrease with the level of seniority—usually an employee will receive between one-half to one-quarter of the options granted to their supervisor. An example of a sample option/ share distribution plan is given in Table 25.2.

The chart in Figures 25.3 and 25.4 provide some representative equity ownership amounts segmented by position and industry. These are an ownership percentage granted at the time of hire. Usually the grant will be restricted in some manner, including subjecting it to vesting. While there is no segment for clean technology itself, usually a company can identify with either an information technology sector or a life sciences sector.

26

Securities

Common Securities Issued by Startups

There are a number of securities that can be used with a startup company. At various points, a company may choose to issue equity or debt securities. A security is simply a fungible financial instrument that represents value. The entity issuing a security is called the issuer. What qualifies as a security depends upon the regulatory structure of each country. Securities laws govern the raising of capital for business purposes.

General Information about Issuing Securities

While issuing securities in your company may seem like a very simple concept, even successful companies have found themselves forced to clean up problems due to improper stock issuances. These problems can range from minor ones that cause time and energy to clean-up to major problems that bring about fines from the Securities and Exchange Commission.

Proper stock and option issuances are crucial for companies as they raise funds from sophisticated investors, consider mergers and acquisitions with other organizations, and contemplate doing an initial public offering. For this reason, it is important that companies exercise due care with their securities issuances.

COMMONLY ISSUED SECURITIES BY STARTUPS

- Common stock
- Preferred stock
- Convertible preferred stock
- Warrant
- Stock option
- Debt instruments

Common Stock

Common stock is an ownership interest in a corporation, entitling the holder to a portion of the corporation's earnings and assets upon the corporation's liquidation or dissolution. If a company only chooses to issue a single class of equity, common stock will be the class. Generally, the founders, management, and employees will be issued common stock of the corporation (it is more typical for investors to be issued shares of preferred stock). Upon liquidation of the company, the common stock will typically be the lowest claim priority, following secured and unsecured creditors, bondholders, and preferred stockholders.

Common stock is the primary form of equity. Common stockholders are typically given voting rights for the election of directors of the corporation's board. Common stock is closely aligned with the corporation's success and failure. When a corporation is profitable, common stockholders are entitled to unlimited appreciation in the value of their common stock. Alternatively, common stockholders may obtain nothing upon a corporation's dissolution once all other creditors and preferred stockholders are paid with priority.

Common stock is advantageous to a corporation because there is no obligation to repay the principle equity invested in the company for the stock, there is no obligation to pay dividends, investors obtain a right to share in the growth of the corporation and investors may influence the management of the corporation through voting for directors. Common stock may be disadvantageous because it dilutes management's interest in the corporation and imposes great risk upon investors.

Preferred Stock

Preferred stock represents a particular class of stock of the corporation with certain rights that common stock does not have. These rights may include liquidation preference, decision-making management control, dividends, antidilution protection, participation, veto provisions, and others. Preferred stock is an alternative equity instrument to common stock and is typically used for venture capital and angel investments. Preferred stock will usually have priority in dividend payment and liquidating proceeds over common stock. Preferred stock may be voting or nonvoting.

Convertible Preferred Stock

Convertible preferred stock represents a subset of preferred stock providing the owner the right to convert the preferred shares into common stock. This subset of preferred stock is the typical equity security issued to venture capital firms and other private equity investors. In most cases, convertible preferred stock will convert automatically into common stock when the company undertakes an initial public offering (IPO) or is acquired.

Warrants

Warrants provide a holder with the right to purchase shares of stock (common or preferred) in a company at a predetermined price, generally at a price above the current price or value of the stock. Warrants represent long-term purchase options that will typically be valid over a period of several years or indefinitely. Oftentimes, warrants will be issued in conjunction with preferred stock or bonds to increase the attractiveness of the stocks or bonds.

Stock Options

Stock options represent the right of the holder to purchase a share of stock at a specific price within a specific period of time. Stock options to purchase common stock of the company are often used for employees and management of high technology or green startup companies as incentive compensation. Many companies will offer these options as part of employee compensation, usually via the stock option plan.

Debt Instruments

Generally, debt instruments, such as notes, bonds, and debentures are typically entitled to receive payment before preferred or common stockholders. Debt instruments may be secured by assets of the issuer or may be unsecured. Debt instrument holders generally do not participate in the management decisions of a business, although they may impose certain affirmative or negative obligations on the business. Debt instruments can be long or short term and carry fixed or variable interest rates. Debt instruments are sometimes preferred by the company because they create a predictable payment schedule to investors, do not lessen management's interest in the growth and voting power in a business, and generally involve less risk for investors than equity investments. Debt may be disfavored, however, because it potentially restrict operations, limit the use of working capital, and tie up assets through pledges of collateral on the debt.

FUEL CELLS: WHAT ARE THEY AND WHY THE HYPE?

The idea of a fuel cell has become the "holy grail" for reducing society's dependence on fossil fuels. In fact, the idea of a fuel cell first arose over 150 years ago when scientist William Robert Grove developed the "Grove cell." Since that time, millions of dollars of investment have been put into the concept of creating a vehicle powered by hydrogen that emits water as its sole byproduct.

The concept itself is fairly complicated (particularly for a lawyer to describe ... although the grandfather of the fuel cell Mr. Grove himself was a lawyer turned scientist, so their is hope), but the basics of a fuel cell involve the creation of energy from hydrogen (H_2) molecules. Through a chemical reaction, the hydrogen molecules are stripped of their electrons. The positively charged hydrogen atoms are passed through an electrolyte membrane and the negatively charged electrons are passed along a wire to create an electric current. The result is that the now positively charged hydrogen atoms combine with oxygen from the air to create water molecules as the exhaust.

Scientists continue to be intrigued by the potential of fuel cells but have yet to sufficiently address certain challenges involved with the distribution, storage, and cost of the hydrogen and related technologies necessary to power fuel cells. New applications including stand-alone generation and new technologies have led to a resurgence of optimism for fuel cells.

Issuing Stock

In general, companies should be aware that both federal and state laws regulate the offer and sale of securities, including stock, options, and warrants. Within the company, it is generally the board of directors that will approve all offers and issuances of securities, oftentimes requiring additional approval from the stockholders.

When deciding how to finance a startup, various strategies can be used to obtain a favorable debt and equity combination. Aside from business strategy, the laws regulating securities are complex, with a mix of federal and state regulations. Competent securities counsel is crucial to the successful financing of a startup. Determining what approvals are necessary will usually be found in the corporate laws of the state of incorporation and the company charter and bylaws. Many companies will prepare a summary sheet for the board of directors that detail the authorized securities outstanding and the required approvals for the issuance of any additional securities.

Prior to the issuance of any securities, the company should be certain that sufficient numbers of securities are authorized under the articles or certificate of incorporation. Following such a determination (or authorizing the shares by filing a proper amendment to the articles with the state), the company will then need to obtain the proper approvals by the board of directors and/or the stockholders of the company.

YOUR STOCK ISSUANCE CHECKLIST

- Has this issuance been approved by the board of directors?
- Has this issuance been approved by the stockholders, if necessary?
- Is this issuance valid under federal law?
- Is this issuance valid under the state law of the state of incorporation?
- Is this issuance valid under the state law of the state where the stockholder will reside?

Federal Regulation of Securities

Most early stage, private companies will be primarily concerned with ensuring that any security issuance is exempt from registration. Before making a sale of stock to a friend, family member, relative, or another individual or company, be certain that you remain compliant with a federal exemption. This is particularly important as the company continues to grow. For startups considering certain more advanced funding sources (e.g., raising money from the public markets via an initial public offering), the company will start to be concerned with different federal securities provisions. But at this point, work with your attorney to ensure each issuance qualifies for some exemption (usually any properly structured issuance will be).

The Securities Act of 1933 (which will often-times be referred to as simply the Securities Act) and the Securities Exchange Act of 1934 (often referred to simply as the Exchange Act) are the two primary sources of federal law regulating securities transactions. The Securities Act regulates all issuances of securities by a corporation. The Exchange Act regulates the resale of securities and generally is not applicable until the corporation has made its initial offering of stock to the public. The principal federal securities regulatory agency is the Securities and Exchange Commission (SEC).

The Securities Act prohibits all sales of securities unless a registration statement containing certain required information has been filed with the SEC and has been declared effective, or unless either the security or the transaction is exempt. Registration with the SEC is expensive and time-consuming and, as a practical matter, startup companies are unable to utilize the registration process. Accordingly, startup companies must structure their securities offerings so that one of the exemptions from the registration requirements of the Securities Act will be available.

Issuing Options

For many entrepreneurs and early employees of startup companies, one of the main reasons they join the venture is the lure of receiving stock or options to purchase stock that may one day skyrocket in value. Stock grants and stock options represent important tools to attract and motivate talented employees. Managing equity compensation is a complex issue to juggle for most entrepreneurs.

HOW DOES A STOCK OPTION WORK?

GreenStartup, Inc. grants one of its employees an option to purchase 100 shares of stock of the company. GreenStartup believes its stock is currently worth $1.00 per share. Therefore, GreenStartup sets the exercise price or strike price at $1.00 per share. This means the employee may give the company $100.00 and will receive 100 shares, but GreenStartup doesn't ask for the money now. The stock option can be exercised for 10 years. Therefore, the employee doesn't have to exercise until he or she has the cash to exercise (at $1.00 per share) or believes the value of the stock exceeds $1.00 per share.

For some employees, this is the best of both worlds—*the employee has the right to buy the stock for $1.00, but is not obligated to buy the stock*. The employee keeps his or her $100.00 but knows he or she can purchase the stock at any time until the option expires. GreenStartup is able to grant an option to the employee which may motivate the employee, but has not had to issue actual stock to the employee. Fast-forward to several years later ... GreenStartup has gone "gang-busters" and now GreenStartup has had experts value its stock at $10.00 per share. The employee is holding an option to purchase the stock at $1.00 per share. When the option is exercised, the employee will pay $100.00, but will receive stock that is valued at $1000, a gain of $900.00.

Instead, if the GreenStartup stock had decreased to $0.10 per share, the employee can just hold the option (and not exercise the option) until it expires and hope that its value increases. The employee keeps the $100. If the employee had purchased the stock for $1.00 per share, it would now only be worth $0.10 or $10, a loss of $90. This is why options continue to be attractive to startups and their employees.

What Is the Difference between Stock and Stock Options?

Stock grants are the simpler of the two to explain. If you receive a grant of stock, you receive a portion of ownership (shares) of the company. At the time you receive the stock grant, you will actually own the stock, have certain voting rights in the company, and may (unless restrictions are applied) be able to transfer your stock to others. If you receive stock options, you are not receiving stock. Instead, stock options are contracts that allow you to buy a certain number of shares of the stock of the company at a certain price within a set period of time. The price that you can buy the stock at is referred to as the exercise price or the strike price. In order to purchase the shares of stock set forth on your stock option, you will pay the company the exercise or strike price in exchange for the shares. This exchange is referred to an exercise of your option. Usually, you may choose to exercise all or a portion of your option.

Why Would You Grant Stock or Options to Employees?

The grant of stock or options may serve multiple purposes. Generally, options and stock are designed to align the interests of the employee with the company by providing a direct reward in the event the company is successful and the stock price appreciates. Ownership of stock or options to purchase stock of the company can be used to offset a lower salary than what could be earned working for a larger, more established company.

There are several major types of stock plans, and the structure of individual plans may vary widely even within a single type of plan. For this reason, some companies may utilize a number of different stock plans, tailoring each to specific circumstances and specific employees. Because each type of plan has strengths and weaknesses, a company should carefully examine its incentive goals and the characteristics of its employees before adopting any plan. It is important to realize that issuances of securities to employees, like securities issuances to any other person, are subject to federal and state securities regulations. Consequently, neither an issuer nor an employee may sell unregistered securities without either registering the security or qualifying under an exemption.

Different Vesting Approaches

There are numerous ways to structure vesting restrictions on options (or stock), including the use of time-based vesting, milestone-based vesting, and a combination approach.

- *Time-based vesting (straight line)*

  Options are released from vesting in equal amounts each month over a particular time period (say monthly, quarterly, or annually over a number of years, usually between two and five years). As an example, an employee has straight monthly vesting over a three-year period. After being employed for 12 months, the employee leaves the company. At this point, the departed employee would only have one-third of the original option grant vested—meaning that the employee could then exercise the option to purchase one-third of the total option grant. The rest of his or her options (two-thirds of the original amount issued) would automatically terminate.

- *Time-based vesting (cliff)*

  No vesting for a particular time period (for instance, the first six months or first year or until financing occurs). Then, once that initial period is completed, the company will then release a certain portion from vesting. Once the cliff period has passed, the rest will typically vest on a straight line basis afterwards (monthly, quarterly, or annually). As an example, an employee's options vest over three years. The vesting will be a one year (12 months) cliff, followed by straight line monthly vesting over the remaining two years. If the employee departs after six months, he or she will have no options vested and unvested options would automatically terminate. However, if the employee departs after 18 months, then one-third will have vested after the cliff and six more months of vesting would have occurred. So the employee could exercise the option to purchase shares equal to 50% of his or her total option grant.

- *Milestone based vesting*

  Options will become vested upon the achievement of particular milestones, rather than based on time periods. As an example, a certain portion of unvested options will vest when (1) the company receives at least $1 million in funding (to incentivize fundraising efforts), (2) the company reaches $250,000 in annual revenue (to incentivize sales), and (3) when the company releases its second generation product (to incentivize product development).

- *A combination-based approach*

  Options will vest on a combination of milestone-based and time-based vesting. As an example, half of the options will vest monthly over a three-year period and the other half will vest based on achievement of certain milestones.

27

Raising Money

Raising Money from Friends and Family

Friends, family, neighbors, and colleagues might be some of the first sources you consider in the funding process for loans and the purchase of the stock of your new company. Garnering investments from the people close to you offers a few advantages: you may not need to expend valuable time and effort making connections and establishing relationships with prospective investors, and friends and family are more likely to offer favorable rates on loans. Plus, your friends and family members also might be more flexible and understanding if their investment falls through.

At the same time, remember that taking an investment from anyone, including a family member or friend, must comply with securities laws and should not be done without understanding the risks of fundraising.

YOUR FRIENDS AND FAMILY FUNDRAISING CHECKLIST

- Is the potential investor accredited?
- If the potential investor is *not* accredited:
    - Is the potential investor a sophisticated individual who has invested in private companies and understands the finances of a startup?
    - Does the potential investor understand the risks of investing in a startup?
    - Is the potential investor able to invest in an illiquid security where they may not be able to get their money back for several years?
    - Can the potential investor afford the investment?
- Will you issue debt or equity to the potential investor?
- Will the potential investor invest in the business or the founder?

Accredited Investors

Many new entrepreneurs have two questions about the concept of accredited investors: First, what is an accredited investor and second, why does it matter?

An accredited investor is someone the SEC deems is likely to be sophisticated and has the financial means to protect themselves when they make potentially risky investments.

To qualify as an accredited investor, an individual must have at least a million dollars in assets (which as of 2010, no longer can include the value of your primary residence) or have an income of $200,000 annually (or $300,000 if the investor is married).

Now why does it matter if an investor is accredited or not? The SEC imposes lower regulatory burdens if a company's stockholders are all accredited investors. Therefore one of the primary reasons why it matters is that it oftentimes takes additional time and money to raise money from unaccredited investors. For example, if you raise money from only accredited investors, you may not need to make any filings with the SEC or the state securities commission. All you would need to do is properly prepare the investment documents and take the investor's money. That's it.

However, what is required if any of your investors are unaccredited? Oftentimes there are more requirements you must meet to comply with the federal and state securities laws. You may be required to prepare a document that includes certain key information about your company, the risks of investing in a private company, and key financial data. You may need to file that information several weeks or months before you can sell the securities. Plus, you may only be able to raise limited amounts of funds. On top of all of this, having an attorney help you to prepare the documents and make the filings can be very expensive. You could add several months of time and several thousands of dollars in costs to raise money from an unaccredited investor.

So should you avoid taking money from unaccredited investors? Perhaps. It is certainly simpler to raise money only from accredited investors. But that may not always be an option, so you may need to consider the most efficient way to raise money from friends and family. If your business chooses to raise funds from unaccredited investors, be sure to work with advisors and attorneys that can help you in the process. Failure to comply with securities laws can allow an investor to get their money back (called a rescission right).

Unaccredited Investors

It is possible for your business to sell to investors that are not accredited. However, not all unaccredited investors are appropriate to invest in a startup or small business. Before deciding to take funds from friends or family members, you should be certain of two key qualities about this prospective investor:

- Are they sophisticated enough to understand the risks of this investment?
- Can they afford to make this investment (and maybe have their funds tied up for years)?

These questions are important since investing in a startup company is not at all like buying stock in IBM or General Electric. Family and friends must understand that the money they put into your business may be lost. Over 50% of new businesses fail in the first few years of operations, and your business is subject to similar risks. So ensure that any prospective investor understands the risks and can afford to take those risks. Additionally, investments into a startup company are typically illiquid, meaning it is difficult to extract those funds from the business. If someone wants to invest in your business, you should educate them that their funds may be tied up for eight to ten years, or lost forever.

Financial relationships with friends and family members carry extra emotional burdens. You may not want your loved ones' financial futures riding on the success of your startup, or their constant input on how to run your business. In order to minimize any potential strain on these relationships, it is important to be as upfront as possible about your realistic expectations for growth and the inherent risks of investing. Remember, other investors in future rounds of funding (such as angel investors and venture capital firms) will look closely on how you valued the equity sold to your friends and family, and may be turned off if they fear SEC complications or inexperienced decision-making.

Selling Securities (Debt or Equity) to Unaccredited Investors

If you have determined that your family member is not accredited, first, be sure to understand the securities law requirements necessary for your friends or family to invest. You will need to ensure that you apply with all state and federal securities laws for the issuance.

The business will be required to take additional steps for the financing that may not be required otherwise such as preparation of investor documentation and filings with the state. In particular, these filings may require the business to wait a certain number of days after making the filing before selling the securities. Be sure to work with an attorney familiar with securities laws as oftentimes the rules are difficult to understand without experience in the field.

Structuring the Investment

A company generally has two options to consider for an early stage investment: debt or equity. In the case of early stage investments into startup companies, the investment tends to be either in the form of the sale of promissory notes (a loan) or the sale of stock (ownership of the company).

For more information on convertible promissory notes, see

- Convertible Note Financing on page 357.
- Sample Term Sheet for Convertible Note Financing on page 359.

Loan

Let's say you've decided to take a loan from a family member. There are usually two different approaches an individual can take. Sometimes a founder will take a personal loan from a family member. In other circumstances, the company will take a loan from a family member. Here, we'll review in more detail the loan from a family member or friend directly to the company.

There are typically two approaches to this type of loan. First, you may enter into a very simple loan that everyone agrees will be paid off at some point in the future. Let's say you decide to take a loan for $20,000 from your great aunt and agree to pay 5% interest and repay the principal balance in one year. This is a simple loan arrangement. In other cases, you may want to have the option to convert the loan into equity of the business. For example, let's say you want to take the same loan for $20,000 from your great aunt, but you also plan to raise an additional $100,000 in the next year and intend to sell stock at that time. You can actually decide to give your great aunt the choice to convert her loan into equity or you can make it automatic. This is called a convertible promissory note.

At least for companies that anticipate doing a full-fledged financing in the near term (say, six to twelve months), issuance of convertible notes is considered to be more company-favorable, and is the easier, faster, and cheaper approach of the two. These are

oftentimes referred to as a "bridge"—since the idea is that the notes will "bridge" the company until a future equity financing.

In this case, the investment is in the form of a promissory note that converts into equity on the terms set in a future "qualified financing" (where the qualified financing typically is defined by having a minimum amount—say $2 million—of total investment). The note will either convert at a discount to the price per share set in the qualified financing (usually between 10% and 30%), will have warrant coverage (usually in the neighborhood of 20% of the dollar amount invested by each investor), or occasionally both. This discount and/or warrant coverage gives the angel investors some additional ownership in exchange for taking the early risk. Sometimes the investor will receive a cap on the valuation for the next round—say a maximum $4 million pre-money—even if the actual pre-money is higher.

Equity or Stock

In other situations, you will want to sell stock or an ownership stake in the business to a family member or a friend. In this case, you are granting them a right to vote on certain key decisions and giving them a portion of the profits of the company in the event you are able to distribute them. The advantage over a loan is that there is typically no interest on the equity and that you don't have to repay the invested amount like with the loan.

For more information on issuing stock or equity, see

• Issuing Stock on page 323.

This approach is sometimes called a "priced" round—when you sell common or preferred stock of the company to an investor—because you are setting a price on the business (e.g., 5% for $20,000 means the business is intrinsically valued at $400,00). A priced round is the more involved route, because it requires that the company and the investors agree on a valuation during a time in the company's life cycle when pegging a valuation is inherently difficult, and involves drafting and negotiating a somewhat more comprehensive and complicated suite of investment documents. The investment can be made in the form of common stock or preferred stock.

If the company does not contemplate raising additional funds from angels, venture, or other institutional investors, and will only be relying on additional small investments, the equity/stock approach may well be more appropriate, as these initial investors will be participating in the upside on terms that are agreed to early in the life of the company.

Raising Money from Angels and Angel Groups

Angel investors (or simply angels) are affluent investors (typically accredited) who provide capital for startups early in the development process, usually in exchange for an equity stake. (The term originally came from individuals who would invest in Broadway theater productions as unnamed donors, and became commonly referred to as "Angels" for their roles in saving productions that had overrun their budgets.) Many successful startups in their infancy have been bankrolled by private sales of debt or equity securities to angel investors. Angels typically look for businesses that have solid management teams and strong growth potential in industries that they know well.

INSIGHTS ON ANGEL INVESTING FROM ANGEL CAPITAL EDUCATION FOUNDATION (ACEF)

How do I know my business is right for an angel group investment?

Angel investment is the "right" source of funding for only a small proportion of entrepreneurial businesses. When considering yourself for investment by an individual angel or angel group, ask yourself these key questions:

- Am I willing to give up some amount of ownership and control of my company?
- Can I demonstrate that my company is likely to realize significant revenues and earnings in the next three to seven years?
- Can I demonstrate that my company will produce a significant return for investors?
- Am I willing take the advice from investors and accept board of directors decisions I may not always agree with?
- Do I have an exit plan for the company that may mean I'm not involved in three to seven years?

When should I approach an angel group?

In general, the best time to seek angel funding is when:

- Your product is developed or near completion.
- You have existing customers or potential customers who will confirm they will buy from you.
- You've invested your own dollars and exhausted other alternatives, including friends and family.
- You can demonstrate that the business is likely to grow rapidly and reach at least $15–30 million in revenues in the next three to seven years.
- Your business plan is in top shape.

What criteria do angel groups use to select entrepreneurs?

No two groups are exactly alike, but generally groups expect to at least see the following:

- A strong management team with experience and proven skills.
- Unique product or service distinguished by an identified competitive advantage and large market.
- Your personal financial investment in the company and investments from your friends and/or family.

- A clear picture of the market for your product or service and realistic plan for market penetration.
- An exit strategy for the investor that is reachable within five to seven years.
- The potential for a strong return on investment.

Source: http://www.angelcapitaleducation.org/dir_resources/for_entrepreneurs.aspx

Angel Investors

According to the Center for Venture Research at the University of New Hampshire and the MIT Entrepreneurship Center, angels invest in nearly 50,000 ventures each year representing annual investments of more than $23 billion.

Angels will oftentimes invest larger sums of money than your friends and family, but these greater sums come with somewhat higher levels of expectations. Angels want to get their investment returned, so will usually only consider funding companies that have valid exit strategies in order to ensure potential liquidity opportunities in a company's life cycle. While it is unusual for a venture capital firm to invest under $1 or $2 million, the majority of angel investments will fall under this level—providing a company with a source of funds to further develop the product and grow the company, without necessitating a $5 million outside investment.

Locating Angel Investors and Angel Groups

Finding the right angel or angels to invest in your startup can be a difficult, time-consuming process. Some angels organize themselves into groups to share research and pool capital, which makes them easier to find. The Angel Capital Association is a good source for finding a group near you. These groups can be found in most large cities, but the bulk of them reside in Silicon Valley, Seattle, Boston, Austin, Denver, New York City and other similar cities or regions with a track record of developing new ventures. Bear in mind, however, that most angels (especially the prominent ones) don't belong to a group. Some angel groups will charge a company to present to their association. In those instances, you should do some diligence to ensure the group is reputable before paying the participation fees.

CLEANTECH-FOCUSED ANGEL ASSOCIATIONS

- Energy Angels (Mid-Atlantic) (http://energyangels.angelgroups.net/)
- Northwest Energy Angels (Pacific Northwest) (http://www.nwenergyangels. com/)
- CalCEF Clean Energy Angel Network (Northern California) (http://www. calcefangelfund.com/)

The best way to meet individual angels is through industry contacts. Though cold calling occasionally works, angels will pay more attention to investment opportunities recommended to them by someone they respect and trust. Your board, advisors, accountant, lawyer, or entrepreneurial friends might possess the contacts you need. Once you've established contact with an angel, you need to present your company in the best light possible. Angels will look for a strong executive summary and management team, a "need-to-have" product or service, industry contacts to support your claims, and the aforementioned exit strategy. Angels will also be attracted to a legally sound business plan free of any significant downstream problems (such as intellectual property conflicts or securities issues).

Structuring Angel Investments

There aren't any generally accepted standards for dealing with angels, so deal terms vary greatly. Some will require intricately structured arrangements rivaling those of venture capital firms, while others (especially those investing very early in the process) will be content with simpler agreements. Angels without significant investing experience may not even know exactly what terms they want. While you might sometimes want to wait for the angel to draft the agreement on their terms, you can always draft a model agreement. Many angels will appreciate this savings of time and expense. The most typical structures are a stock investment (either common or preferred stock) or a convertible note financing. More information on both structures is available in the sections that follow.

For more information on structuring an angel investment, see

- Raising Money from Family & Friends on page 329.
- Convertible Note Financing on page 357.
- Sample Term Sheet for Convertible Note Financing on page 359.

Since angels are primarily concerned with getting a return on their investment, most will not demand an active role in the business. This means that unlike most venture capital firms, angels will seldom insist on board representation or veto rights over employee decisions. Usually, angels will only require the right to veto significant changes to the business plan, management salary levels, and the amount of equity available for employee incentive programs. Most angels demand less equity than venture capital firms, lessening your amount of equity dilution. Angels are also free from some of the restrictions that bind venture capital firms. For instance, angels will sometimes allow founders to cash out partially by selling some stock directly to investors during a funding round. Venture capital firms will almost never allow such a transaction, as they fear it will cause the founder to be less committed to the enterprise.

One of the risks of accepting investments from individual angels, rather than through an angel group or investment firm, is that you may have less knowledge about the potential investor. Since venture capital firms rely on their reputation to attract founders into doing deals, they must protect this reputation by holding themselves to certain accepted standards.

IDENTIFYING ANGEL INVESTORS

- *AngelSoft* (angelsoft.net)
  Consolidated platform to submit business plans to 450+ venture capitalists and angel groups. A free version and a premium version for $250/month. Up to three applications may be outstanding at any time.
- *Angel Capital Association* (angelcapitalassociation.org)
  North America's professional alliance of angel groups provides information on the more than 265 angel associations across the country.

- *vFinance, Inc.* (vfinance.com)
  Offers paid searching tools to find angel investors based on net worth on the individual, industry that the angel will invest in, and location of the investor. Customized searches for $1.00 to $2.50 per contact.

Raising Money from Venture Capital

In deciding which transaction is in the best interest of your company, many of the same considerations discussed above will affect your decision as well as the needs of any future investors, as the various transactions have different tax, liability, and post-transaction structural consequences.

HOW DOES THE VENTURE CAPITAL PROCESS WORK?

From a company's standpoint, here is how the whole venture financing transaction and relationship looks:

- The company starts up and needs money to grow. The company seeks venture capital firms to invest in the company.
- The founders of the company create a business plan that shows what they plan to do and what they think will happen to the company over time. The business plan should include how fast the company will grow, how much money it will make, who the key managerial leaders will be, and other relevant information.
- The venture capitalists look at the plan, and if they like what they see they invest money in the company. Very often the most important aspect of the plan is a clear articulation of who is running the company and the market size. Venture capitalists typically deeply value leadership and management success when dealing with startup companies that all begin to look alike. Good management is a key differentiator.
- The first round of investment is typically called the "Series A" round and the company will receive cash in exchange for equity ownership, which is usually given in the form of preferred stock. Over time a company may receive three or four (more or less depending on the needs of the company) rounds of funding before going public, getting acquired, or going out of business.

Can Your Business Raise Money from Venture Capitalists?

One of the pitfalls for some green entrepreneurs is falling in love with the SunPower, eBay, Google, or Yahoo!, startup funding path. Let's just call it, the Venture Capital Model.

That is, take an interesting technology, get a couple of angels to invest early, then find a venture capital firm to kick in several millions along the way until you can get acquired or go public. While this model worked well for some notable successes (Apple, Google, and Microsoft), not every successful idea has grown from venture investments.

For help determine if venture capital is right for your business, go to

- Venture Capital & Clean Technology on page 241.
- Venture Capital "Fit" Test on page 361.

Nationwide, institutional venture capital firms typically make only between 2500 and 3500 investments annually (and only about one-third of those are initial investments in startups, while many of those are follow-on investments). Green businesses make up a small portion of those approximately 3000 investments (approximately 10%). And with over one million small businesses started each year, the chances of getting one of those coveted VC investments are extremely small.

Even still, green business and technology entrepreneurs are oftentimes surprised to learn that their technology just isn't the type that traditionally gets funded by an institutional investor. Perhaps the company's market isn't large enough or the investment required is too great for VCs, to be interested. Yet many technology entrepreneurs will spend 100% of their time looking to raise a Series A round from venture capital firms. For many companies, the Venture Capital Model may be either unlikely or impossible. Or, perhaps the Venture Capital Model won't work for the company until after several years of growth or development.

Venture Capital Financing

There are a number of examples of highly successful companies that utilized venture financing: Amazon.com, America Online, Amgen, Apple Computer, Cisco Systems, Compaq, DEC, Federal Express, Genentech, Google, Intel, Lotus, Netscape, Oracle, Seagate, Sun Microsystems, Tesla, 3Com, and Yahoo!. Public perception is that VCs only fund high tech companies. However, the truth is that VCs will fund a variety of companies that fit their investment profile and provide returns consistent with their internal metrics. While firms may invest in industries outside of green or high technology fields, it is still typical that a firm will focus its investments into certain fields, industries, or technologies. This focus is the result of the ability of the firm to understand the technology, market, and potential of any investment, as well as to allow the firm and its partners to offer its portfolio companies ongoing value as an outside advisor.

Likewise, the perception exists that venture capital firms only invest in mid-stage or advanced-stage companies. While there are many firms that will focus their investments on mid-stage and advanced-stage companies, firms do choose to invest in early stage companies. Matching the typical investment stage and technology or market focus of the venture capital firm is integral to obtaining funding from a venture firm.

Returns on Investments

In return for the money it receives, the company gives the VCs stock in the company as well as some control over the decisions the company makes. The company, for example, might give each VC firm a seat on its board of directors. The company might agree not to spend more than $X (say $250,000) without the VC's approval. The VCs might also need to approve certain people who are hired, loans that are made, and other key decisions.

In many cases, a VC firm offers more than just money. For example, it might have good contacts in the industry or it might have a lot of experience it can provide to the company. The value an experienced VC may add to a startup company often may transcend mere financing.

One big negotiating point that is discussed when a VC invests money in a company is, "how much stock should the VC firm get in return for the money it invests?" This question is answered by choosing a valuation for the company. The VC firm and the people in the company have to agree how much the company is worth. This is the "premoney valuation" of the company. The VC firm then invests the money in the company and creates a "post-money valuation." The percentage increase in the value determines how much stock the VC firm receives. A VC firm might typically receive anywhere from 10% to 50% of the company in return for its investment. The original stockholders are diluted in the process. If the situation exists where the stockholders own 100% of the company prior to the VC's investment, then following an investment by the VC firm in exchange for 50% of the company, the original stockholders shares would now represent the remaining 50% ownership in the company.

After several rounds, or series, of financing, the company and the investors will usually be looking for liquidity for their investment. Private companies without a market to buy or sell their shares have few opportunities for an investor to "cash out" their stock. Therefore, most VC-backed companies will look for some type of a liquidity event such as an initial public offering or an acquisition event. At or after either event, the VC firm and company will look to end or scale back the relationship. But, to be fair, reaching such a point often will involve three to seven years, multiple rounds of financings, and a substantial amount of time to find and finish such a transaction. In order to satisfy its investors, a VC will ultimately need to be able to extract its investment (plus a healthy return) to return the funds to its investors. For example, in many clean technology companies, it isn't uncommon that a group of institutional investors have invested between $50 and $100 million before an initial public offering or some type of merger or acquisition event.

Most VCs will look at a highly successful return on their investment if they are able to return 10 times or more of their investment back to the fund. The odds of this happening are low, so having one "home run" in the investment portfolio can pay off the numerous low performing investments. VCs are still happy with returning two to three times the investment on a company. If a VC has 10 portfolio companies it has invested in with one big winner (10 times), one or two medium winners (two to three times), one or two breakevens and the rest losers, the fund could wind up a success for the investors and the venture partners. Venture capital firms operate in a very risky game where they hope to find one Google or Apple and avoid investing in too many Webvans (a famous web-grocery delivery company that once had about $800 million in venture capital, but ended up with $830 million in losses and just $40 million on hand when it closed up shop) or Kozmo.com (a small-goods delivery service that raised more than $250 million only to be forced to liquidate in 2001).

RESEARCHING VENTURE CAPITAL FIRMS AND ANGEL INVESTORS

- *National Venture Capital Association* (nvca.org)
  The trade association of the venture capital industry provides a list of the member-venture capital firms and links to their Web sites.

- *vFinance, Inc.* (vfinance.com)
  Offers paid searching tools to find venture capitalists based on the size of deal, industry that the VC will invest in, and location of investor. Customized searches for $3.00 per contact. Offers paid searching tools to find angel investors based on net worth on the individual, industry that the angel will invest in, and location of the investor. Customized searches for $1.00 to $2.50 per contact.
- *Capital Hunter* (capitalhunter.com)
  Provides a database of venture capital and active private equity investors and key executives, and downloads of funding event details. Subscription of $59.00 per quarter.
- *MoneyTree Report* (pwcmoneytree.com)
  Provides a database of venture capital transactions that can be customized by location, investment stage, industry, and keywords. Public access allows for searching the current quarter's results only. Free.
- *Grow Think Research* (growthinkresearch.com)
  Provides database of venture capital activity detailing U.S.-based companies, funding transactions, executives, directors and advisors, and investors. Subscription of $499.00 per month or individual searches for $5.00 per record.
- *TheFunded Web site* (thefunded.com)
  TheFunded.com is a community of founders and CEOs to research, rate, and review funding sources worldwide. Offers reviews, information and actual term sheets from various funds. Free for entrepreneurs.

Other subscription-based services include Dow Jones VentureSource, Venture Xpert Web, and Capital IQ.

Obtaining a Business Loan

While some startup companies seek equity financing instead of debt financing, entrepreneurs may still find it helpful to consider the various bank loans available. The Small Business Administration has programs to assist businesses obtain loans that they may not be able to otherwise obtain.

Bank Loans

Bank loans can be classified according to whether the loan is short term or long term, and whether the loan is secured or unsecured. Short-term loans are generally used to finance the company's inventory needs, accounts payable, and general working capital. Interest rates are typically lower on short-term loans than on long-term financing. Long-term loans typically require a larger amount of collateral to secure the financing far into the future. They are usually used to finance fixed assets, such as the company's property, plant, and equipment.

Loans may also be secured or unsecured. Unsecured loans are simply promises to pay a debt. If the borrower defaults on the loan, the lender's only recourse is to sue the borrower. In such a situation, the lender will not have any priority claim to a particular piece of the borrower's property. Consequently, businesses have a difficult time obtaining unsecured loans unless they have a strong credit history. Secured loans are also promises to pay a debt, but the promise is "secured" by the property of the debtor (called "collateral"). If the borrower defaults on the loan, the creditor can recover his losses by seizing the property that was collateral for the debt.

For companies that are unable to obtain banking loans from traditional large financial institutions, some startup companies will choose to work with *community or local banks*. These banks may offer a complete banking relationship and be willing to provide financial products with more varied minimum amounts, payment plans, or interest rates. Startups also may benefit from *microloans*, which have begun to be an important tool for newly established small businesses. These loans can range from several thousands of dollars to up to $35,000. Microloans are funded by the Small Business Administration through grants to nonprofit community lenders that oversee the lending process to business borrowers. The unique feature is that the lending and credit decision is made locally by the community lender. Each community lender will have individual credit and lending requirements, but the maximum term of these loans is six years. The microloans will require the borrower to provide a personal guarantee and some form of collateral. Additionally, the community lenders will require the borrower to complete a business planning and training program prior to issuance of the loan. But the tradeoff is that microloans are easier to obtain than a traditional bank loan.

Specific Loan Programs for Startups

Some examples of the types of loans available to entrepreneurs and startup businesses are as follows:

- *Working capital lines of credit.* Under a line of credit, a party may borrow funds as needed, up to a specified maximum amount. The line of credit is used to fund the working capital and cash needs of the business. To secure the loan, the bank will sometimes use the company's accounts receivable or inventory as collateral. The term of the loan may vary and is often renewable. Borrowers will pay interest on the outstanding balance of the line of credit.

- *Short-term commercial loans.* Short-term loans are usually given for a specific expenditure, such as a piece of equipment. Interest is paid on the lump-sum of the loan and is often a fixed rate, so businesses usually do not face much risk of rising interest rates. Short-term loans may be as short as 90–120 days, or may extend from one to three years. The loan is typically secured by collateral, such as accounts receivable, inventory, or a fixed asset of the business. Most loans to startup companies and new small businesses are short term, and the lending

agency will review the company's cash flow and credit history before providing funds.

- *Long-term commercial loans.* Long-term loans (those with terms longer than one to three years) are more difficult for new businesses to obtain because the risk that the new business will default increases with the length of the loan. The length of the loan usually ranges from five to seven years, although loans secured by real estate may extend much longer. Long-term loans are generally used for business expansions and to fund major plant and equipment purchases. Lenders usually require that the loan be secured by the asset being acquired. In addition, the lending agency will review the company's business plan and cash flow to determine whether the company will be able to repay the principal and interest over the term of the loan. Lenders may also require insurance to protect the collateral.

- *Small business credit cards.* Small business credit cards offer an alternative to working capital lines of credit. They provide a quick source of limited funds when cash-flow is tight. The interest rates are typically only slightly less than the interest rate on individual consumer credit cards, and may not have very high spending limits. Small business owners who are considering this option should also be aware that they may have personal liability for the credit card, at least until the business has an established credit history of its own and the business owner can negotiate a new arrangement. If the credit history of either the business owner or the business itself is less than flawless, the credit card company may also require that the business deposit a specified amount of cash as collateral for securing a credit line.

- *Letters of credit.* Businesses engaged in international trade frequently use letters of credit as a method for making payments. In these situations, the buyer and seller will arrange a contract for the sale and shipment of goods. The buyer will then deposit money (or take out a loan) at his local bank in the amount of the letter of credit. The buyer's bank will issue the letter of credit to the seller's bank in the foreign country. The letter of credit will specify certain documents that must be presented in order for buyer's bank to transfer the funds. Such documents frequently include a commercial invoice, the bill of lading for the shipment of the goods, and insurance documents. The seller's bank will then notify the seller that a letter of credit was opened in his favor and that the goods may be shipped. Once the goods have been sent, the seller will present the requisite documents to his local bank for approval. If the documents conform to the letter of credit, the issuing bank will transfer the money to seller's bank and the money will be deposited in the seller's account.

  Letters of credit are beneficial for both parties to the transaction for a couple of reasons. First, by requiring certain documents in the letter of credit, the buyer is offered some protection that the goods were sent on a particular date and that they were shipped in a particular condition. Second, the seller is offered protection that he will be paid, since the buyer's bank is required to honor the letter of credit upon presentment of conforming documents. The flipside of these advantages, however, is that parties must be exceptionally precise when filling out the letter of credit and presenting the required documents. Even if the goods arrive on time and in the required condition, the issuing bank will not be obligated to make a payment if the documents do not conform to the letter of credit or if they are not received on time. Thus, the parties must pay very close attention to detail and must be as accurate as possible in their terminology.

Small Business Administration-Backed Loans

The Small Business Administration (SBA) does not grant loans. Instead, they will often-times act as a guarantor for obtaining a loan from a banking institution. In general, to qualify for a loan backed by the SBA, the business must be owner-operated, a for-profit company, and structured as a sole proprietorship, corporation, LLC, or professional part-nership. Businesses including retailers, manufacturing companies, construction compa-nies, and professional services business such as doctors, dentists and veterinarians can all qualify for SBA-backed loans.

Loans backed by the SBA are attractive for certain businesses because they are targeted at companies that have been unable to obtain a traditional loan. SBA-backed loans will typically require much small down payments, payback terms that are extended and can be used for initial startup capital. SBA-backed loans can generally be approved for the purchase of an existing business or franchise, purchase or expansion of real estate or fixed assets, or for use of working capital. The loan process is similar to the process described above, except that the bank will submit an application on your behalf to the SBA for approval.

TYPICAL BANK LOAN APPLICATION PROCESS

Obtaining a business or commercial loan from a bank will vary depending on the specific rules and policies of the bank. However, the following represents a common application process for obtaining a bank loan.

- *Interview.* Most often a bank will schedule an initial interview to help pro-vide you information on the process and to gauge if your business is a viable candidate for a loan. This can sometimes be an informal meeting at a bank branch or a telephone conversation.
- *Submission of loan application.* This is the submission of an application that will oftentimes include items such as financial statements, lists of assets, business plan, personal asset records of the founders, and so on.
- *Closing of the loan.* Depending on the type of loan, this could involve a pro-cess whereby you sign the loan documents at the branch or elsewhere, or in the event it relates to assets or real estate, could require the involvement of a title company.

28

More about Fundraising

Developing Your Fundraising Strategy and Plan

The following provides details on raising outside investment, and is applicable to both angel investors and venture capital. Prior to beginning any fundraising plan, you should talk with advisors and investors to determine if your business is an appropriate candidate for outside investment from institutional or angel investors.

YOUR 10-STEP FUNDRAISING CHECKLIST

- Make a funding self-assessment.
- Establish a fundraising team.
- Determine the funding needs of the business.
- Analyze the marketplace for investor targets.
- Create the necessary fundraising tools.
- Solicit interest and meetings from targeted investors.
- Make presentations to prospective investors.
- Receive feedback from prospective investors.
- Business due diligence.
- And finally, the *"Yes. Here's a term sheet."*

Step 1: Make a Funding Self-Assessment

The decision to raise money from outside investors is a significant one. Regardless of whether the fundraising process is successful, raising money from outside investors will take substantial time away from the business, will be a distraction, and has no guarantee of success. Therefore, before making the decision to pursue outside funding, determine whether you are prepared to engage in the fundraising process, whether your business can support the time required, and if there are other funding alternatives that may be a better fit.

One of the first steps in the fundraising process is to do an initial assessment of whether financing makes sense for the business. Many business plans received by professional investors will be immediately rejected because the company is not a good "fit" for that angel investor or venture fund or for financing in general. This could mean the company is serving a market that is too small or is a niche market. It could be that the technology has few barriers to entry. Or, it could be a company that will have fairly low gross margins.

Consider taking the *"Venture Capital Fit Test"* in this chapter to determine if your business fits the "standard" criteria for venture funding.

NOT A GOOD FIT FOR VENTURE CAPITAL ...

Many very successful businesses or businesses with the potential for great success do not fit the venture capital funding model. For example, a business plan to provide technology consulting services may not be a typical business model funded by venture capitalists even though it can be a successful and profitable business.

Here are some examples of markets, products and technologies that tend *not* to be "fits" for venture capital funding:

- A *market* such as retail, mining or banking, may not be a market the venture firm will invest in or has contacts in and knowledge of.
- A *product or service* that is highly technical, extremely complicated and difficult to explain to a lay person may not be a technology that the venture capitalist is willing to try and convince one's partners to invest in.
- A technology that will only serve a *small niche* and will involve an extensive process of education, selling and distribution may not have the risk-reward profile for a venture firm.
- A business designed to sell a *low-cost product at a low margin*, taking market share from the industry leader may not be a model that a venture firm will find fits its strategy.

Green businesses should also consider that given the broad range of businesses that qualify as "clean technology" to ensure that they recognize that their business still needs to fit into these broader venture capital criteria.

Companies may find that they are not a company that is a good fit for venture funding. In some cases, this will mean that the company's model and goals do not align with the venture capital funding model. In other cases, a company will recognize that they need to attempt to grow the business, further develop the product, or increase the experience level of the management team. In both cases, the company should focus their efforts on growing the business through other funding sources until the business more closely matches the business model funded by venture capital firms.

Step 2: Establish a Fundraising Team

For many companies looking to begin fundraising, the chief executive officer is the logical choice to lead the fundraising efforts. However, other companies may establish another individual within the company to lead the fundraising process such as the chief financial officer or a founder of the company. In any case, the company will generally need to select one individual to be the lead for the process and set the tone and goals for the efforts. This lead individual will hold initial meetings, make personal contacts and handle much of the "selling" process for the entity. Expect that during the solicitation and negotiation phase of the process, the fundraising lead will devote a minimum of 20% of one's time to these efforts, and potentially much more if the process requires.

Generally, the company will begin the fundraising planning and strategizing with the entire management team. The management team will typically help with general budgeting and funding strategy evaluations, as well as preparation of information and materials to provide to potential investors. However, as the process moves from preparation to solicitation, many companies will task one individual or team to be responsible for managing the fundraising process (oftentimes the CEO and an administrative assistant) to coordinate contacts, timelines, provide materials, follow-up, and scheduling. As the process moves forward from the solicitation phase into evaluating term sheets and legal due diligence, companies may then decide to involve their outside counsel to a greater extent.

In later fundraising rounds for a company (following a seed/angel round or a Series A round), current venture capital investors of the company may play a role in the process making additional introductions and solicitations. However, in most cases, the follow-on investment process will continue to be initiated and managed by the company itself.

The team should also consider setting a basic timeline, high level discussions of the financial needs of the company, funding goals, and the acceptable level of dilution from a financing, and an early discussion of the networks of individuals familiar with the venture community to generate early contacts and leads in the process. The plan should involve a discussion of the preparation of the key tools needed for solicitations. The management of the company should also discuss the prospect that fundraising may not be successful, and a maximum amount of time the company will devote to fundraising before pursuing other alternatives or revising the business model.

How long will it typically take to close an initial venture capital investment in a company?

- Under 30 days—1% of respondents
- 30–60 days—18%
- 60–90 days—45%
- 90–120 days—26%
- More than 120 days—10%

Average 80 days to complete from the receipt of the business plan to closing of the investment. Note: Days measured from the receipt of the business plan to closing of the investment.

Source: Profit Dynamics Inc.

Step 3: Determine the Funding Needs of the Business

One of the frequent questions you will hear from potential investors is: *"How much money are you trying to raise and what will you use it for?"*

The fundraising process is predicated on the fact that your business needs capital to grow. Determining how much capital you will need, when you will need it, and where that capital will be spent needs to be addressed prior to soliciting investors. Generally, your business plan should set out the needs of the business for each year of a five-year period, with an emphasis on the next 15 to 18 months of the business. In addition, the business plan should identify where the funds will be used, with a fair degree of specificity.

Venture capital firms do tend to have minimum investment levels in order to ensure that the partners of the firm can service the company on the board, through strategic guidance, and through networking. As such, it is useful to see the average investment levels for venture capital deals. The general rule of thumb is that venture firms prefer their investments to be within the range of $2 million to $6 million for an initial investment.

To better understand the issues, let's take an example at play. Assume that your business plan projections hold that the business will need $15 million over the next five years. In the first year, however, you are only projecting that the business will require $1.5 million of cash. Should you set out to raise $1.5 million, $15 million or something in between?

There are numerous perspectives on the debate. The two extreme positions of this debate are:

- *Raise as much as you can* (1) to avoid having to raise funds again when the management least needs to be distracted from building the business and (2) to allow the business to invest quickly and grow rapidly to tackle the small market windows.

- *Only raise what you need* since (1) few VCs are willing to go "all-in" in a startup still developing its product; (2) unused money sitting in a startup's bank makes VCs unhappy; (3) more investment dollars increase dilution of the founders; and (4) raising the full complement of funds will set the company on a specific path and may not allow the management to modify the business model from a "home-run" company to a "solid-double" company (to use the baseball terminology).

The experts who say to raise as much as possible may encourage an entrepreneur to raise the funds needed for the first three years or perhaps even first five years. And while this approach may increase the dilution you face (since you will be effectively raising the funds at a lower valuation), these experts may argue that investing heavily in a solid business idea and team early on can create a much larger business than a more methodical approach (effectively increasing the total payout for all, even if the relative percentage owned by the management of the company decreases). If your product or service represents a major opportunity, and it will require a rapid implementation and development strategy, this full "go-to-market" strategy may fit within your funding request.

On the other hand, raising $10 million for a first-time entrepreneur or for a new idea represents a much more significant investment for a VC than $2 million (or even $500,000). Most VCs will be much more willing to shell out the additional $8 million after a year or two of success, but would have doubts about handing over a blank, $10 million dollar check to most entrepreneurs. Therefore, some suggest that an entrepreneur pursue a more measured approach to fundraising—raising only what you will need for a 15-month aggressive development period. While this type of an approach will require the company to raise additional rounds later, it is generally easier to solicit larger investments (likely to be at higher prices per share) as the company continues to make positive progress. In the hypothetical example above, raising only $1.5 million is likely to be lower than many VCs will prefer for an initial investment round, and therefore the requested amount may need to be higher to attract the interest of certain VCs.

The average dollar amount of a first round investment by venture capital funds

- $4.1 million in Q3 2009
- $5.1 million in Q2 2009
- $4.5 million in Q1 2009
- $4.6 million in Q4 2008

Average dollar amount of a first round investment by venture capital funds of $4.58 million.

Source: PWC MoneyTree.

The timetable to receive funding from a VC will usually take three to six months, but can be faster depending on the investment cycle of the venture firm. However, you should adequately prepare the business to have sufficient capital to continue to operate in the event the investment process takes a longer period to complete.

Step 4: Analyze the Marketplace for Investor Targets

Before beginning your fundraising process, the first step is identifying likely investor targets. In most cases, through a quick review of the portfolio companies of any venture firm (oftentimes listed on the firm's Web site), you will notice an identifiable investment strategy. Focus your efforts on firms that have a strategy that most closely aligns with your company. Look at comparable companies and find their investors to identify who you should also target. And, if you have contacts at the portfolio company, ask for a referral since those types of referrals are usually successful at generating meetings and interest.

HOW TO TARGET INVESTORS FOR YOUR BUSINESS

According to one investment banker in the clean technology space, "Investors in clean technology companies are much less 'geographically-focused' than in sectors such as software. Cleantech investors are attempting to match the fund's background and experience with the right business." Look at the following characteristics of firms to develop a list of prospects:

- *Location.* It is usually preferable for an investor to be located in the company's hometown, but as indicated above, this isn't a limiting factor for some clean-tech investors. Otherwise, look to see if the investor has invested in any other companies located in your town or region. If you are from out of the investors' home town, most VCs will encourage you to find a local VC or investor and then approach them about the possibility of coinvesting.
- *Industry focus.* Identify which investors are investing in other companies in your industry or utilizing similar technologies. If you are a solar company look for funds investing in energy generation.
- *Investment stage.* While a firm may invest in businesses at various stages of company development most firms have a preference in their strategy. If you are an early stage company, focus your efforts on companies that are investing in either seed or Series A investments.
- *Size of investment.* If your business model requires a substantial investment or will need significant follow-on investments in a short period, look at firms that make investments that match those needs.
- *Activity level.* Only eight venture firms investing in early stage companies did more than nine deals in 2006, according to Entrepreneur.com. No firms investing in late-stage companies did more than seven deals in 2006. Look to see if a potential investor is actively funding new companies or is only involved in follow-on investments.
- *Industry reputation.* Word of mouth is perhaps the best source of inside information about various firms, particularly from portfolio companies. If you do not have contacts in the field, many discussion boards will offer some insights into the reputation of a firm or firms or look at TheFunded.com.

How to Find Out Who Invests in Your Technology, Industry, or Area?

Target your search for venture capital firms by finding out the investors of other startups in your industry or field of technology. A great, free tool is the MoneyTree "Custom Search" function. This search can be used to identify specific companies that received funding in the most recent quarter and their investors.

If you are a Texas-based clean technology company and wanted to find out who had invested in other Texas-based clean technology companies in the third quarter of 2009, simply input those parameters (Texas,

What is the fastest a company has ever received a first-round investment, from the time they received the business plan until the deal closed?

- Under 30 days—41% of respondents
- 30–60 days—39%
- More than 60 days—20%

Average "fastest" deal time from respondents is 40 days.

Source: Profit Dynamics Inc.

Industrial/Energy) and submit the report. The report would tell you that Glycos Biotechnologies, Inc., Ingrain, Inc., and Kosmos Energy LLC fit these search criteria.

If you then clicked on Glycos, you'd find that the company "develops metabolic engineering technologies." And Glycos received a $5 million early stage investment from DFJ Mercury and Draper Fisher Jurvetson. If you clicked on Ingrain, you'd find that the company is a clean technology company that "provides computational rock physics data." And Ingrain received $15 million from Energy Ventures and TPH Partners LLC. Finally, if you selected Kosmos Energy, you'd find they are a later-stage investment from Warburg Pincus LLC and may not fit your criteria.

Using this information, you would know of three potential investors that invest in clean technology companies located in Texas. Comb your personal network to see who you know at those investment firms or try to reach out to connections at these companies to find an introduction to these investors. Visit the MoneyTree Web site (www.pwcmoneytree.com) from PricewaterhouseCoopers and the National Venture Capital Association for more information.

RESEARCHING VENTURE CAPITAL FIRMS AND ANGEL INVESTORS

- *National Venture Capital Association* (nvca.org)
  The trade association of the venture capital industry provides a list of the member-venture capital firms and links to their Web sites.
- *vFinance, Inc.* (vfinance.com)
  Offers paid searching tools to find VCs based on the size of deal, industry that the VC will invest in, and location of investor. Customized searches for $3.00 per contact. Offers paid searching tools to find angel investors based on net worth on the individual, industry that the angel will invest in, and location of the investor. Customized searches for $1.00 to $2.50 per contact.
- *Capital Hunter* (capitalhunter.com)
  Provides a database of venture capital and active private equity investors and key executives, and downloads of funding event details. Subscription of $59.00 per quarter.
- *MoneyTree Report* (pwcmoneytree.com)
  Provides a database of venture capital transactions that can be customized by location, investment stage, industry, and keywords. Public access allows for searching the current quarter's results only. Free. (More information above)
- *Grow Think Research* (growthinkresearch.com)
  Provides database of venture capital activity detailing U.S.-based companies, funding transactions, executives, directors and advisors, and investors. Subscription of $499.00 per month or individual searches for $5.00 per record.
- *TheFunded Web site* (thefunded.com)
  TheFunded.com is a community of founders and CEOs to research, rate, and review funding sources worldwide. Offers reviews, information and actual term sheets from various funds. Free for entrepreneurs.

Other subscription-based services include Dow Jones VentureSource, Venture Xpert Web, and Capital IQ.

Step 5: Create the Necessary Fundraising Tools

There are a number of different tools that will be important for the fundraising process. Investors will prefer different sets of materials to review—some will only want to see an executive summary business plan over email while others prefer you to mail a full business plan. In order to be able to provide materials that an investor requests, you should prepare the tools, materials, and resources that will be used in your fundraising process.

What information would you like to receive with the initial contact?

- 2–3 page Executive Summary—56%
- 10–15 page "Mini" Business Plan—22%
- Full Business Plan—14%
- Information by telephone only—7%
- Cover letter describing the company—1%

Source: Profit Dynamics Inc.

Be certain to prepare materials that look professional, are accurate, and tell the story of your business. A company will only have one chance to make your initial introduction and word-of-mouth will spread about companies that are unprepared for the process. Invest the time up front to truly prepare and vet materials prior to sending them to any potential investor targets.

SAMPLE FUNDRAISING TOOLS

- A prepared and practiced *"elevator speech"*
- Two to three page *executive summary business plan*
- 10–15 page *"Mini" business plan*
- *Full investor-focused business plan*
- *One page letter*, with a two-paragraph introduction into the business
- *Management team biographies* and resumes
- *List of references* (each of whom have a full set of company materials)
- *List of customers* (if you have any), likely customer targets and testimonials of current customers where available
- Set of *company financials* (current and projections)
- 15–20 slide *PowerPoint investor presentation*
- Complete set of *due diligence materials* (company records, documents, and files)

WHAT TO INCLUDE IN YOUR POWERPOINT INVESTOR PRESENTATION?

- A memorable introduction and theme to the presentation
- Company background (name, location, structure, stages of growth, names of presenters)
- Opportunity/problem addressed
- Marketplace
- Revenue sources
- Management team
- "Go-to-market" strategy
- Marketing
- Competition

- Competitive advantage
- Risks
- Financial summary
- Timeline
- Capital needs and funds utilization
- Why should the investor put money into your business?

Limit your presentation to 15–20 slides and be sure to avoid any errors, typos, or mistakes.

WHAT TO INCLUDE IN YOUR COMPANY FINANCIALS?

- Bottom up numbers with assumptions clearly spelled out
- Provide numbers that look five years out (and two years back, if applicable)
- Break out expenses by department
- Break out revenue by product category
- Avoid the common problem of underestimating expenses or overestimating revenue
- Be able to explain numbers—investors will "drill down"
- Avoid using the phrase, *"This market is so large that we plan to take just 2% of this market to be a success."* Investors have heard it before and want a company with specific targets in the market and an aim to be a market leader rather than taking a small niche of a larger market.

Be sure to have your numbers checked by an accountant and by an individual familiar with making investor presentations.

Step 6: Solicit Interest and Meetings from Targeted Investors

The popular phrase says, "You only get one chance to make a first impression." And so this goes with investing. Most professional investors will speak with hundreds (if not thousands) of entrepreneurs each year. You are hoping to stand out … in the right way.

Email makes it easy to mass-mail your business plan to every venture capital firm out there. Unfortunately, an unsolicited, untargeted email blast is not necessarily the best approach to building a base of interested investors. The goal of these initial solicitations is to obtain a face-to-face meeting with a potential investor.

To understand how best to solicit venture capital firms (or other professional investors), it is important to understand how the process of choosing to invest works. From the company's perspective the key player in this investment "drama" is known as your "champion." The champion is generally a general partner of the fund (or sometimes an associate, but nearly

What's the most common way you have found the companies you have actually invested in?

- Referral by another VC—34%
- Direct contact by the entrepreneur—30%
- Referrals by intermediaries—17%
- Referrals by accountants and attorneys—13%
- Investor events (such as industry forums and venture capital conferences)—7%

Source: Profit Dynamics Inc.

always someone with a certain level of clout within the fund). When a company gets funded, there is nearly always one firm partner who "champion's" the investment and is willing to stick one's neck out to get the deal financed. Generally, in the investment process, the CEO of the company will first meet with a contact at the firm. The hope is this meeting is with a person who is the firm's "specialist" in your particular industry, who has at least some availability of time to have another company under one's wing, and has clout with the copartners of the firm. Ultimately, the hope is this first meeting will be with your potential champion or with a person who will introduce you to your champion.

The champion is the company's advocate and main point of contact during the process. They look at your business and your team, and see that with financing and a bit of input from the investor-director, your business could produce big returns. The champion will drive the internal process within the fund, setting up meetings, bringing in experts, finding additional investors to invest in the deal if necessary and ultimately making the pitch to the other partners to make the investment.

How would you prefer the initial contact by the entrepreneur (or their representative) be made to you?

- E-mail—42% of respondents
- Postal mail—29%
- Telephone—17%
- Fax—8%
- Referral—4%

Source: Profit Dynamics Inc.

As you can probably tell, the champion really is the lynchpin in this entire process—and that is why a targeted solicitation is so crucial. The hope is you are not only finding the right set of venture capital firms to solicit, you're looking for the individual at the firm who will become your champion.

One certainty is that VCs will each have unique preferences, so the importance of matching their preference is important. Some prefer an introduction to come by referral. Others are comfortable with an unsolicited email or telephone call. Most VCs appear to favor receiving an executive summary business plan of a couple of pages answering the key questions on the business. This gives them the opportunity to judge for themselves following a high-level review.

Utilize the resources available to you to identify these preferences and use your referral network if it is available. Not all VCs will be interested (that is a given) and many might not respond to your initial contact. Be diligent, follow up and consider alternative approaches to build a connection to your points of contact.

Should We Hire a Placement Agent or Finder to Help with Fundraising?

Referrals by agents can be successful (17% of VCs in the Profit Dynamics survey listed referrals by intermediaries as the most common source of deals). However, these referrals are only as good as the agent and their personal networks. Before hiring any agent, be certain that you obtain information on prior placements including the names of the companies and the investment firms. You should request a list of references as well.

Securities laws limit your ability to pay commissions (or similar types of compensation) for sales of your company's securities. Only a licensed broker-dealer may receive commissions under securities laws. Here are some questions you may consider asking of any potential agent:

- Are you a registered and licensed broker-dealer?
- Can you provide any references to venture capital firms who can offer information about your services and experience?

- Can you provide a list of companies and venture firms where you were successful in raising funds?
- Can you provide a list of companies where you were not successful in raising funds?

Step 7: Make Presentations to Prospective Investors

The first hurdle in the fundraising process is getting an initial meeting with a potential investor. Obviously, even getting a meeting is a challenge (perhaps 1 out of a 100 chance for an unsolicited submission to a venture capital firm)—so one of the first keys to the process is ensuring that you take full advantage of the meeting. Landing a meeting is certainly a good step in the right direction, but initial meetings are just that, initial. The initial meeting represents an opportunity to sell the vision for the company to the investors and, hopefully, land a follow-up meeting. Depending on the investor, that initial meeting could be formal or informal.

For a firm that is interested in an investment, there will generally be a series of meetings designed to be a two-way communication between the venture capital firm and the startup company. Once you have been invited to make a presentation to the investor, you and your team will be asked to showcase the company and provide background on the company's vision, the management team, the broader market, and the details behind the approach you plan to utilize.

WHAT TO *EXPECT* IN YOUR INVESTOR MEETINGS

- A meeting with *one or several individuals* from the venture capital firm.
- A formal investment pitch will usually be *1 to 2 hours long* (although an informal discussion before being asked to provide a formal pitch could be between a half hour and an hour—or more depending on the firm).
  - Prepare a presentation that will take *40–45 minutes to go through without interruption*, but be aware that you'll likely be interrupted quite a bit.
  - Be sure to leave *at least 20 minutes for questions* at the end, even if it means going through some slides or information somewhat quickly at the end.
  - Immediately *stop the presentation for a question.*
- Prepare a PowerPoint slide presentation of *no more than 15–20 slides*. (Discussed above is a sample of slides you should consider including.)
- Prepare *answers to 15–20 of the most likely questions* you expect to receive.
- Be prepared to discuss *assumptions in creating your financial models*—and expect to receive questions about valuations even at an initial meeting.
- *Be flexible.* It is better to get off schedule if the investor is engaged and asking questions. The VC can review your prepared materials after you leave, but can't engage you with questions then.

And a couple practical points …

Ask your contact what types of technology is available. Should you bring your laptop? A CD? Flash memory drive? Projector? WiFi? Bring several hardcopy printouts of your materials to provide to your primary contact and any additional parties that may join in the presentation.

WHAT TO *AVOID* IN YOUR INVESTOR MEETINGS

- Don't have any *misspellings or errors* in your presentation materials.
- Don't look *unprepared*. Practice the presentation several times to varied audiences—particularly to individuals without an industry or technical background in your technology.
- Don't get *stumped by a question*. Prepare an answer to the 15–20 most-likely-to-be-asked questions.
- Don't *fudge something you don't know*. Offer to find out the answer and then quickly follow-up once you've answered the question.
- Don't present *proprietary information* in the investment presentation. You may need to reveal such information at a later point, but early on you should avoid offering such information since VCs will not generally sign nondisclosure agreements.
- Don't ask the VC to *sign an NDA*.
- Don't bring *team members* to the meeting who will not be participating or presenting.
- Don't forget to *ask questions* about the VC as well. The VC wants to know you are as interested in them as they may be in you.
- Don't schedule *investor presentations back-to-back*. Take time in between presentations to reflect on the prior presentation and modify items as appropriate.
- Don't *change the presentation simply because one VC recommended* you to add or change a particular portion. This process is highly subjective, so note trends of comments, but don't overhaul the presentation based on the concerns of one person alone.

Do I Need a Venture Capital Firm to Sign my NDA?

The short answer is no. In an ideal world, you would want everyone to sign a nondisclosure agreement if they will discuss the proprietary aspects of your technology. But common practice within the venture capital community is to refrain from signing NDAs. The reason is that the VCs (and their attorneys) believe that there is some risk for VCs to sign an NDA when they see so many similar business plans on a daily basis—many of which may overlap in some capacity. Therefore, avoid the embarrassment of asking a VC to sign a nondisclosure agreement—industry practice is that VCs don't sign NDAs.

Step 8: Receive Feedback from Prospective Investors

Even though a company soliciting investors will hear "no thanks" or "send us an update on your efforts" during the solicitation process, these negatives can be very helpful in eventually securing funding. One of the most important things you can receive from potential investors is feedback. Obviously, most companies that ultimately receive funding will hear "no" or "we're not sure" many more times than "yes." Therefore, understanding why an investor just isn't sold is extremely helpful.

First, expect to receive numerous refusals in the process—this generally is to be expected in the process (this means you are meeting with lots of potential investors, which is crucial). But, secondly (and most importantly), use any outright "no" or even a lukewarm "send us an update on your fundraising progress" to find out information to help you tailor your pitch and focus your business plan. Here are a few questions you may ask:

- What is the most/least interesting proposition we offer?
- Where do you see the needs to build out our team?
- What would you need to see in six months to invest?
- What should we reconsider in our business plan?
- Can you recommend another firm that might be a good/better fit (be careful with this one)?

A "No," a "Yes," or a "Maybe"?

Many entrepreneurs struggle to interpret the message from a VC they just met with. Oftentimes the meeting goes well and the entrepreneur senses that investor seemed engaged. Then things seem to unexpectedly cool down. Calls are less frequent. Emails go unanswered. What gives? If a venture capital firm partner is interested, the process will likely move fairly briskly. Generally, you can expect to hear things like:

- "I want to get you in front of the firm partners. Are you and your team available to present next week?"
- "What is your timing—I think we want to get involved?"
- "What do you need from us to move forward?"
- "Let me have some of our people do some more analysis and diligence work."
- "Can we get a list of customers and references to interview?"
- "When can I come and meet the rest of your team?"

If the reception you receive is more lukewarm, you can expect that the VC is probably not entirely convinced by some portion of your business plan.

When you've gotten in the door for a face-to-face meeting with a venture capital firm, you've obviously done something right. Generally, the company has at least "intrigued" someone at the firm if they've arranged for a meeting. By inviting you to a face-to-face meeting, you've passed the initial sniff test (an interesting business plan, a good management team, or a well-respected referral). And once you've passed that test, you are "fund-able."

HOW CAN I TELL IF THEY ARE *REALLY* INTERESTED?

- VC calls you (or quickly calls you back).
- VC sets up a follow-up meeting.
- VC wants to meet other members of the company and team.
- You receive an email or document with a number of questions.
- VC provides a checklist of items they need to review.
- You are asked for a list of references or customers.
- VC asks you to speak with a technical specialist or expert.

What does "fund-able" mean? A select few business plans are no-brainers for an investor (think Marc Andreessen's potential new company). The rest of the business plans a VC sees will have some warts—perhaps an unseasoned management team, a large potential competitor, a complex technology or a market that isn't a multibillion dollar space. In each case, there are some positives and negatives to weigh for any potential investor. Getting your foot in the door for meetings or discussions makes it likely that you are "fund-able"— simply that you fall within a broad category of companies that the VC believes is worth consideration or further investigation.

But a "fund-able" company isn't a "funded" company. If you are still in the "fund-able" category after the meeting, the VC may still not be ready to commit to writing you a check but might not want to refuse the investment just yet. Perhaps you'll find another investor and will come back to this VC to fill out the round. Perhaps you'll have a large customer sign up in the next couple months. Perhaps you'll have a technological breakthrough. These "perhaps" may keep you in the realm of "Keep us updated."

How should you handle a lukewarm reception? First, don't wait on a firm or firms to get back to you with a final answer. As discussed in the box below, an outright "no" is unlikely. Be polite and try to gather as much feedback as you can to help in the process, but don't alienate a potential investor by pushing too hard for a yes or no response. In the meantime, continue to setup meetings, build relationships and focus your message. The process of building the business shouldn't be put on hold. Momentum is a tough thing to regain once it is lost. Second, be persistent in your efforts to try to keep prospective investor updated. Let a potential investor know when you've reached certain goals you discussed in your presentation, you've hired new talent into the organization, or you've received positive press. Don't just hope for a potential investor to change its mind at some point down the road; give the firm or firms a reason to change it.

One of the biggest mistake an entrepreneur can make in the fundraising process is wasting valuable time "waiting on" prospective investors to change their mind or give you a final "no." Continue to build momentum and act on your business plan. Meet with many potential investors and keep communication lines open with all potential investors (even those that may have said "no" outright).

WHY DON'T THEY JUST TELL US "NO" IF THEY AREN'T INTERESTED?

Once an entrepreneur has had an initial meeting with a potential investor, the entrepreneur may expect to hear the following:

> *"Yes, we're interested in investing in you. Let's move forward."*
> or a
> *"No, but thanks for coming in. Good Luck."*

Then, in both scenarios the entrepreneur can move forward or move on, rather than be in limbo.

Unfortunately, limbo may be more likely than not as an entrepreneur may not get an outright rejection and instead may hear something like:

> *"We're interested if you can find a lead investor."*
> or
> *"Let's see if you can build up some customers/hit some development milestones/make some significant sales and then come back to us."*
> or

> *"Like the idea, keep us posted on the fundraising process."*
>   or
> *"Looks to be a bit too early stage. Come back to us when you've got some additional traction."*
>
> Unfortunately, none of these responses are particularly helpful to the entrepreneur and the startup company. They aren't a yes or a no. They are more like a "maybe."
>
> *So why don't they just tell you "no?"*
> - It is no fun to tell someone "no."
> - What if they are wrong and you turn out to be the next Google, Apple, Genentech or Cisco? How bad would they look as the VC firm who passed on that …
> - Sometimes (but not very often) things do change and the business does become a much more attractive candidate.

Step 9: Doing Business Due Diligence

If the process continues to go well, the VC will begin its process of further validation of the business. The VC is looking to make a substantial investment of time and money into the business and wants to be certain about the underlying assumptions the business is based upon.

Much like a job interview, expect that the VC will do its own homework on the company—speaking with your references, talking to your current or prospective customers, researching into the management, and consulting with contacts within the industry. Ultimately, the goal is to not to have anything surprise them. Therefore, talk with your references and customers before you give names to the VC. Let the contacts know what to expect and why they'll be receiving calls from potential investors. Provide your references with a copy of the business plan or certain materials to ensure they understand the business and can advocate on your behalf.

Be prepared to have your financial projections and current financial statements thoroughly reviewed by representatives of the potential investor. These requests for information may seem time-consuming, but it is important to help minimize the time this research and review will take. In some cases, use your attorney, your accountant, or other contacts to ensure the process goes smoothly.

Step 10: And Finally, "Yes. Here's a Term Sheet."

The finish line is in sight: A term sheet.

As the initial meetings and business diligence processes come to a conclusion, the final step in the solicitation process will culminate with a presentation to the firm partners (usually instigated by your champion). Following this presentation, the partners will discuss the business and make a decision to invest.

Find out when the partners of the VC firm have their weekly meeting (traditionally, Silicon Valley firms have scheduled these meetings on Monday mornings). It may be a good idea to follow up with your champion the following day, after the fund has their all-partner meeting. Utilize this information to see where the partnership stands on your potential investment.

And then begins the process of turning that "yes" over the phone or in a meeting, into a financing contract and cash in the bank. The first step will be receipt of a term sheet from the investor—nearly always drafted by the lawyers for the venture capital firm.

Once you've received the term sheet, the work of the company is far from complete. The company must now drive the process to completion to secure the funding. One mistake many entrepreneurs make is failing to realize that once a term sheet has been received, the company must manage the process to close the investment—including final negotiation of the term sheet and working with legal counsel to finalize the investment contracts. The company should work closely with the venture capital firm and partner, but don't assume that the VC will oversee the deal for the company. Traditionally, the company should utilize its advisors and counsel to set a timeline and manage completion of the transaction.

Convertible Note Financing

Many emerging companies raise money in the first instance not from institutional investors, but from "angels"—friends, family, and high net worth individuals—in a seed or angel round. In doing so, they need to decide whether to structure the seed round as a "bridge" (convertible debt) financing or a "priced" (preferred or common stock) financing. This section will examine convertible note financings, and the differences with "priced" financings.

DETERMINING IF A "BRIDGE" FINANCING IS APPROPRIATE

- Are you planning to raise additional funds from venture capital or other institutional investors?
- Are you planning to raise these funds in 6 to 18 months?
- Have you received positive inquiries from potential institutional investors?

If you answered "*yes*" to each of these questions, a bridge round may be appropriate. Ultimately, a bridge round is an effective tool when it is a "bridge" to a later financing event. If you aren't planning on such an event or are unlikely to reach it, then consider a priced round.

If a company finds an early stage investor who wants to invest capital into the business, how should the company decide what percentage of the company that investment is worth? It is tough to value a company without much to go on. As a result of this uncertainty, a company will need to decide whether to issue convertible debt that will convert at a future point (usually when a larger financing occurs) or issue equity interest based on a current estimate of the value of the company. And a growing number of investors are considering a hybrid approach—the convertible note with a cap on the future pre-money valuation in the conversion financing event.

Background on Convertible Notes

Convertible notes have many different names—a loan, a bridge loan, bridge notes, convertible promissory notes, or just plain notes. Ultimately, they all mean basically the same thing: a debt instrument issued by the company to an investor. The company will issue an investor a promissory note in exchange for a cash investment. The investor will receive a promissory note that earns interest, and, in the event the company is able to sell shares of its stock, will automatically convert into stock down the road. Startups like convertible notes because they are easy to explain, easy to prepare, don't require a lot of effort to value the company, and convert into equity down the road.

The convertible note concept only works when the company is planning toward a future equity financing—usually between 3 and 12 months after the convertible notes are issued, but sometimes as long as 18 to 24 months. Sometimes, convertible notes are used where a company is actively pursuing venture capital or angel funding or in the case where indications are that a venture capital firm or firms are interested in funding a company, but the process is going slower than expected or the venture capital firm and the company are unable to agree on the valuation.

Typically, the transaction uses a straightforward promissory note, which may be secured or unsecured, has a due date 3 to 12 months later (or longer depending on the terms agreed upon), and has a low interest rate. The promissory note provides that if a company does an equity financing above a certain amount (say at least $2 million) before the due date of the note, the principal and interest under the note automatically convert at the equity closing into whatever is being issued (this is usually referred to as series next preferred or series A preferred, if this represents a first round of funding), at a discount to the price per share as cash investors (this discount may range between 10% and 50%, depending on the length of time until conversion, or may increase over time by 5% to 10% each month the note is outstanding). If there are multiple lenders with notes outstanding, usually the notes will all rank equally as to payment and security priority. In some cases, the bridge lenders will take warrants in lieu of a discount or in addition to that discount. Warrants are simply stock options that allow the investor to purchase shares of common stock at a certain price within a set time frame. And some investors have negotiated a cap on the pre-money valuation for the notes in the equity financing (the conversion).

Benefits of a "Bridge" Financing

Some experts encourage early stage startups to raise money involving seed financings with a convertible note with a discount (sometimes with a cap on the future pre-money valuation). Others encourage the company to go with a priced round. So what are the pros and cons of doing a bridge round?

Pros:
- *Doesn't require a valuation.* In early stage investments, the valuation is difficult to settle on or much lower than the company would like.
- *Preparing financing documents for a bridge financing are simpler.* Simpler documents results in a quicker turnaround time (faster money in the door) and lower legal fees.

Cons:
- *Misalignment of interests.* Strangely enough, debt investors may actually want a *lower* premoney valuation for your eventual Series A round because it would *increase* their ownership percentage. This may be why caps in pre-money valuations are in vogue.

- *Unfavorable terms on the convertible notes.* While the terms of the notes may not require a valuation, investors may insist on terms that are unfavorable to the company such as founders' personal guarantees, heavy penalties in an event of default, grants of security interests in the company's assets, and others.

Discuss the pros and cons of the priced versus bridge round with your mentors, other entrepreneurs, and your attorney. There isn't a single "right" approach, but the key is determining the best approach for your business.

Sample Term Sheet for Convertible Note Financing

<table>
<tr><td colspan="2">MEMORANDUM OF TERMS FOR THE PRIVATE PLACEMENT OF SECURITIES
OF GREENSTARTUP, INC.</td></tr>
<tr><td colspan="2">This term sheet summarizes the principal terms of the proposed financing of GreenStartup, Inc. (the "Company"). This term sheet is for discussion purposes only; there is no obligation on the part of any negotiating party until a definitive note and warrant purchase agreement is signed by all parties. This term sheet is subject to the satisfactory completion of due diligence. This term sheet does not constitute either an offer to sell or an offer to purchase securities.</td></tr>
<tr><td>Amount to be raised</td><td>$250,000</td></tr>
<tr><td>Type of security</td><td>Convertible promissory notes (the "Notes").</td></tr>
<tr><td>Warrants</td><td>Warrants to purchase securities issued in the Company's next equity financing having an aggregate exercise price equal to 20% of the principal amount of the Notes.

-or-

Warrants to purchase common stock having an aggregate exercise price equal to 20% of the principal amount of the Notes.</td></tr>
<tr><td>Interest rate</td><td>Prime plus 2% per annum.

-or-

6% per annum.</td></tr>
<tr><td>Maturity</td><td>Principal and accrued interest shall be converted on or before December 31, 2012 into equity securities issued in the Company's next equity financing in an aggregate amount of at least $5,000,000 (including conversion of the Notes) (the "Next Equity Financing").

If the Next Equity Financing does not occur on or before December 31, 2012, principal and accrued interest shall be payable upon demand of the Holder.

-or-

If the Next Equity Financing does not occur on or before December 31, 2012, principal and accrued interest shall be due and payable on such date.

-or-

If the Next Equity Financing does not occur on or before December 31, 2012, principal and accrued interest shall be payable in four equal quarterly installments.</td></tr>
</table>

Conversion discount (Only used if the Company will provide the Note Holder the ability to convert at a discount)	Each Note will convert at a 10% discount to the price in the Next Equity Financing. -or- Each Note will convert at a 10% discount (the *"Conversion Discount"*) to the price per equity security paid by investors in the Next Equity Financing, as adjusted as follows: For each full month that the Notes are outstanding, the Conversion Discount shall be increased by 2.5% up to a maximum of 35%, as set forth in the table below.

Month 0 = 10.0%	Month 6 = 25.0%
Month 1 = 12.5%	Month 7 = 27.5%
Month 2 = 15.0%	Month 8 = 30.0%
Month 3 = 17.5%	Month 9 = 32.0%
Month 4 = 20.0%	Month 10 = 35.0%
Month 5 = 22.5%	Month 11+=35.0%

Conversion cap (Only used if the parties agree on a cap on the conversion pre-money valuation)	The Notes will convert at a maximum price per share equal to $4 million divided by the number of fully-dilute shares of the company in the Next Equity Financing.
Subordination	The Notes will be subordinate in right of payment to all current and future indebtedness to banks and other financial institutions. -or- The Notes will be subordinate in right of payment to certain indebtedness to banks and other financial institutions. -or- None of the above. *(No subordination.)*
Security interest	The Notes will be unsecured. -or- The Notes will be secured by all of the Company's assets. -or- The Notes will be secured by certain assets of the Company, including computer equipment, network equipment, and electronic storage equipment.

Investors	Principal Amount of Note
Jane Angelita	$100,000
John Familia	$75,000
Joe Entrepreneur	$25,000
ABC Ventures LP	$50,000
	$250,000

Closing date	The closing of the sale of the Notes will occur on or before March 31, 2012.

Venture Capital "Fit" Test

Select 1 to 5 for each of the categories below. Total your score below.

1. Size of your market: _____

1	Under $500 million
2	$500 million to $1 billion
3	$1 billion to $3 billion
4	$3 billion to $5 billion
5	Over $5 billion

2. Revenues in five years: _____

1	Under $10 million
2	$10 million to $20 million
3	$20 million to $35 million
4	$35 million to $50 million
5	Over $50 million

3. How much investment do you require now? _____

1	Under $500,000 or over $20 million
2	$500,000 to $1 million; $10 million to $20 million
3	$1 million to $2 million; $7.5 million to $10 million
4	$2 million to $3 million; $5 million to $7.5 million
5	$3 million to $5 million

4. Product gross margins: _____

1	Under 40%
2	40% to 50%
3	50% to 60%
4	60% to 70%
5	Over 70%

5. Industry: _____

1	Other
2	Consumer products and services; retailing/distribution; healthcare services
3	IT services; networking and equipment; computers and peripherals
4	Telecommunications; semiconductors; media and entertainment; medical devices and equipment
5	Industrial/energy; software; biotechnology

6. Location of company: _____

1	Other
2	Within a 2 hour drive of any of below
3	Austin; Chicago; Denver; Philadelphia; San Diego; Seattle; Washington DC
4	Boston; New York City; Southern California
5	Silicon Valley

7. Management team business experience and success: _____

1	Very low
2	Low
3	Medium
4	High
5	Very high

8. Costs to bring the product to market: _____
   (e.g., sales, marketing, distribution, consumer education, etc.)

1	Very high
2	High
3	Medium
4	Low
5	Very low

9. How extensive are the barriers to entry you have?   _____
   (e.g., intellectual property, proprietary information, lead to the game)

1	Very low
2	Low
3	Medium
4	High
5	Very high

10. Current commitment to Business   _____
    (e.g., full-time founders, office space, development products, angel investors)

1	Very low
2	Low
3	Medium
4	High
5	Very high

Bonus points (1 point for each):   _____

- Former CEO of a venture-backed company
- Received VC Funding at prior company
- 1 point for every five people you know personally at any venture capital firm

TOTAL: _____

If you scored

45 or higher	*Excellent fit.* Start sending out business plans.
40 to 44	*Good fit.* Try to expand your network to increase your fit.
35 to 39	*Okay fit.* Look to improve areas that are a 3 or below and expand size of network.
Under 35	*May not be a fit.* Determine whether the business is not currently at a stage ready for VC funding or whether the business model does not fit with the typical VC model.

In addition, some venture capital firms will *not* consider any company that does not score *at least a 4* on each of the first 4 questions.

For instance, your product may need to participate in a market at least $3 billion in size, a plan to reach sales of $50 million in five years, and a product with gross margins above 60%.

29

Strategies for Managing Startup Intellectual Property

General Intellectual Property Strategies

Effectively managing intellectual property (IP) assets is a difficult proposition for a new startup given the fiscal constraints nearly every new company will face.

> **YOUR INTELLECTUAL PROPERTY MANAGEMENT CHECKLIST**
>
> - Incorporate IP costs into your budget and business plan.
> - Develop a trade secret policy.
> - Utilize nondisclosure agreements (NDAs).
> - Understand and utilize copyright protections.
> - Ensure your employees and consultants enter invention assignment agreements.
> - File for provisional patents.
> - Don't miss the patent filing deadlines.
> - Develop an international patent strategy.
> - Research competing trademarks before you pick a name or brand.

Incorporate IP Costs into Your Budget and Business Plan

There are certain costs of doing business, including the costs of protecting IP. In particular, the cost of filing a patent ranges from $10,000 to $40,000 (or even more in some instances). It is important for a company to properly budget for future costs of IP. If your company decides to file a provisional patent, you should ensure that you will have the funds available to file the full patent 12 months later.

A potential investor may not require you to hold the patent at the time you are raising funds, but may want to ensure that you have properly protected the IP as a trade secret and that you intend to use the funds raised to file a patent.

Develop a Trade Secret Policy

Trade secrets include any information (formula, technique, pattern, physical device, program, idea, process, compilation of information, or other information) that (1) provides a business with a competitive advantage (that is generally known and not readily discoverable) and (2) where the individual or company takes reasonable steps to protect the secret and maintain these protections and prevent improper acquisition or theft. The scope of trade secret protection can be incredibly broad and, therefore, can be a key tool for startups. From day one, the founders of the company should put a trade secret protection policy in place. This policy will limit the disclosure of key information internally and externally to prevent information from being spread inappropriately.

Many startups forget that some of their employees will depart the company, so be certain that when this happens you are diligent in conducting an exit interview and reminding the employee of his or her ongoing obligations. With that in mind, be sure that your employees have signed confidentiality agreements that include their time as employees and their obligations after they depart. Not all employees or consultants need to know all matters—be selective in who knows what within your company (although this is difficult when your startup is only a few people with everyone pulling various oars!) Also, with the extensive use of emails and email attachments, you should be careful about inadvertently forwarding confidential documents.

KEY ASPECTS OF A TRADE SECRET PROTECTION POLICY

Company confidential documents and materials

- Label any proprietary or confidential documents and materials as "confidential" or "proprietary."
- Use password protections on key files and programs.
- Require all confidential or proprietary documents to be locked in file cabinets, desks, or company safes.

Employees

- Provide written copies of your trade secret policy and offer regular reminders of the importance of the policy.
- Limit disclosure of proprietary or confidential matters only to employees that will need such information.
- Utilize exit interviews for departing employees to remind of ongoing confidentiality restrictions and responsibilities.

Hiring employees or consultants

- For hiring procedures that will involve the disclosure of confidential information, utilize nondisclosure agreements prior to commencement of hiring process.
- Include confidentiality restrictions in all new hire and consultant agreements (include language that the new hire will not utilize confidential or proprietary information in subsequent jobs and will not reveal information to future employers).

Third party correspondence and meetings

- Utilize nondisclosure agreements in all meetings where confidential or proprietary information may be exchanged.
- Have visitors sign nondisclosure agreements when visiting or touring your facilities.

Utilize Nondisclosure Agreements

Non-disclosure agreements (NDAs) are an important tool for technology companies. You should prepare a mutual and one-way NDA early on in your development (or ask your attorney for a simple form of these agreements). As discussed below, not all individuals will sign an NDA (your attorney will not need to sign an NDA nor will most venture capitalists sign one). However, you should put in place a policy within your organization that all discussions with outsiders that involve discussions or observations of proprietary information will require an NDA to be in place. Also note (as discussed below) that signing an NDA is not, in itself, enough to provide trade secret protection. Be sure that you consider other ways your trade secrets may be released including at technical presentations and conferences, or through published papers and articles. When in doubt, keep your key secrets a secret.

An NDA is designed to protect the company from unauthorized disclosure of company confidential or proprietary information. However, if the company attempts to enforce the NDA, a court may find that the company did not take "reasonable precautions" required under trade secret laws. Remember, an NDA alone might not be sufficient to protect your trade secrets if the company fails to take other precautions to protect its confidential information.

WHO DO I NEED TO SIGN A NON-DISCLOSURE AGREEMENT?

Potential business partner or collaborator?
Most definitely. You should have mutual NDAs in place before discussions begin.

Potential hire?
Yes, if confidential matters are a part of the interview.

New employee?
You should have a policy in new hire documentation whereby new employees will sign confidentiality restrictions (among other restrictions).

Prospective investor (other than venture capital firm)?
Potentially. Most sophisticated investors and regular angel investors will be hesitant to sign NDAs to review and investment opportunity. However, in the event that you intend to discuss key technical aspects of the company, you should consider it.

Venture capital firm where I am sending my business plan?
Most will not since they see so many similar business plans and presentations that could create problems for the firm. You can try, but know that industry practice is that VCs don't and won't.

Your accountant or auditor? Usually will be covered in an engagement letter. Otherwise, probably yes, particularly in the case where they will come into contact with proprietary information or data.

Your attorney (or potential attorney)? Probably not. Usually covered in an engagement letter. Attorneys (even potential attorneys) are also bound by ethical standards limiting their ability to share confidential information. Some attorneys may sign NDAs.

Understand and Utilize Copyright Protections

Copyright protection represents another inexpensive tool that startups can utilize to protect certain aspects of their technologies. You should use the copyright notice for all documents produced by the company (© with the year the information or document was created or published, as well as the name of owner of the copyright which will typically be the name of your company). These copyright protections exist at the time your information is created.

Additionally, green and high tech companies should utilize these protections for their Web sites and for any software they create or design. For companies that have Web sites (both sites created solely for informational purposes and for sites created by e-commerce or other web technology companies), you should also place copyright notes on each web page. Additionally, copyright protections should be used for any software or programs the company produces, including source code and full software packages. In the case of software code or programs, you may also want to file your code with the U.S. Copyright Office to assist with enforcement actions where others utilize your code without permissions. These filings do have a small fee associated with them.

Ensure Your Employees and Consultants Enter Invention Assignment Agreements

While it may seem counterintuitive, your company may not automatically own something invented by your employees or consultants while employed or retained by your company. Therefore, it is very important that you put agreements in place that will ensure your company keeps rights to the inventions your company and its employees, or consultants make.

From day one, your company should be certain that each founder has executed an invention assignment for future inventions as well as properly assigns their rights to any IP that is being contributed to the company. Don't delay on making these assignments—one of the worst things to happen to a promising startup will be a founder departing the company prior to assigning key IP to the company and the remaining team is left without rights to key technologies.

Additionally, all new green and high-tech startups should put standard employment and consulting agreements in place for all employees and consultants. Be sure to properly follow state laws for proper assignment of your IP. Some states have specific language that must be included in any agreement assigning IP rights. Remember, each state has differing laws, so if you have employees located in different jurisdictions (at virtual offices or working from home), be certain that you've carefully drafted assignment language to avoid any issues down the road. Software companies should be sure that copyright is included in the assignment provisions to keep rights to any software code written by your employees or consultants.

Finally, be aware of unique problems that sometimes result from developing IP with a third party. Startups should be careful when collaborating with a university, a government agency, or other companies or organizations. If you work with a university or university personnel, be certain that the agreement does not permit the university to hold any rights to the current inventions or any future improvements or enhancements. Likewise, research conducted with funds from government grants oftentimes permit the government to have rights to the IP, permit additional research to be done on the developed technologies, or may even forfeit your rights to the IP if not properly disclosed. Be certain to review any grant agreements closely for the IP provisions and insist on narrowly tailored language with respect to ownership done at government facilities. Finally, remember that any work done with a partner company or organization could subject you to joint ownership of the developed technology. Without a specific agreement, IP will likely be co-owned by the partner organizations or companies.

The key to preventing ownership problems is to be certain that all agreements are drafted very carefully and have clearly identified provisions of assignment. Early on, your startup should attempt to create standard agreements that have carefully crafted assignment language for the particular state where the work will be done. Be certain to closely review all agreements for any provisions that may limit any rights you hold to your current and future intellectual property.

"MAGIC" LANGUAGE TO ASSIGN INVENTIONS

Be aware that certain states (including California, Washington, Illinois, and others) require that certain disclosures be made by the employer in the assignment agreement for the assignment of the patent rights to be valid. If such disclosures are not made, the assignment may be held to be invalid.

If you did not secure a valid assignment and the employee has refused to sign an invention assignment, the company still may hold "shop rights" which grants a nonexclusive license to use the invention developed by the employee, but this right may not be further assigned and would only be nonexclusive—which may scare away certain potential investors.

Patenting for Startups

Patents potentially represent one of the most valuable assets of a green business, but also may represent a substantial expense and use of time for a business that may have neither the money or the time available. This section is designed to provide you with strategies to manage the patenting process for an early stage business.

Filing for Provisional Patents

Patents are a key differentiator for high tech startups. In many cases, investors, customers, and potential competitors will look more favorably on a company with a strong patent portfolio in the key technologies of the field. Therefore, it is important to consider key steps to take as an early stage company with respect to protecting patent rights domestically and abroad. The expense of filing a full patent application can be upward of $40,000. Therefore, it is important to find ways to maximize the value for patents.

A provisional patent application (PPA) offers a company or individual to make a less complicated filing for $75 (up to $150 for certain inventors and companies). From the date of this filing, you have one year to submit a full application. This timing may allow your company to protect itself while it raises sufficient funds to create its patent portfolio. Companies have begun to utilize this process in order to attain a "patent pending" label on the invention and subsequent improvements. With the relative ease and lower fees of a provisional patent, a company may consider filing a provisional patent on both the initial invention and then on subsequent improvements on the invention.

Once the one-year period nears expiration, the company can make a determination as to a full patent filing. As a result, it is important to lay out a schedule for filing a provisional patent application that you can overlay with your budgeting. If you know you have to raise funds in the next six months, then perhaps filing a PPA makes sense since you are relatively sure that you be taking money from investors to pay for an IP strategy. If you plan to launch your product in two years and know that cash won't begin flowing until the launch, then a $40,000 expense a year from now may create some cash-flow problems without a product ready for launch. And, in any event, a "patent pending" product still is not a full patent and doesn't provide any absolute protections.

To file a PPA, you must (1) pay the required fee; (2) attach a cover sheet (one page in length); (3) submit a description of the invention with detailed information on the invention, the process for its creation, and details on its use (the description is an identical requirement as with a full patent filing); and (4) provide illustrations (if necessary). Filing a provisional patent allows the holder to claim the date of filing of the PPA as the original filing for purposes of priority, but permits the holder to wait up to one year for a full filing. However, just like in the case of a full U.S. patent application, your company will need to file an international patent application within one year of the provisional filing.

Don't Miss the Patent Filing Deadlines

While not all companies are prepared to file for patents early in their formation, you should be aware of certain actions that will prevent you from filing in the United States or abroad. In the United States, you will be unable to file a patent if your invention (1) has been "offered for sale" for greater than one year; (2) has been used publicly for greater than one year; or (3) has been published. Some companies have been barred from filing where they have offered the product, even though they have not made any sales prior to the one year deadline.

Outside the United States, most countries do not grant patents on the basis of the "first-to-invent," but operate on a "first-to-file" basis. Therefore, public use or offers for sale will not bar you (or others) from filing a patent on your product. As such, you should consider filing as soon as possible if your first filing will be international. Once you have filed a U.S. patent application, you will typically have one year from that filing date to file for an international patent.

Develop an International Patent Strategy

You must apply for a patent in each country where you want protection (except that you are able to file a single application for protection throughout the European Union). For the United States, you have a one-year period after an invention is sold or made public to file for a patent application. Elsewhere around the world, if your invention is made public or sold before you have filed a patent application, your invention might no longer able to be patented in that country.

Be careful with sales, presentations, and other disclosures to international parties if you are planning to apply for protection elsewhere in the world.

Additionally, make sure you have a patent and IP strategy that considers international implications. It is expensive to protect your IP worldwide. So plan accordingly and consider which markets are going to be targets for your product, and what costs the company can spend to apply for the necessary protections in those markets. These may be tough choices for any early stage company, but it is *better to make a choice than have a choice made for you*. Discuss with an IP attorney to see what strategy makes sense for you.

Trademarking for Startups

Choosing a name for the business or a product is oftentimes a challenging process given all of the potential considerations: other company names, domain names, international translations, double meanings, existing trademarks, and so on. This section is designed to help you understand key considerations for trademarks both domestically and abroad.

Research Competing Trademarks before You Pick a Name or Brand

Trademarks are an important part of developing your brand and brand awareness among potential customers. Selecting the right mark is important because the day you begin to utilize that mark, you've begun the process of brand identity.

Where many companies get into trouble is being forced to change their mark after selling a product or using that mark for branding and advertising. Imagine a company that spent money on advertising, has established a Web site, and began making sales, only to receive a letter from another company claiming infringement of their mark. This company would have invested time and money in establishing the brand only to see it slip away due to incomplete diligence at the start.

How can you prevent this from happening to your company? One course of action is to hire a third party to do a search once you've selected a mark—oftentimes your attorney can recommend a service to assist you—but this approach can be expensive. Before you hire a third-party service, you should take some initial steps to ensure your mark distinctive. Start with a thorough search on the Internet, searching your word choice and similar derivations of those words. Once you've come up with a list of possible names, begin looking at trademark databases maintained by the various government agencies (a list of databases is available in the Trademark section of this chapter). Consider discussing your choices with a Trademark attorney or a third-party search service. Once you've selected a mark or series of marks, only then you should consider filing.

You should also understand the international implications of selecting a particular mark—especially if you consider international markets to be a key marketplace for the sale of your products.

> **FREQUENTLY ASKED QUESTIONS ABOUT TRADEMARKS**
>
> *What are the benefits of registering a trademark?*
> - You reserve a mark you intend to use in the future.
> - You make your mark appear in searches others do before they adopt a mark thereby preventing others from adopting your mark inadvertently.

- You provide yourself additional remedies when someone else uses your mark or a mark likely to be confused with it.
- You make your rights in a mark federal and nationwide.
- You enable your use of ® next to your registered mark. ® is the symbol for showing that the U.S. Patent and Trademark Office has confirmed that you have the exclusive right to use this mark in connection with your goods or services. Using the ® gives notice to others that you have this right.

When should I apply to register my mark?

- For a green or high tech company, where product adoption is the key, you should consider protecting your mark as soon as you select it. Although using a mark without registering it confers rights in the United States, you need to register it to enforce those rights if someone else adopts your mark.

May I use ™ next to my mark?

- You may use ™ as soon as you have ascertained that no one else in the United States has rights in the mark. It may be good to check with your attorney before making this determination to avoid potential infringement. However, it is not necessary to file an application to register your mark before you use the ™ symbol.

Which country should I begin with?

- Most companies will first register in the country or countries they plan to sell their product into.

What does it cost to file marks in the United States?

- To register a mark in the United States, the registration process (including legal fees) generally costs between $2500 and $5000 (but remember, it can cost more if there are objections from the Trademark Office).
- Also remember that, in many cases, these costs are spread out over the entire registration process which oftentimes takes between one and two years.

FREQUENTLY ASKED QUESTIONS ABOUT INTERNATIONAL TRADEMARKS

Which countries should I register in?

- Most U.S. companies will first register in the United States. The Paris Convention rights will allow you to apply to register in most countries of the world for six months and still obtain the United States filing date.
- You may also consider filing a European Community Trademark ("CTM"). Most countries, other than the United States, are "first-to-file" countries (meaning that the first to file an application for a mark, as opposed to the first to use a mark in commerce, has superior rights). The CTM provides coverage

in 25 countries (Austria, Belgium, Cyprus, Czech Republic, Denmark, Estonia, Finland, France, Germany, Greece, Hungary, Ireland, Italy, Latvia, Lithuania, Luxembourg, Malta, Netherlands, Poland, Portugal, Slovak Republic, Slovenia, Spain, Sweden, and United Kingdom). As of the writing of this book, Switzerland and Norway are not in the Community and require separate applications.

- If you are a U.S. company and envision your product being sold in other countries in the Western Hemisphere, you may also file in Canada, Mexico, and in a number of countries within South America.

With so many countries throughout Asia, where should I begin?

- You may consider a modest filing program in Asian countries including Japan, China (PRC), Taiwan, South Korea, Singapore, and Hong Kong.
- Many companies will also look at filing in Australia and New Zealand due to their use of English.

What does it cost to file marks internationally?

- Foreign applications cost $1500 to $3000, depending on the country.
- A Community Trademark application costs from $4000 to $6000.
- In many countires, as with U.S. filings, these costs are spread out over the entire registration process which oftentimes takes between one and two years.

30

Service Partners

Finding and Working with a Lawyer

When many people think of attorneys, they think of the trial attorneys popularized in fiction. However, experienced entrepreneurs know that attorneys also play a crucial role in business. Legal counsel provides crucial input and advice at nearly every stage of your company's growth, from the initial formation of your company, to negotiating a lease for office space and hiring employees, to selling or licensing your product or services, to taking on financing, selling the business, or listing on a publicly traded market.

HOW TO USE LEGAL SERVICES EFFICIENTLY

- Discuss fees before a major transaction and on a regular basis.
- Keep your lawyer informed.
- Involve your attorney at board meetings.
- Have efficient meetings.
- Prevention is the best cure (involve your attorney before issues become problems).
- Give a "heads up" of issues (business or legal) that may be coming up.
- Centralize communications with your attorney and law firm.
- Utilize paralegals and junior associates.
- Don't try to do key legal work yourself.

Researching and Identifying an Attorney

There are many ways of coming up with an initial list of attorneys and firms to consider. One of the most effective ways is by referral. Talk to other business founders and see who they've worked with. If they like their attorney, then you may want to consider working with their attorney. As you start to plan your business, take note of attorneys that represent companies in your industry; that work with local chambers of commerce and business organizations; and that give presentations or lead business-related workshops. Solicit references from your friends and business acquaintances for corporate attorneys that they have worked with and liked. It will not take you long to create a short list of potential attorneys and firms. Your next task will be to get further information on these candidates, and interviews are a good way to learn more about a potential attorney or firm.

You should spend time with all qualified potential counsel. In these meetings, you will want to see if the attorney has the breadth and depth of legal experience that you will need to help you reach your entrepreneurial goals. Ask if the attorney has experience with companies in your line of business, and what type of work the attorney has done for those clients. Also important is the attorney's personality and the working environment of the attorney's firm. Is this a person or group of people you can imagine yourself coming to with legal problems? Would you be able to work easily with them? Are they proactive, responsive, and approachable?

Many attorneys who work with entrepreneurs have business experience themselves. You will also want to ask about this experience too, since, as noted above, corporate attorneys can also become trusted business advisors. A few other considerations for your interview include the attorney's communication skills and use of technology in their delivery of legal services and the attorney's connections in the community, since a well-connected attorney may be a source of valuable introductions for your business.

INFORMATION ABOUT ATTORNEYS ONLINE

- Avvo.com
- Martindale.com
- Lawyers.com
- Lawyers.Findlaw.com

Startup Counsel

For most startups, your attorney will likely serve as your general counsel during the early stages of the business (until it makes sense to hire an employee to provide legal support on a full-time basis). Your startup counsel should help with preparation of securities issuances and key employment documents such as offer letters, confidentiality agreements and invention assignments; management of the stock option process including issuance of proper paperwork and tracking of share exercises; offering strategy assistance with respect to your intellectual property; and coordination of financing events including negotiating both debt and equity deals for you. Many of the organizational milestones during your first year will require some level of legal guidance, and a well-trained lawyer can help you sidestep potential missteps.

In short, while corporate attorneys are necessary for many tasks associated with running your business, they can be much more than mere legal technicians helping with isolated transactions. A strong, continuing relationship with your business counsel can be a major asset for your company.

Intellectual Property and Patent Counsel

Companies that intend to file for intellectual property protections should consider retaining appropriate intellectual property or patent counsel, particularly if the company is inexperienced in the patent or trademark filing process. Patent Attorneys will typically have a scientific or technology background through education or experience that will allow them to draft patent applications and otherwise assist with issues before the Patent Office. Trademark Attorneys are generally specialists in assisting with and managing the process of obtaining trademarks. In each case, these intellectual property attorneys can assist with a single filing or can manage your initial and ongoing filings, which may include working with local counsel in various countries for international filings.

Specifics for Intellectual Property and Patent Counsel

The process to be a registered *Patent Attorney* requires that the attorney have a particular educational background and achieve a passing score on the U.S. Patent and Trademark Office registration examination. Candidates for the examination are required to have scientific and technical training, including a bachelor's degree in a field of natural science or technology, such as biology, chemistry, computer science, engineering, and physics, or technical training including certain engineering certifications, work experience, or sufficient relevant coursework. Patent Attorneys may represent clients before the Patent Office, may prepare, file, and prosecute patent applications for their clients before the Patent Office, and may give patentability opinions.

When selecting patent counsel, you should attempt to find an attorney that has a scientific or technical background that is related to the field or fields of your products. In many cases, you can obtain referrals from other companies operating in your industry. A patent counsel will either utilize an independent patent search firm or may, in some cases, have resources at the firm to provide this service. An independent search firm may be preferable as the results of the search are often relied upon by potential investors.

Likewise, filing for trademark protections are not always an intuitive or an easy process. Therefore, in many cases hiring an *attorney that specializes in trademark protections* can assist with the process and help you to manage your ongoing obligations. Check with your corporate attorney for recommendations and cost estimates.

Using Legal Services Efficiently

Legal fees are a cost of doing business, but you can find ways to minimize fees by developing an efficient relationship with your attorney. Ask your attorney to help prepare form contracts that only require minimal negotiations. Ask for an "initial review" of a contract to see if any red-flags are raised in the review before requiring a full review and markup. Call the attorney directly when an issue arises and ask if they can quickly assist with the issue. Before you try and save money doing it yourself, try and use other methods to control the legal fees (there are several listed below). Don't be short-sighted in your approach—it will often cause well-intentioned companies more heartaches than they ever imagined.

Keeping Fees and Expectations Reasonable

It pays (literally) to create an efficient and effective relationship with your attorney. Why? Because your lawyer will typically charge by the hour. As a result, you should consider some of the following steps to help manage the relationships with your outside counsel.

Discuss fees before a major transaction and on a regular basis. No one likes arguing over legal bills—it isn't fun for you and it isn't fun for the attorney. Many of the situations over billing arise because the client had an expectation as to how much something should cost, but the actual cost was higher. Why does this happen? In many cases, a client asks the attorney to undertake a project without ever asking for an estimate on fees. Research into a question or issue may seem to be a simple process; but issues can be more complicated and necessitate a specialist. Therefore, ask for estimates up front for any sizeable transaction and review the bill afterwards. If the bill looks off base with the original estimate, call the billing attorney to discuss the differences. Most good attorneys view the relationship with a startup client as a long-term preposition. Therefore, it is in everyone's interest to meet expectations on quality and, to the extent possible, on price. If the client has a better sense as to the cost a particular service or transaction will cost, then the likelihood of miscommunications decrease substantially. Your attorney will appreciate these discussions since it leads to fewer painful conversations down the road.

Keep your lawyer informed. Sometimes a client will decide to try and minimize legal bills by cutting off any communications with their attorney unless they have a specific legal need to discuss (and usually a rapid turnaround time). This means limiting any interactions until you plan to raise money from an angel investor or you have a founder depart the company and need a separation agreement or you get a "cease and desist" letter in the mail. But the reality is context behind the transactions helps ensure effective and efficient work. So include your attorney on emails with updates on your product development, your sales wins, and your new hires. Your attorney should not bill you for the time he or she spends reading these updates (and if they do, find yourself another attorney). By providing this information, you may find that your attorney is better able to inform you of important information. When your attorney sees an email about an important new hire, perhaps your attorney will shoot you a quick reminder to add the new hire to your directors and officer's insurance; or if you email about a new customer in China, your attorney may remind you of the tax or IP implications. Keeping your lawyer informed of nonlegal matters may add some billable time to your monthly legal bills, but it will also eliminate larger bills down the road. For example, imagine if your attorney hadn't reminded you about paying local Chinese taxes and then you got slapped with a fine or worse.

Involve your attorney at Board Meetings. Many of the most important discussions of your new business will be taking place during your regular board meetings. For some of these discussions, you'll need a seasoned legal advisor to walk you through key issues and help you avoid pitfalls. Therefore, you'll probably want to have your attorney present in person or via telephone. But what about the costs—doesn't this really start to add up if my attorney is at a two hour board meeting every month or every quarter? First, these costs are well-spent in most cases. Having your attorney present will give your investors and other board members additional confidence in the organization (and hopefully you too). But if costs still seem like a major problem, talk to your attorney. In some cases, you can front-load the meeting with discussions that would need counsel's presence and spend the later time focused on issues that are less likely to involve legal issues. In addition, some attorneys have a separate fee structure for meeting attendance that may help keep your bill lower or may agree to waive fees for board meeting attendance.

Have efficient meetings. If you need to meet with your attorney or want to schedule a time to talk over a conference call, give your attorney a synopsis of the issue or question in advance of the call or meeting, or send over an agenda before the meeting. Your attorney should know quite a bit, but if he or she doesn't know the answer, you've made your meeting an inefficient one. Or, if you are planning to have a meeting to discuss numerous hiring issues, your attorney may ask you if a colleague who is an employment lawyer can be asked to join you for the first part of the meeting.

Make your attorney ask you to provide them something only once. A disorganized or forgetful client can make a relationship more costly than it needs to be. If your attorney asks you to send the employment records of an employee to prepare a separation letter, send them. If you forget and they have to ask again, you may have just paid twice for your forgetfulness.

Prevention is the best cure. Most attorneys are happy to "turn off the meter" for certain types of meetings or discussions. Perhaps you can schedule a lunch every quarter. You may call these "relationship meetings" or a "business update meeting." In any case, go out to lunch or coffee with your attorney (and ask to have a nonbillable lunch—but you can take turns buying lunch). Spend the lunch updating your attorney on the status of the business and letting them know what could be coming down the pipe. These meetings are a great way to get advice on the key issues you may soon face. Going global? Find out how your attorney's other clients did it. Thinking about licensing technology from a local university? Find out what your attorney knows about the tech transfer office. Use these nonbillable lunches as a chance to pick your counsel's brain on trends, potential issues, and to highlight challenges that will be coming up. Some people view using their attorney like going to the doctors—"you won't see me at the doctors unless there is a good chance I could die or lose a limb because of the injury." Instead, think about getting regular checkups. These are an investment in your business and will prevent larger expenses down the road (like having to deal with a lawsuit, which is a surefire way to get stuck with a huge legal bill!).

Giving a heads up. Are you having second round meetings with VCs and expect term sheets in the next month? Have you retained investment bankers and are planning an acquisition in the next few months? Are you looking into partnering with a European company to sell your products into the EU? In each of these cases, you are likely to have a need for legal counsel for these transactions. Rather than letting your attorney hear about the deal when they receive the signed term sheet, provide a heads up so the attorney can consider how to staff the transaction and can perform certain pre-activities when the deal approaches a level of certainty.

Dealing with the last minute. Your attorney never wants to tell you that they don't have time to address your problem or manage your deal. But if you send them documents at noon and want them returned by five that same day, you could have problems getting your documents returned. Or, it could be that the partner will need to do this project oneself rather than delegate to a more cost-effective associate. Give your attorney ample time wherever possible, and where the project is going to have an extremely tight time frame, give as much notice of the tight deadline as possible. This allows for proper staffing (so the highest billing attorney isn't stuck doing a project better suited for a junior associate).

Centralize Communications. In the beginning stages of your organization, you may find that each of the founders will be communicating with your attorney. As the organization grows, this approach probably will become much less efficient and effective. Be sure that you have a protocol in place for the use of legal services. You do not want the entire organization to have the ability to request work to be done by your attorney. In many organizations, a key finance person will manage the relationship with your general counsel and a key engineering or research person will manage the relationship with your patent or IP counsel. If anyone at your company needs to use legal services, you should have the person go through this contact person, and you should put in place a "backup" in the event the primary point of contact in your organization is unavailable.

Paralegals and Junior Associates. Certain tasks are best performed in the hands of a paralegal or a junior associate. This includes managing typical stock option grants (unless you do these in-house), handling certain types of securities filings, and other more routine corporation actions. You should discuss work allocation up front with the lead attorney. Most firms will attempt to use the most cost-effective resources where they can, but if you have a rush project, those plans may go out of the door and the most "available" person, rather than the most cost-effective person will be doing the work. Plan ahead to avoid this type of scenario.

Don't try to do key legal work yourself. To save money, some companies will become amateur lawyers—using an old contract as a guide, failing to get a lawyer's eyes on a term sheet, or not informing their lawyer of a letter from a customer's counsel. Trying to save money on legal services by doing them yourself nearly always costs you down the road. Certain times, a company may be able to handle certain legal items without or with minimal supervision by their attorney (Table 30.1). However, that "self-drafted" contract with a customer could make your potential investor's lawyer nervous and could sink the deal, or could require your lawyer spend hours redrafting the contract and trying to have the old contract voided, or the improper separation letter could raise age discrimination issues.

Remember that there are risks associated in acting without counsel. While you can find some good sample and example forms from books or the Internet, no business is fully

TABLE 30.1

Tasks for the Counsel and the Entrepreneur

Event/Transaction	Importance of Counsel	Counsel's Task	Entrepreneur's Task
Formation of the company	Medium–low	Structural questions and implications on taxes	The state formation of the corporation or LLC
Founders stock and agreements	Medium	Issuance of stock since securities laws are implicated	Determine allocations and vesting
LLC operating agreement	Medium	LLC operating agreements that implicate numerous founders and implicate decision-making authority	Single-member LLC operating agreements
Employee agreements	Medium–low	Decision making on certain employment issues such as noncompetition, invention assignment, trade secrets, severance, and option issuance	Preparing standard agreements
Patents	High	Preparation of a patent or provisional patent application	Research of potential competing patent claims
Trademarks	Medium	Preparation of a trademark application	Doing a basic trademark search and domain name search
Third-party investments	High	Involvement of counsel is highly encouraged (particularly for investments by sophisticated parties)	Preparation of due diligence required by the transaction

"standard" and many attorneys can assist with preparation of many forms of agreements and the entrepreneur can use this reviewed form as a template going forward. While an entrepreneur may be able to handle certain responsibilities without counsel when the organization only involves two or three founders, as the number of parties involved grows, the need for a qualified outside counsel will rapidly increase.

Smart use of legal services may be more prudent and cost effective than trying to refrain from the use of any legal services. Many experienced startup counsels are familiar with these restrictions and limitations, and can help create a cost-effective solution.

Finding and Working with an Accountant

The primary purpose of accounting is to provide you, as a business owner, with useful information for decision-making purposes about your business. The goal of this section isn't to transform you into an accountant or to help you earn your CPA. Instead, the goal here is to help you as a business owner or executive to make the process for your accountant much easier and to help you make business decisions based on the accounting data you have.

YOUR ACCOUNTING/BOOKKEEPING CHECKLIST

- Put in place a financial recordkeeping policy.
- Manage the expense report process.
- Regularly add financial information into your accounting or other financial tracking system.
- Purchase the right accounting and bookkeeping software for your business.
- Consider outsourcing your accounting/bookkeeping.

Accounting helps you manage your cash flow so you know how much of your line of credit you need to utilize each month. Accounting allows you to identify trends in sales revenue and expenditures and identifies which products and services contribute the most to your bottom line. Accounting information tells you if you can expand your operations to that second office, how much money you'll need to borrow to invest in that new product line you've been working on, and, perhaps most importantly, how much you can afford to pay yourself each month. Understanding accounting information gives you the power to understand what drives the profitability of your business and to make the best decisions for your company.

Let's talk about how to set up your accounting system so you can start using your accounting information for decision-making purposes. Accounting information can be stored in hard copy format or electronically through the use of accounting software. You can keep your records in any way that makes sense to you as long as your records accurately reflect your business transactions. For thousands of years, before the advent of computers, businesses kept track of all their accounting information by hand through the use of ledger books. While manually keeping track of your transactions in ledger books is completely acceptable, using electronic accounting software is much more efficient. Today's accounting software is very user-friendly, very reasonably priced, and will literally save you hundreds of hours of time when compared to recording and summarizing all of your transactions by hand.

Keeping Track of Records

Make sure to keep good records associated with every sale, purchase, payment, and receipt your company is involved in. A good record includes the date, amount, and important details associated with the transaction (i.e., who you are paying or who is paying you, what the payment is for, etc.). Make sure to keep multiple records associated with a single transaction together. Take for example, a $500 utility bill: Here, the important records associated with this transaction include the bill you receive from the electric company and a copy of the check you write to the utility company, and any receipt for payment. Staple these records together if you need to.

In the beginning of your business operations it might seem like you don't even have a business to account for. Don't let the absence of revenues fool you into not keeping track of your start-up expenses. These initial expenses are part of your operating costs. If these expenses are unaccounted for, your financial records won't reflect the true costs of your operations. Relying upon incomplete financial information leads to poor business decisions. In addition, many of these expenses are deductible for tax purposes. Failing to keep track of these expenditures means that you won't remember to deduct them on your tax return, which results in you paying more income tax than you are legally obligated to pay. No business owner wants to pay more tax than one has to, so keep track of all expenses you incur, even before you start generating revenues.

Tips for Effective Handling Expense Reports

One of the most common expenses through the life of a business are expense reports—which may be expenses the business founder or owner occurs in buying things for the business or could include the expenses your sales people incur on their trips and travels to secure customers. Because these expenditures can be overlooked or forgotten about, here is a list of key steps you can take to manage expense reports:

- Develop a policy on expense reports and publish it to employees.
- Utilize a standard expense form for ease of use and consistency of information.
- Require all expense reports to be completed and submitted timely.
- Attach receipts for all out-of-pocket expenses (Note: The IRS may deny certain deductions if copies of receipts are not available).
- On the expense report, the date, time, and business relationship should be included for each expense.
- All expense reports should be signed and dated by the submitting employee and the individual approving the transaction.
- Track all expenses related to meals and entertainment in line items separate from travel costs (only a portion of meals and entertainment expenses are allowed to be deducted).
- Have an approval process in place and process expense reports in a manner that is consistent with processing of other approved invoices.

- Retain copies of all paid expense reports consistent with internal record retention policies (that should match applicable tax recordkeeping requirements).

Outsourcing Your Bookkeeping

If you don't want to have to worry about overseeing the bookkeeping work of your part-time employee, consider outsourcing your bookkeeping to a small accounting firm. Bookkeeping fees at smaller firms can be very reasonable, ranging from a few thousand dollars to tens of thousands of dollars per year depending on the level of service and frequency of reports you require.

Search online or visit Web sites such as:

- www.buyerzone.com
- www.osibusinessservices.com
- www.sbsuite.com

Most of these Web sites will provide you with quotes from various bookkeepers in your area.

Inputting Information into Your Accounting System

At the end of every day or every week, set aside time to enter all of your company's transactions by date and event into your accounting system. It's generally a bad idea to wait until the end of the month to record your transactions as thinking about entering an entire month's worth of information can be overwhelming. The more transactions you have, the more often you should record your transactions into your accounting software (Table 30.2). After you enter each individual transaction into your accounting software, the software will "post" the information to your general ledger.

TABLE 30.2

Accounting Software Review's 2010 Software Report

Rank	Software	Price ($)	Overall Rating[a,b]	Ease of Use[a,c]
1	Peachtree Complete	199.99	4.0	4.0
2	Myob business essentials pro	99.00	3.5	4.0
3	Quickbooks Pro	178.95	3.0	4.0
4	Netsuite Small Business Accounting[d]	1188.00	3.0	2.5
5	Cougar mountain	1499.00	3.0	3.0
6	Bookkeeper 2009	39.99	3.0	3.5
7	Simply accounting pro	169.99	2.5	4.0
8	CYMA IV Accounting for Windows[e]	595.00	2.5	3.0
9	Daceasy	499.99	2.5	2.5
10	Bottom line accounting	399.00	2.0	2.0

Source: Accounting-software-review.toptenreviews.com (accessed 1/13/2010).
[a] Maximum of four.
[b] Summary measure of 11 categories.
[c] One of eleven categories.
[d] Price per year.
[e] Price per module.

Once all of your transactions have been recorded (and subsequently posted to your general ledger), you can use the information in your ledger to create financial reports with a few clicks of your mouse. These financial statements summarize your business' activities for whatever period of time you specify. This aggregated and summarized information is what you use to determine how profitable your company was this week (or month, quarter, year, etc.) and to make future business decisions. Financial reports allow you to see your company's operations at the aggregate level (am I making a profit?) and at a very detailed level. This detailed level will provide you with very specific information that you can use to improve your company's operations. For example, you need to be able to generate and understand your financial statements to be able to tell how fast your customers are paying you for purchases made on credit, how much money is being tied up in inventory sitting in your warehouse or on your store shelves, which individual product or service contributes most to your bottom line, and if you'll need to use $5000 or $50,000 of your line of credit next month.

PLASTICS: WHAT'S THE "CLEAN TECH" OPPORTUNITY?

Plastics represent a substantial portion of the U.S. energy usage, and therefore many view this as a logical opportunity for "greening." However, don't be surprised that many consumers are not aware of the environmental ills of plastics. According to a 2007 study:

- Seventy-two percent of U.S. respondents did not know that plastics are derived from oil or natural gas.
- Forty percent of U.S. respondents believe that all plastics are biodegradable.

Tax Recordkeeping

Proper recordkeeping is a key consideration for any startup company, especially in the early stages when you do not have an established set of policies, an experienced accounting staff, or chief financial officer. Recordkeeping for tax purposes should be integrated into an entire document retention policy for the company. Most of the items discussed below will be key documents for accounting purposes.

YOUR TAX RECORDKEEPING CHECKLIST

- Receipts of income
- Payroll and employee information
- Records of purchases
- Business expenses
- Records of company assets

Receipts of Income

Retain all documentation that provides evidence of income. Gross receipts are the income you receive from your business. You should keep supporting documents that show the amounts and sources of your gross receipts. Documents that show gross receipts should include the following:

- Receipts of sales
- Customer invoices
- Shipping records
- Bank deposit records
- Credit card payment records (particularly in the event you are primarily selling goods or services via the Internet)

Payroll and Employee Information

Early on, startup companies should create and utilize standard documents for all new employees that join the company. These records should be kept for a number of purposes, including for tax purposes. In addition, as you begin to provide compensation to your employees and make withholdings from the employee's compensation, you should be carefully tracking and recording these amounts. Documents that show payroll and employee information should include the following:

- Records evidencing hiring
- Records evidencing changes in title, job status, or salary
- Termination documentation
- Cancelled checks
- Paystubs or payroll run reports
- Withholding calculations

Records of Purchases

The purchases described in this section reference items purchased and resold to customers, or goods purchased for manufacturing into finished products to be sold to customers. Since these purchases will likely represent an important portion of your costs of goods and your inventory, the company should keep records that identify the costs of purchase and payment information. Documents that show purchases should include the following:

- Cancelled checks
- Credit card records
- Invoices
- Shipping and receiving documentation and reports

Business Expenses

The business expenses are those that are used for aspects of the business other than those directly related to items to be sold to customers. These include research and development costs, meals and entertainment, electricity bills, and similar costs. These expenses may be in the form of expenses that are submitted by your sales people for reimbursement or could involve more traditional purchases of office supplies. Documents that show purchases should include the following:

- Cancelled checks
- Credit card records
- Receipts
- Invoices

Records of Company Assets

Assets for your company may range from machinery, vehicles or technical research equipment to buildings, software, or furniture. The key with company assets is to retain documentation on the purchase, ongoing improvements and maintenance, and disposal of the assets. Documents that show purchases should include the following:

- Acquisition records (invoices, shipping/receiving information; payment information)
- Maintenance records and costs
- Depreciation reporting
- Disposal records
- Real estate records
- Leases and rental records

TAXES: THE IMPORTANCE OF RECORDKEEPING

You need good records of all business revenues and expenses when you prepare your tax return. You need even better records if you are selected for audit by your state taxing authority or the IRS.

Here are some tax-related record-keeping tips to remember:

- *Identify the source of every receipt.* As a business owner you will receive cash or property from various sources. Keeping track of what the receipt relates to will help you identify if the receipt is taxable or not. For example, receipts from the sale of inventory or services are taxable but receipts from business contributions made by owners are not taxable.
- *Identify the business purpose behind every expenditure.* While you will be keeping track of your business expenditures for accounting purposes anyway, it's a good idea to keep track of the *business purpose* behind each expenditure.

A business expenditure must be both "ordinary" and "necessary" to be deductible for tax purposes. An ordinary expense is one that is considered to be common in your trade or business and a necessary expense is one that is considered to be appropriate for your trade or business. Note that some expenditures are deductible for accounting purposes but not for tax purposes (and vice versa).

- *Keep the records you use to prepare your tax returns.* If you are audited by the IRS or your state taxing authority, you may be asked to provide documentation supporting the revenues, expenses, and tax credits you report on your tax return. As a result, it is important to keep your tax records and all supporting documentation until the relevant statute of limitation expires. The IRS has three years to give you a tax refund or audit your tax return and 10 years to collect any tax due. This time period begins on the date that your tax return is due or the date you file your tax return (whichever is later). Therefore, always keep at least three years of tax return information on hand in case you are audited. A complete set of records makes an audit examination much less painful.

31

M&A and IPOs

Deciding to Pursue an IPO or Merger

Not every company represents a viable candidate for pursuing an initial public offering (IPO) or a merger and acquisition (M&A) transaction. The decision to undertake an IPO or to enter into acquisition negotiations oftentimes will be done over many months, if not years. And a number of key factors must be examined before any company definitively decides to pursue either of these transactions.

> **KEY CONSIDERATIONS FOR PURSUING AN "EXIT" EVENT**
>
> - Strategic considerations
> - Financial considerations
> - Industry considerations
> - Other considerations

There is no "right" time to consider going public or entering into a sale or merger transaction. However, there are certain factors that will begin to drive the decision, one of which is that the company begins to receive significant interest from investors, advisors, or even potential acquirers. Said one startup CEO, "I didn't realize we were a good candidate to be acquired until we started getting calls from companies we someday hoped to become."

Perhaps the decision will be driven by the need for capital for scaling of your technology or product; perhaps you are unable to compete in certain markets due to customer perceptions; perhaps a key investor is not willing to invest in follow-on rounds; or perhaps key members of management have expressed a desire to begin to transition out of the company. In each of these cases, the management team and the board of directors may begin early discussions into the various options for the company.

Whether your company decides to pursue an IPO or a sale of the company, the factors for deciding the proper timing are largely the same. The decision ultimately hinges on: (1) the needs and vision of the founders, employees, and investors; (2) the current status of the company (such as profits, revenue growth, competitors and industry growth); and (3) market conditions in your industry. In making your decision, you should consult with accountants, attorneys and, in certain cases, with investment bankers. The window of opportunity for an IPO can quickly close for your industry or for the IPO market at large, virtually halting all IPO activity until the market warms again. Companies that may not be mature enough for listing on Nasdaq may be better candidates for raising money through an IPO on the Toronto Stock Exchange or the AIM. Fortunately or unfortunately, companies often find that outside factors will primarily drive their choices and options for pursuing various transactions.

Strategic Considerations

The IPO or an acquisition presents your company with new strategic opportunities and challenges. An IPO may increase the publicity and stature of your company, resulting in new customers, employees and joint ventures. Additionally, the publicly traded shares available post-IPO provides the company with a currency of sorts, aside from cash reserves, to make acquisitions of its own. With a sale of the company, a well-positioned acquirer may provide economies of scale, new business contacts, and additional expertise for the target company. (Note: In an acquisition, the buyer is referred to as the "acquirer" or "acquiring company" while the company being acquired is the "target" or "target company.")

Along with these opportunities are challenges, many stemming from the fact that you will inevitably give up some control over the decisions and direction of your company. The IPO process and life as a public company place additional pressures and restrictions on management through SEC filing and disclosure requirements (including disclosure regarding executive compensation, customer contracts, and other previously confidential information), restrictions on publicity, increased fiduciary duties and liability of the directors and officers, increased public scrutiny of management decisions, and the impact of Sarbanes-Oxley. Failure to satisfy these many requirements may expose the company and its directors and officers to securities litigation from the stockholders.

A sale of the company may allow you to avoid the requirements of a public company (if the acquirer is private) but can also result in a total loss of control for the founders or, at a minimum, an additional layer of oversight from the acquiring company depending on the structure of the transaction and the intent of the parties. The acquiring company may implement major changes in personnel, compensation, benefits, policies, or company culture.

Financial Considerations

An IPO and a sale of the company may both provide company stockholders with the ability to "cash out" their shares of stock, but the sale of the company liquidity usually will allow the stockholders to receive cash for their shares more quickly. In an IPO, the founders of the company are typically limited by the investment bank (or the underwriters) for a period of time (known as a "lock-up" period) in which the founders cannot sell their shares. This period, typically six months, leaves the founders subject to fluctuations in the company share price (including market-wide fluctuations).

A sale of the company gives stockholders immediate liquidity (the ability to sell or "cash out" their stock) in the form of cash consideration or in shares of the acquirer under a stock-for-stock exchange—shares which are likely not subject to lock-up provisions like in an IPO. For a cash transaction, the amount of consideration for stockholders is guaranteed; the tradeoff is that the upside is capped since consideration is fixed (unlike in an IPO where there is no ceiling on share price and thus the stockholders' upside). If consideration is in the form of stock in the acquirer, the target stockholders do have some potential upside but that depends on the performance of the acquirer as a whole—not just the performance of the target. Founders may be willing to sacrifice the benefits of immediate liquidity for the opportunity to enjoy all the upside of the company.

Another consideration in your decision should be the transaction costs and, in the case of an IPO, the ongoing compliance costs for a public company. When taking into account the underwriters, filing, legal, and other fees, the IPO alone can cost in excess of $1.5 million even for relatively small offerings (according to Schultheis et al., 2004). Once your company is public, disclosure, legal, and accounting requirements will add additional expense to the company. Studies have shown that with Sarbanes-Oxley, the average annual corporate governance compliance costs for companies with less than $1 billion in revenue is $2.8 million! A sale of the company will also incur transaction costs in the form of legal, accounting, and consulting fees, although many of these expenses will be borne by the acquirer.

Industry Considerations

The sector or industry of the company also affects the choice to pursue an IPO or a sale of the company. For IPO candidate-companies, the business model must make sense to investors who may be unfamiliar with certain technologies, markets, or customer bases. Therefore, certain industries tend to have products that more easily translate into IPO candidate-companies. In industries with active acquirers, companies may be less willing to undertake the investment required to pursue an IPO.

For example, Medical Device companies tend to be less likely to go public through an IPO. Rather, companies in this sector are more likely to see exit events from medical device giants such as Medtronic, Boston Scientific, or Guidant. In certain subsectors, such as neuro/spinal with 10–12 active acquirers and cardiovascular with 4–5 acquirers, the number of attractive acquirers leads companies to pursue an M&A exit over a potential IPO.

Other Considerations

The decision of whether to pursue an IPO or a sale of the company will also depend on the position of your company as well as current market conditions. Companies experiencing rapid growth or that are in an industry currently favored by investors (see the many dot-com IPOs in the late 1990s) are better suited for an IPO while those companies experiencing constant but slower growth might find a sale of the company to be more appropriate. The market will also ebb and flow from year to year, quarter to quarter with various levels of M&A and IPO activity. Recently, M&A have been the more popular route in the United States for venture-backed companies with $31.9 billion in deals for 2006 compared to $3.75 billion in IPOs during that same period according to VentureSource.

OPPORTUNITIES AND NEW TECHNOLOGIES IN SOLAR

- *Thin film.* A thin, flexible product which uses much smaller amounts of silicon (1% or less) but also produces a photovoltaic effect (although currently at conversion efficiencies of 6–11% versus 14–22% for conventional crystalline silicon solar cells).
- *Organic solar cells.* Uses a fairly simple manufacturing process to create organic photovoltaic receptors, but currently has low conversion efficiency (3–5%).
- *Gallium arsenide.* Has conversion efficiency of greater than 30%, but the economics have not been proven to work.
- *String ribbon.* Utilizes thin layers or ribbons of silicon which eliminates much of the waste created through traditional production.

Mergers and Acquisitions

In deciding which transaction is in the best interest of your company, many considerations will weigh on your decision as well as the needs of the acquiring company, as each of the various types of M&A transactions have different tax, liability, and post-transaction structural consequences.

COMMON M&A STRUCTURES

- Direct or forward merger
- Forward triangular merger
- Reverse triangular merger
- Stock acquisition
- Asset acquisition

The M&A Transaction Process

Initial Negotiations

In advance of any M&A transaction, initial negotiations will take place between representatives of both the target and the acquirer to discuss the basic aspects of the deal (estimated purchase price, deal structure, assets to be acquired if an asset acquisition, postclosing corporate organization, etc.). If the parties decide to proceed with a deal, the next step is for the parties to sign confidentiality agreements and perhaps exclusivity agreements. Significant confidential information is exchanged throughout the due diligence and negotiation processes and confidentiality agreements are necessary to protect these proprietary information, especially in case the deal fails. Your lawyer should be able to draft a confidentiality agreement. An exclusivity agreement may also be appropriate and demanded by the acquiring company given the substantial time and cost involved in the diligence and negotiations process. If an acquiring company is going to dedicate the resources to the acquisition, they want to make sure they are not going to get "scooped" on the deal by another company.

Letter of Intent or Term Sheet

The next step of the process is for the parties to draft either a "letter of intent" or a "term sheet" outlining the key terms of the transaction. The parties may want to avoid signing a binding letter of intent or term sheet in case negotiations fail. The level of specificity in the term sheet or letter of intent will vary depending on the transaction but may include terms such as a description of the assets or liabilities to be transferred (if an asset acquisition), the purchase price and form of consideration, employee matters, indemnification provisions, termination provisions, and tax considerations. The more terms that are established at the outset, the more likely that the parties will be able to reach a final agreement. The final agreement, be it a merger agreement, asset purchase agreement, or stock purchase agreement, may mirror the terms of the letter of intent or term sheet but not necessarily so in all aspects.

Due Diligence

After the parties have agreed to terms in a letter of intent or term sheet, the companies can begin the due diligence process. As a first step, the acquiring company (typically through their counsel) will prepare a diligence request list, identifying the documents and information the acquiring company would like to review. The list will vary depending on the specific terms of the transaction but may include: common corporate documents, customer/supplier lists, material contracts, list of real estate and personal property, list of intellectual property and copies of licensing agreements, corporate financing documents, evidence of insurance coverage, list of current or pending litigation, financial information, tax information, employee information (including compensation, benefit plans, etc.), and environmental issues and liabilities. The target company then provides the documents either in hard copy or electronic format for the acquiring company and its counsel to review. The target company may also conduct due diligence of the acquirer if, for instance, the consideration is in the form of acquirer stock, but that due diligence is typically much less extensive.

If problems arise during the due diligence process, they may be resolved either by the company providing additional documentation, adjusting the purchase price or structure, or addressing the problems in the representations and warranties of the target company (discussed more fully below).

Transaction Documents

The merger agreement or asset purchase agreement sets forth the terms of the transaction and the obligations of the parties. The agreement will set the mechanics of the closing, the purchase price and transaction structure, the deliverables of the parties, and the assets and liabilities to be transferred in the event of an asset acquisition.

A large and occasionally heavily negotiated section of these agreements contains the "representations and warranties" of the parties. For the target, representations and warranties will often address the organization and standing of the company within the state of incorporation, capitalization matters, proper authority to enter the agreement, necessary consents (third party or governmental), accuracy of financial statements, compliance with laws, taxes, real and personal property, material contracts to which the company is a party, benefit plans, intellectual property, insurance, customers, personnel, litigation, environmental matters, and other necessary facts. Any exceptions to the representations and warranties are listed in a "schedule of exceptions" or "disclosure schedule" to the agreement. For the acquiring company, organization and standing and authority for the agreement may be sufficient disclosures to make (unless the consideration is in the form of stock in the acquiring company in which case additional representations and warranties may be appropriate). The representations and warranties not only serve as a means of disclosure of relevant information between the parties but are also critical in connection with the indemnification provisions discussed below.

Unless the merger or asset acquisition is structured as a "sign-and-close," where the merger or sale closes immediately following the signing of the agreement, the agreement will contain conditions to the parties' obligations to close at a later date (a "delayed closing"). For example, the acquirer's obligation to close may be dependent on obtaining certain approvals, the nonoccurrence of material adverse events since the signing, the representations and warranties of the target remaining true as of the date of closing, the performance or compliance with the covenants in the agreement, or the signing of employment or noncompetition agreements by designated employees. Many of the same obligations may apply to the acquirer, but again tend to be less extensive. If the conditions have not been satisfied, the other party can walk away from the deal. Given this possibility, the precise wording of these provisions is very important so as to avoid giving the other party too much discretion in whether to close the deal.

With a delayed closing, the agreement will also likely contain a termination provision allowing a party to terminate the agreement in the case of mutual agreement by the parties, failure to close before an agreed upon date, a material breach by the other party or a governmental order enjoining, restraining, or prohibiting the transaction. Another important section of the agreement contains the indemnification provisions. Often the target company and its stockholders will indemnify the acquiring company in the case of a misrepresentation or a breach of a warranty in the agreement (highlighting the importance of the target company's representations and warranties). To cover these claims, a portion of the purchase price may be held in escrow for the period of indemnification.

Merger

The first type of merger transaction is the statutory merger or consolidation in which two entities combine to become one "surviving" entity. There are typically three forms that the statutory merger can take, but all result in the acquiring company holding an equity stake in the target. The first is a direct or forward merger, where consideration is paid to the target company stockholders and then the target company merges into the acquiring company, with the acquiring company as the surviving entity (Figure 31.1).

The remaining two forms of merger structures are "triangular mergers," in which the acquiring company forms a subsidiary that either merges into the target (reverse triangular merger) or the target merges into the subsidiary (forward triangular merger). The result is the target becomes a wholly owned subsidiary of the acquiring company (Figures 31.2 and 31.3). One advantage of a triangular merger is that the acquiring company may be able to limit any liability stemming from the merger solely to the assets of the subsidiary (the assets of the parent company cannot be touched). Additionally, there may be tax advantages in structuring the deal this way (more on this below). All three forms of statutory merger require the approval of stockholders of both the target and the acquiring company.

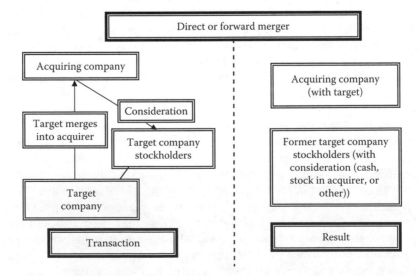

FIGURE 31.1
Diagram illustrating a direct or forward merger.

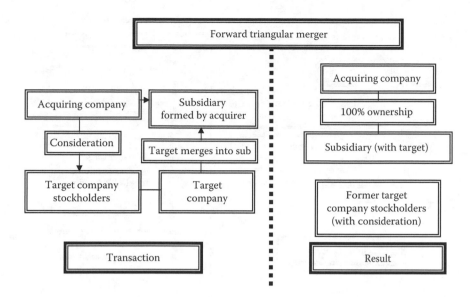

FIGURE 31.2
Diagram illustrating a forward triangular merger.

Stock Acquisition

A second transaction type is known as a stock acquisition where the acquirer agrees to purchase all outstanding shares of the target company's stock (Figure 31.4). This is actually an agreement between all the stockholders of the target company and not necessarily the company itself (although the company is often a party as well). Like in a merger, the result is that the acquiring company holds an equity stake in the target. Even if the acquirer is not

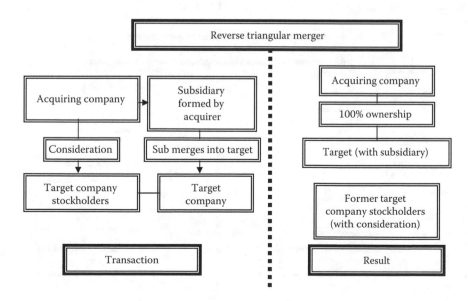

FIGURE 31.3
Diagram illustrating a reverse triangular merger.

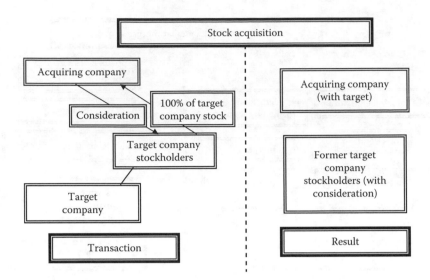

FIGURE 31.4
Diagram illustrating a stock acquisition.

able to acquire all of the target company's stock, they may be able to acquire a majority and then, depending on relevant state laws, acquire the remaining stock in what is called a "second-step merger."

Asset Acquisition

A third transaction type is an asset acquisition where the acquiring company acquires some or all of the target company's assets but not an equity stake as in the previous two transactions (Figure 31.5). One advantage of an asset acquisition is that the acquiring company may choose only those assets it wants to purchase and also limit the liabilities it assumes

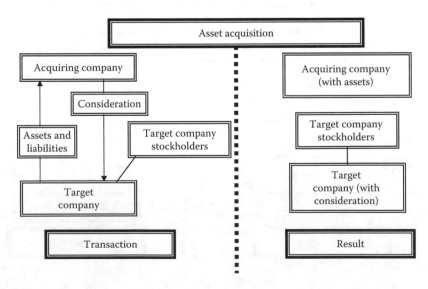

FIGURE 31.5
Diagram illustrating asset acquisition.

(with some exceptions). This may be disadvantageous to the target company since it will retain all unassumed liabilities.

These various transactions will have different requirements for board and stockholder approval, consents by third parties and tax implications (namely, whether the target company or target company stockholders will need to immediately recognize income from the transaction). You should consult with an attorney and an accountant in considering these differences. The structure of the transaction will often heavily depend on the needs and preferences of the buyer as well.

Initial Public Offerings

This section will provide an overview of the timeline, key events, and issues in the IPO process.

YOUR IPO TIMELINE

- Selecting the managing underwriter
- Organizational meeting
- Registration statement preparation
- Due diligence
- Road show
- SEC comments
- IPO

Pre-IPO Stage

Selecting the Managing Underwriter

Selecting the managing underwriter or underwriters for your IPO will depend on (1) the size of the offering, (2) the needs of your company, and (3) the experience and specializations of the various investment banks vying to serve as an underwriter. Companies will often select more than one underwriter and up to four or five to serve as the managing underwriter (if more than one, then "co-managers"), especially in the case of larger offerings or if specialized or local expertise is needed. In the case of co-managers, one of the firms will serve as the lead underwriter, but it is not unheard of to have co-lead underwriters. The managing underwriters will form a "syndicate" of other investment banks to share in marketing your shares to investors and to share in some of the risk.

In making the selection, your company may choose to consider several investment banks all at once, allowing them to compete for your business in a "beauty contest" or "bake off" where each bank will present to the company. The alternative is to deal with the investment banks one at a time until you find your underwriter. Underwriters will encourage this latter approach but that is largely because they would rather not compete with others to get your business. Given the narrow market window, the beauty contest approach may be more time efficient and allow you to get your shares to market sooner and provide a better comparison of the underwriters available.

Once the underwriters are selected, your company counsel will usually participate in the negotiation of the underwriter agreement which will cover all aspects of the offering. Then the *real* work begins.

Organizational Meetings

Once the company has selected a managing underwriter, the next step of the IPO process is to hold an "organizational meeting" with the company management, the underwriters, the company counsel, and other key participants in order to set a timetable for the transaction and to designate responsibilities among the parties.

Registration Statement Preparation

If not before, then soon after the organizational meeting, the company should begin preparing the registration statement to be filed with the SEC under Form S-1. The registration statement contains a "prospectus," which is a detailed description of the business, management, finances, and other required information. The prospectus is part selling document, presenting the company in a positive light to investors, but also part disclosure document, identifying the risks in the investment. Identifying the risks is critical to allow the underwriters, directors, and officers to limit liability. Given this delicate balance, the company should rely on experienced lawyers and advisors to take the lead role in drafting the registration statement, with active participation from the underwriters and company management.

After filing the registration statement, the SEC will review the statement to ensure compliance with the applicable form and then will respond with comments to the company regarding any deficiencies in disclosure. The company then amends the registration statement per the comments and the SEC reviews the amendments, again issuing comments if appropriate. This process continues until the registration statement is declared "effective."

The underwriters and the syndicate will distribute to potential investors a "preliminary prospectus," (also known as a "red herring" because of required red print on the cover) stating that the prospectus is merely preliminary and is incomplete. It is advisable to wait until at least one round of comments have been received since the printing of the preliminary prospectus can be quite expensive and the company will want to avoid multiple printings if significant changes are required by the SEC.

Due Diligence

While the registration statement is being prepared, the company will also engage in the "due diligence" process, which is a legal and business review of a company (organizational documents, board actions, stockholder agreements, financial documents, environmental compliance, contracts, etc.). This process is generally performed by the management, company's counsel, and the underwriters and their counsel in order to (1) ensure the accuracy and completeness of information contained in the registration statement and (2) to assist the underwriters in properly valuing the company. As in a sale of the company, the company's role in the diligence process will be gathering all of the relevant documents for the lawyers and underwriters to review.

Waiting Stage

Road Show

After the registration statement has been filed, the company will work with the managing underwriters to prepare a presentation for potential investors as part of the "road show"—a series of presentations by the management team in various cities before large groups of investors and one-on-one presentations over the course of two to three weeks and arranged by the underwriters. Increasingly, companies may perform part of its road show virtually, through the use of Internet-based presentations.

SEC Comments

Typically, the company will wait to embark on its road show until it has responded to SEC comments to the registration statement, as any material changes to the preliminary prospectus would require the company to recirculate the prospectus reflecting the changes at a substantial printing cost and potential delay. While waiting for approval, it is important to limit company publicity (namely public statements of company officials) as this could result in a violation of the securities laws (typically called the "quiet period"). Up until the moment the SEC declares the registration statement effective, the company, upon the advice of the underwriters may decide to delay the offering to a later quarter if market conditions have changed. In extreme cases, the IPO may even be terminated.

Initial Stock Offering

After the company has responded to the "rounds" of SEC comments and made all required amendments, the SEC will declare the registration statement effective and then the company and underwriters can decide on a final offering price. At this point, the final prospectus can be printed and distributed to investors and your company is now public. At this stage, the company will undertake efforts (usually with the help of the underwriters) to sell the public stock to potential investors.

Appendix: Additional Resources

Green Entrepreneur Handbook Web Site

- For updates, links, more up-to-date information and answers to your questions, visit: http://www.greentrepreneur.org/

Green Magazines and News

- E Magazine (http://www.emagazine.com/)
- Ecologist (http://www.theecologist.org/)
- Mother Earth News (http://www.motherearthnews.com/)
- Our Planet (http://www.ourplanet.com/)
- Recycling Today (http://www.recyclingtoday.com/)
- Sun & Wind Energy (http://www.sunwindenergy.com/)
- Sustainable Industries (www.sustainableindustries.com/)

Green Web Sites, Blogs, and Online Resources

- Eco Geek (www.ecogeek.com)
- Energy Tax Incentives (http://energytaxincentives.org/business/renewables.php)
- EnviroWeb (http://www.smallbiz-enviroweb.org/Resources/SmallBizWebPubs.aspx)
- Greener World Media (www.greenbiz.com)
- Grist (www.grist.org)
- Natural Capitalism Solutions (www.natcapsolutions.org)
- Renewable Energy World (www.RenewableEnergyWorld.com)
- The Lazy Environmentalist (www.lazyenvironmentalist.com)
- Tree Hugger (www.treehugger.com)

Government and Association Resources

- Biomimicry Institute (www.biomimicryinstitute.org)
- Database of State Incentives for Renewables and Efficiency (www.dsireusa.org/)
- Pew Center on Global Climate Change (http://www.pewclimate.org/)
- Small Business Guide to Clean Air Regulations (www.cleanair.org/Air/SmallBusinessGuide.pdf)
- The Department of Energy/Energy Efficiency and Renewable Energy (www.eere.energy.gov)
- The National Renewable Energy Laboratory (www.nrel.gov)

- The Rocky Mountain Institute (www.rmi.org)
- U.S. Green Building Council (www.usgbc.org)
- United States Environmental Protection Agency (www.epa.gov)

Startup Web Sites, Magazines, and News

- All-Biz Network (www.all-biz.com/)
- AllBusiness.com (www.allbusiness.com/index.jsp)
- Business 2.0 (money.cnn.com/magazines/business2/)
- Business Owners' Toolkit (www.toolkit.cch.com/)
- Entrepreneur (www.entrepreneur.com/)
- Fast Company (www.fastcompany.com/)
- MoreBusiness.com (www.morebusiness.com/)
- SBA Resources (www.sba.gov/hotlist/)
- Service Corps of Retired Executives (SCORE) (www.score.org)
- Wired (www.wired.com/)

Tech Magazines and News

- CIO (www.cio.com/)
- C-NET (news.com.com/)
- ComputerWorld (www.computerworld.com/)
- First Monday (www.uic.edu/htbin/cgiwrap/bin/ojs/index.php/fm/index)
- MIT TechReview (www.techreview.com/)
- SlashDot (www.slashdot.org/)
- ZDNet (www.zdnet.com/zdnn/)

Startup-Focused Blogs

- TechCrunch (www.techcrunch.com/)
- GigaOM (gigaom.com/)
- ReadWriteWeb (readwriteweb.com/)
- Mashable (mashable.com/)
- VentureBeat (venturebeat.com/)
- Startup Nation (www.startupnation.com)
- OnStartups.com (onstartups.com)
- VentureWire (www.venturewire.com)

Fundraising Resources

- The Funded (www.thefunded.com)
- National Venture Capital Association (www.nvca.org)

- European Private Equity and Venture Capital Association (www.evca.com)
- Venture Economics (www.ventureeconomics.com)
- vfinance.com (www.vfinance.com)
- Private Equity HUB (www.pehub.com)
- MoneyTree Report (www.pwcmoneytree.com)

References and Reading List

Adams, R. 2002. *A Good Hard Kick in the Ass: Basic Training for Entrepreneurs*. New York, NY: Crown Business.

Bagley, C. E. and Dauchy, C. E. 2003. *The Entrepreneur's Guide to Business Law* (2nd edition). Mason, OH: West Educational Publishing.

Berkery, D. 2007. *Venture Capital for the Serious Entrepreneur*. New York, NY: McGraw-Hill.

Bolles, R. N. 2007. *What Color is Your Parachute? 2008: A Practical Manual for Job-Hunters and Career-Changers*. Berkeley, CA: Ten Speed Press.

Bradford, T. 2006. *Solar Revolution: The Economic Transformation of the Global Energy Industry*. Boston, MA: MIT Press.

Brandt, S. C. 1967. *Entrepreneuring: The Ten Commandments for Building a Growth Company*. Friday Harbor, WA: Archipelago Publications.

Carson, R. 1962. *Silent Spring*. Boston, MA: Houghton Mifflin.

Collins, J. 2001. *Good to Great: Why Some Companies Make the Leap ... and Others Don't*. New York, NY: HarperCollins.

Collins, J. and Porras, J. 1997. *Built to Last: Successful Habits of Visionary Companies*. New York, NY: HarperBusiness.

Cooney, S. 2009. *Build a Green Small Business: Profitable Ways to Become an Ecopreneur*. New York, NY: McGraw-Hill.

Croston, G. 2008. *75 Green Businesses You Can Start to Make Money and Make a Difference*. Irvine, CA: Entrepreneur Press.

Croston, G. 2009. *Starting Green*. Irvine, CA: Entrepreneur Press.

Davidow, W. H. 1986. *Marketing High Technology: An Insider's View*. New York, NY: Macmillan, The Free Press.

Drucker, P. F. 2006. *Innovation and Entrepreneurship*. New York, NY: HarperCollins.

Etsy, D. and Winston, A. 2006. *Green to Gold: How Smart Companies Use Environmental Strategy to Innovate, Create Value, and Build Competitive Advantage*. New Haven, CT: Yale University Press.

Friedman, T. 2008. *Hot, Flat, and Crowded: Why We Need a Green Revolution—and How It Can Renew America*. New York, NY: Farrar, Straus, and Giroux.

Gartner, W. B., Shaver, K. G., Carter, N. M., and Reynolds, P. D. 2004. *Handbook of Entrepreneurial Dynamics: The Process of Business Creation*. Thousand Oaks, CA: Sage Publications.

Gilbert, J. 2004. *The Entrepreneur's Guide to Patents, Copyrights, Trademarks, Trade Secrets, & Licensing*. New York, NY: Berkley Books.

Godin, S. 2006. *Small Is the New Big: And 183 Other Riffs, Rants, and Remarkable Business Ideas*. New York, NY: Portfolio Hardcover.

Gosden, F. F. Jr. 1989. *Direct Marketing Success: What Works and Why*. New York, NY: Wiley.

Heath, C. and Heath, D. 2007. *Made to Stick: Why Some Ideas Survive and Others Die*. New York, NY: Random House.

Hess, K. L. 2001. *Bootstrap: Lessons Learned Building a Successful Company from Scratch*. Carmel, CA: S-Curve Press.

Hill, B. E. and Power, D. 2002. *Attracting Capital from Angels: How Their Money—And Their Experience—Can Help You Build a Successful Company*. New York, NY: Wiley.

Hirshberg, G. 2008. *Stirring It Up: How to Make Money and Save the World*. New York, NY: Hyperion.

Horn, M. and Krupp, F. 2008. *Earth: The Sequel: The Race to Reinvent Energy and Stop Global Warming*. New York, NY: W.W. Norton & Co.

Jones, V. 2008. *The Green Collar Economy: How One Solution Can Fix Our Two Biggest Problems*. New York, NY: HarperOne.

Kaplan, J. 1996. *Startup: A Silicon Valley Adventure*. New York, NY: Penguin.

Kawasaki, G. 2004. *The Art of the Start: The Time-Tested, Battle-Hardened Guide for Anyone Starting Anything*. New York, NY: Portfolio Hardcover.

Koester, E. 2009. *What Every Engineer Should Know About Starting a High-Tech Business Venture*. Boca Raton, FL: CRC Press.

Komor, P. 2004. *Renewable Energy Policy*. Bloomington, IN: iUniverse, Inc.

Lerner, J., Hardymon, F., and Leamon, A. 2004. *Venture Capital and Private Equity: A Casebook*. New York, NY: Wiley.

Livingston, J. 2007. *Founders at Work: Stories of Startups' Early Days*. New York, NY: Apress.

MacVicar, D. and Throne, D. 1992. *Managing High-Tech Start-Ups*. Boston, MA: Butterworth-Heinemann.

Makower, J. 2009. *Strategies for the Green Economy*. New York, NY: McGraw-Hill.

McDonough, W. and Braungart, M. 2002. *Cradle to Cradle: Remaking the Way We Make Things*. New York, NY: Farrar, Straus, and Giroux.

McQuown, J. H. 2004. *Inc. Yourself: How to Profit by Setting up Your Own Corporation* (10th edition). Franklin Lakes, NJ: Career Press.

Mohr, J. 2001. *Marketing of High-Technology Products and Innovations*. Englewood Cliffs, NJ: Prentice-Hall.

Moore, G. A. 2002. *Crossing the Chasm: Marketing and Selling Disruptive Products to Mainstream Customers*. New York, NY: Harper.

Nesheim, J. 2000. *High Tech Start-Up: The Complete Handbook for Creating Successful New High Tech Companies*. New York, NY: Simon & Schuster.

Pernick, R. and Wilder, C. 2007. *The Clean Tech Revolution: The Next Big Growth and Investment Opportunity*. New York, NY: HarperCollins.

Pernick, R. and Wilder, C. 2008. *The Clean Tech Revolution: Discover the Top Trends, Technologies, and Companies to Watch*. New York, NY: HarperCollins.

Pressman, D. 2006. *Patent It Yourself* (12th edition). Berkeley, CA: Nolo Press.

Rich, S. R. and Gumpert, D. E. 1987. *Business Plans that Win $$$: Lessons from the MIT Enterprise Forum*. New York, NY: Harper & Row.

Roam, D. 2008. *The Back of the Napkin: Solving Problems and Selling Ideas with Pictures*. New York, NY: Portfolio Hardcover.

Roberts, E. B. 1991. *Entrepreneurs in High Technology: Lessons from MIT and Beyond*. New York, NY: Oxford University Press.

Savitz, A. W. and Weber, K. 2006. *The Triple Bottom Line: How Today's Best-Run Companies Are Achieving Economic, Social and Environmental Success—and How You Can Too*. New York, NY: Wiley.

Schultheis, P. J., Montegut, C. E., O'Connor, R. G., Lindquist, S. J., and Lewis, J. R. 2004. *The Initial Public Offering, A Guidebook for Executives and Boards of Directors* (2nd edition). Washington, DC: Wilson Sonsini Goodrich & Rosati.

Seireeni, R. and Fields, S. 2008. *The Gort Cloud: The Invisible Force Powering Today's Most Visible Green Brands*. White River Junction, VT: Chelsea Green Publishing.

Stathis, M. 2005. *The Startup Company Bible for Entrepreneurs: The Complete Guide for Building Successful Companies and Raising Venture Capital*. Worthing, UK: AVA Publishing.

Swanson, J. A. and Baird, M. L. 2003. *Engineering Your Start-Up: A Guide for the High-Tech Entrepreneur* (2nd edition). Belmont, CA: Professional Publications.

Van Osnabrugge, M. and Robinson, R. J. 2000. *Angel Investing: Matching Start-Up Funds with Start-Up Companies*. San Fransisco, CA: Jossey-Bass.

Viardot, E. 1998. *Successful Marketing Strategy for High-Tech Firms* (2nd edition). Boston, MA: Artech House.

Wilmerding, A. 2006. *Term Sheets & Valuations—A Line by Line Look at the Intricacies of Venture Capital Term Sheets & Valuations*. Boston, MA: Aspatore Books.

Index